UNITEXT - La Matematica per il 3+2

Volume 113

More information about this series at http://www.springer.com/series/5418

Lorenzo Peccati · Mauro D'Amico
Margherita Cigola

Maths for Social Sciences

 Springer

Lorenzo Peccati
Department of Decision Sciences
Bocconi University
Milan, Italy

Margherita Cigola
Department of Decision Sciences
Bocconi University
Milan, Italy

Mauro D'Amico
Department of Decision Sciences
Bocconi University
Milan, Italy

Additional material to this book can be downloaded from http://extras.springer.com.

ISSN 2038-5714   ISSN 2532-3318   (electronic)
UNITEXT - La Matematica per il 3+2
ISSN 2038-5722   ISSN 2038-5757   (electronic)
ISBN 978-3-030-02335-5   ISBN 978-3-030-02336-2   (eBook)
https://doi.org/10.1007/978-3-030-02336-2

Library of Congress Control Number: 2018958490

Preface

This textbook is the natural evolution of some teaching notes prepared for the course on "Quantitative Methods - Module 1, Mathematics" in the "Bachelor in International Government" program at Bocconi University. While many books are available on statistics for the social sciences, few try to cover mathematics for the social sciences. This textbook represents such an attempt, the aim being to present some mathematical tools of value in the social sciences through a number of examples.

These mathematical tools are:

- Linear algebra;
- Differential calculus;
- Integral calculus;
- Dynamic systems.

The style of presentation is informal, and most of the examples are calibrated with reference to the application areas.

We are grateful to the referees, who examined the first version of this textbook and made insightful suggestions, and to our colleagues Ross MacMillan and Massimo Marinacci for useful suggestions regarding the contents.

The sole responsibility for any error, whether in the English or in the mathematics, resides with the authors.

The solutions of the exercises provided in the textbook are available at the following link http://extras.springer.com.

Milan, Italy   Lorenzo Peccati
July 2018   Mauro D'Amico
                                                             Margherita Cigola

Contents

1 **Linear Algebra** 1
   1.1 Vectors and Matrices 1
        1.1.1 Introductory Examples 1
   1.2 Operations on Vectors. 4
        1.2.1 Vector Representation in $\mathbb{R}^2$ and in $\mathbb{R}^3$ 4
        1.2.2 Vector Addition 5
        1.2.3 Scalar Multiplication 7
        1.2.4 Linear Combination of Vectors 8
        1.2.5 Linear Dependence/Independence of Vectors 10
   1.3 Matrices 16
        1.3.1 Types of Matrices 16
        1.3.2 Operations on Matrices: An Inventory 18
        1.3.3 Transposition 19
        1.3.4 Matrix Addition 20
        1.3.5 Multiplication of a Scalar by a Matrix 21
        1.3.6 Multiplication of Matrices 22
   1.4 Determinants 41
        1.4.1 Notion and Computation 41
        1.4.2 How to Invert a Matrix Using Determinants 50
        1.4.3 Determinants and "Viability" of a Leontief System 51
   1.5 Rank of a Matrix 52
   1.6 Statistical Applications of Linear Algebra. 58
   1.7 Linear Applications. 67
   1.8 Linear Algebraic Systems 71
        1.8.1 A Special Case: Cramer's Systems 72
        1.8.2 The General Case 74
   1.9 Applications to Networks 86
   1.10 Some Complements on Square Matrices 87
   1.11 Exercises 93

2   Differential Calculus   99
    2.1   What's a Function   99
           2.1.1   Intervals   100
           2.1.2   Easy Functions.   101
           2.1.3   Elementary Functions   101
           2.1.4   Continuous Functions   110
           2.1.5   An Annoying Detail and a Tribute to L.D. Landau   112
           2.1.6   The Small o Algebra   114
           2.1.7   Some Rankings   115
    2.2   Local Behavior and Global Behavior   118
           2.2.1   Local Behavior, Derivative and Differential   118
           2.2.2   Notation for Derivatives   125
           2.2.3   Derivative and Differential: What's the Most Important
                    Notion   126
           2.2.4   The Computation of Derivatives, also of Order >1   128
    2.3   What's a Function of a Vector   152
           2.3.1   Graphic Representation of a Function of $n \geq 2$
                    Variables   155
           2.3.2   How Big is a Vector?.   157
           2.3.3   Derivatives of Functions of a Vector   158
           2.3.4   Unconstrained Extrema for Functions $f : \mathbb{R}^n \to \mathbb{R}$ 163
           2.3.5   Constrained Extrema   170
           2.3.6   The General Case.   178
    2.4   Exercises   192

3   Integral Calculus   197
    3.1   Integrals and Areas   197
    3.2   Fundamental Theorem of Integral Calculus   198
    3.3   Antiderivative Calculus   201
           3.3.1   Integration by Parts   202
           3.3.2   Integration by Substitution   203
    3.4   An Immediate Application: Mean and Expected Values   204
           3.4.1   Expectation and the Law of Large Numbers   206
           3.4.2   Density Function and Distribution Function   207
           3.4.3   Discrete Distributions   209
    3.5   Frequency/Probability Density Functions: Some Cases   210
           3.5.1   A Special Distribution   211
    3.6   People Survival   214
    3.7   Exercises   217

4   Dynamic Systems   221
    4.1   Introduction   221
    4.2   Local Information: The Motion Law   224
           4.2.1   Discrete Time   224

	4.2.2	Continuous Time	226
	4.2.3	Motion Law of a DS	228
	4.2.4	Autonomous Systems	229
4.3		Extracting Info from a Motion Law	232
4.4		Classic Approach	233
	4.4.1	Linear Discrete Systems	234
	4.4.2	About Some Special Discrete Systems	246
	4.4.3	Continuous-Time Systems	249
	4.4.4	Continuous Systems: Separable Equations	250
	4.4.5	Continuous Systems: Linear Differential Equations of the First Order	253
	4.4.6	An Interesting Socio-demographic Model/1	254
	4.4.7	Linear Continuous Systems	260
4.5		Numerical Approach	282
	4.5.1	Discrete Systems	282
	4.5.2	Continuous Systems	283
4.6		Qualitative Approach	285
	4.6.1	Equilibria: Notion and General Systems	285
	4.6.2	How to Find Equilibria for Autonomous Systems?	287
	4.6.3	Nature of an Equilibrium Point	289
4.7		A Newcomer: The Phase Diagram	290
	4.7.1	Notion	290
	4.7.2	Equilibria in a Phase Diagram	293
	4.7.3	Behaviors Revealed by a Phase Diagram	295
	4.7.4	Continuous Systems	307
4.8		Some Politically Relevant Applications	309
	4.8.1	Consensus Diffusion	310
	4.8.2	Arab Springs	315
	4.8.3	Growth and Demographic Trap	318
4.9		Military Applications	321
	4.9.1	War	321
	4.9.2	Guerrilla	322
4.10		USA Against USSR: An Old Story	324
	4.10.1	An Electoral Application	329
4.11		Epidemics	332
	4.11.1	The B Model	332
	4.11.2	The R Model	333
	4.11.3	The RI Model	334
	4.11.4	The RIV Model	335
4.12		Exercises	337
References			341
Index			343

About the Authors

Lorenzo Peccati was born in 1944 and is an Emeritus Professor of Mathematics at Bocconi University. He was the Editor of the European Journal of Operational Research for 10 years. He has authored over 140 publications in national and international journals as well as several books.

Mauro D'Amico was born in 1963 and is an Instructor in Mathematics at Bocconi University. He has an extensive experience in teaching both university and business school courses. In addition, he has written several textbooks in Italian.

Margherita Cigola was born in 1961 and is an Associate Professor of Mathematics at Bocconi University. Her research focuses on optimization theory, and she has an extensive experience in teaching applied mathematics.

Chapter 1
Linear Algebra

1.1 Vectors and Matrices

1.1.1 Introductory Examples

Example 1.1.1 The Government can choose the value of some variables, under its control, for instance:
(1) — Official Discount rate;
(2) — Amount of T-Bills to be issued;
(3) — Tax rate to be applied.
We can construct the Table 1.1 collecting decisions:

Table 1.1 Decisions table

Variable	Decision
Discount rate	$4\% = 0.04$
T-Bills issued	$10 \text{ billions} = 10^8$
Tax rate	$40\% = 0.4$

Once it is clear that the first number is the discount rate, the second one is the amount issued of T-bills, the third one is the tax rate, we can breakdown the table and reduce it to:

$$\begin{bmatrix} 0.04 \\ 10^8 \\ 0.4 \end{bmatrix}$$

we will call such entity a (column) *vector of dimension 3* or, inside out, a (column) *3-vector*. "Column" refers to the fact that we have stacked the components. An equivalent representation would be:

$$\begin{bmatrix} 0.04 & 10^8 & 0.4 \end{bmatrix}$$

© Springer Nature Switzerland AG 2018
L. Peccati et al., *Maths for Social Sciences*, UNITEXT - La Matematica
per il 3 + 2 113, https://doi.org/10.1007/978-3-030-02336-2_1

and we would call it (row) *vector of dimension* 3. The *dimension* we have indicated refers simply to how many numbers we have listed.

Example 1.1.2 The election results in 4 States of 3 parties are represented in the Table 1.2 below.

Table 1.2 Election results

State\Party	Democratic	Republican	Ultraliberal
Seaside	10000	12000	400
Nice Hills	5000	4000	300
Mountain	500	500	50
Island	4000	3000	220

The same breaking down process, used in the previous example, can be applied and we get this entity:

$$\begin{bmatrix} 10000 & 12000 & 400 \\ 5000 & 4000 & 300 \\ 500 & 500 & 50 \\ 4000 & 3000 & 220 \end{bmatrix}$$

We will call it *matrix of type* (4, 3), as 4 is the number of rows and 3 is the number of columns. An intriguing aspect is that we can meaningfully represent this matrix both as a collection of 3 column vectors:

$$\begin{bmatrix} 10000 \\ 5000 \\ 500 \\ 4000 \end{bmatrix} ; \begin{bmatrix} 12000 \\ 4000 \\ 500 \\ 3000 \end{bmatrix} ; \begin{bmatrix} 400 \\ 300 \\ 50 \\ 220 \end{bmatrix}$$

portraying the performances of the 3 parties in the 4 states or as a stack of row vectors describing the performances of the 3 parties for each of the 4 states:

$$\begin{bmatrix} 10000 & 12000 & 400 \end{bmatrix} ; \begin{bmatrix} 5000 & 4000 & 300 \end{bmatrix}$$

$$\begin{bmatrix} 500 & 500 & 50 \end{bmatrix} \quad ; \begin{bmatrix} 4000 & 3000 & 220 \end{bmatrix}$$

Both of these aspects will turn out to be relevant:

$$\begin{cases} \text{a matrix as a sequence of columns} \\ \\ \text{a matrix as a stack of rows} \end{cases}$$

We consolidate the ideas emerged in the two previous examples.

Definition 1.1.1 Let $x_1, x_2, \ldots, x_n$ be real numbers[1] ($x_s \in \mathbb{R}$ for $s = 1, 2, \ldots, n$). An array:

$$\begin{bmatrix} x_1 \\ x_2 \\ \cdots \\ x_n \end{bmatrix} \quad \text{or} \quad \begin{bmatrix} x_1 & x_2 & \cdots & x_n \end{bmatrix}$$

is called *vector of dimension n* or, for short, an *n-vector*. Another way to denote a vector $\mathbf{x} \in \mathbb{R}^n$ is:

$$\mathbf{x} = [x_r] \text{ with } r = 1, 2, \ldots, n$$

As, in several cases, the variation range of r is defined, its specification can be omitted.

The first one is a *column* vector, while the second a *row* vector. By default vectors are column vectors. The real numbers $x_1, x_2, \ldots, x_n$ are called *components* of the vector. The set of all (real) vectors is denoted with $\mathbb{R}^n$.

A good idea[2] consists in using a single (bold) letter to indicate a vector, e.g.[3]:

$$\mathbf{x} = \begin{bmatrix} x_1 \\ x_2 \\ \cdots \\ x_n \end{bmatrix}$$

Indices of x denote components (x_3 is the third component of $\mathbf{x}$). Sometimes we will index vectors with symbolic exponents: $\mathbf{x}^1$ and $\mathbf{x}^2$ will be the symbols for "the vector # 1" and "the vector # 2". The transformation of a vector from column to row or the reverse is called *transposition* and is denoted by "T". For instance, if:

$$\mathbf{x} = \begin{bmatrix} 1 & -2 & 3 & -4 \end{bmatrix}$$

we can write:

$$\mathbf{x}^{\mathrm{T}} = \begin{bmatrix} 1 \\ -2 \\ 3 \\ -4 \end{bmatrix}$$

---

[1] The set of real numbers (correspoding to the points of a straight line) is usually denoted with $\mathbb{R}$. $x \in \mathbb{R}$ means that x is a real number.

[2] We will use systematically this convention.

[3] We are familiar with the = sign between numbers:

$$2 + 3 = 5$$

hopefully recognized by each reader. We will start to use the equal sign between vectors and matrices. In the case of vectors, = means that all the corresponding components are equal.

The special vector with null components:

$$x_1 = x_2 = \cdots = x_n = 0 \text{ or } \begin{bmatrix} 0 \\ 0 \\ \vdots \\ 0 \end{bmatrix} \text{ or } \begin{bmatrix} 0 & 0 & \cdots & 0 \end{bmatrix}$$

is called a *null vector* and is denoted by $\mathbf{o}_n$. Since generally, the value of n is obvious in the context, we will denote it simply by $\mathbf{o}$. The name is determined by the fact that the letter "o" is somehow similar to the number 0 and the null vector plays similar roles to those of the number 0.

1.2 Operations on Vectors

1.2.1 Vector Representation in $\mathbb{R}^2$ and in $\mathbb{R}^3$

2-vectors can be represented on a Cartesian plane as points or arrows. The vector $\begin{bmatrix} a \\ b \end{bmatrix}$ corresponds to the point (a, b) with abscissa a and ordinate b in a Cartesian plane or with an arrow starting from the origin and ending at (a, b) as Fig. 1.1 shows.

Fig. 1.1 Vectors in the plane

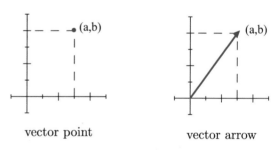

vector point   vector arrow

We can try to extend these representations in a three-dimensional space. For instance, a vector in $\mathbb{R}^3$ could be seen as a segment connecting the origin with any point (a, b, c) as illustrated in Fig. 1.2.

Fig. 1.2 Vector 3-D

But we do not insist too much.

1.2.2 Vector Addition

Example 1.2.1 The fiscal system of some State collects taxes from three categories:

$$\begin{cases} \text{Private people} \\ \text{Companies} \\ \text{Institutions} \end{cases}$$

The amounts collected (in billions) in the first year of observation (2013) are the components of this 3-vector:

$$\mathbf{x} = \begin{bmatrix} 4 \\ 5 \\ 1 \end{bmatrix}$$

The amounts collected in the second year of observation (2014) are:

$$\mathbf{y} = \begin{bmatrix} 4.5 \\ 5.2 \\ 1.6 \end{bmatrix}$$

The vector of the total amounts collected over two years is:

$$\mathbf{z} = \begin{bmatrix} 4 + 4.5 \\ 5 + 5.2 \\ 1 + 1.6 \end{bmatrix}$$

We can write the following vectors for the total fiscal revenues:

$$\begin{bmatrix} 4 + 4.5 \\ 5 + 5.2 \\ 1 + 1.6 \end{bmatrix} = \begin{bmatrix} 4 \\ 5 \\ 1 \end{bmatrix} + \begin{bmatrix} 4.5 \\ 5.2 \\ 1.6 \end{bmatrix} = \begin{bmatrix} 8.5 \\ 10.2 \\ 2.6 \end{bmatrix}$$

or as:

$$\mathbf{z} = \mathbf{x} + \mathbf{y}$$

the mechanism works because the numbers of the components of the two addenda are equal.

Let us formalize this:

Definition 1.2.1 Given two vectors $\mathbf{x} = [x_r]$, $\mathbf{y} = [y_r] \in \mathbb{R}^n$, we call the *sum*[4] of the two vectors the n-vector $\mathbf{z} = [z_r]$ and we write:

---

[4]For many people "sum" and "addition" are synonimous. In Mathematics the operation which combines two objects (in this case: vectors) is called "addition", while "sum" is the name reserved for the result. If you add eggs, salt and cheese you make an "addition", the omelette (hopefully) you get is the "sum".

$$\mathbf{z} = \mathbf{x} + \mathbf{y}$$

if:

$$z_r = x_r + y_r \text{ for every } r = 1, 2, \ldots, n$$

It is easy to prove that this operation has of five (rather exciting) properties:

- (**Closure of $\mathbb{R}^n$ w.r.t. addition**): the sum of two vectors in $\mathbb{R}^n$ is in $\mathbb{R}^n$.
- (**Commutativity**): the order of the addenda does not matter:

$$\mathbf{x} + \mathbf{y} = \mathbf{y} + \mathbf{x}$$

- (**Associativity**): given three vectors $\mathbf{x}, \mathbf{y}, \mathbf{v} \in \mathbb{R}^n$ it is

$$(\mathbf{x} + \mathbf{y}) + \mathbf{v} = \mathbf{x} + (\mathbf{y} + \mathbf{v})$$

- (**Existence of the neutral element**): there exists a unique element $\mathbf{e}$ in $\mathbb{R}^n$, such that $\mathbf{x} + \mathbf{e} = \mathbf{x}$ for every $\mathbf{x}$. It's the null vector

$$\mathbf{0} = \begin{bmatrix} 0 \\ 0 \\ \ldots \\ 0 \end{bmatrix}$$

  therefore $\mathbf{e} = \mathbf{0}$.
- (**Existence of the opposite**): for every $\mathbf{x}$ in $\mathbb{R}^n$, there exists a unique $\mathbf{y} \in \mathbb{R}^n$, such that $\mathbf{x} + \mathbf{y} = \mathbf{0}$ (the neutral element). It is:

$$\begin{bmatrix} -x_1 \\ -x_2 \\ \ldots \\ -x_n \end{bmatrix}$$

  reasonably denoted as $-\mathbf{x}$.

An interesting geometric representation of the sum of two vectors will turn out to be useful later.

Let $\mathbf{x} = \begin{bmatrix} 1 \\ 3 \end{bmatrix}$ and $\mathbf{y} = \begin{bmatrix} 2 \\ 2 \end{bmatrix}$. We can represent $\mathbf{x} + \mathbf{y} = \mathbf{y} + \mathbf{x}$ on a Cartesian plane getting the Fig. 1.3: the symmetry of the parallelogram is implied by commutativity.

Fig. 1.3 Parallelogram sum

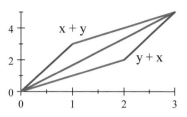

1.2.3 Scalar Multiplication

As usual, we start, with an example.

Example 1.2.2 The Government of a European country has decided to spend in 2015 for Defense, Education, Public Health and Salaries, the amounts (in billions of €) collected in the following 4-vector:

$$\mathbf{x} = \begin{bmatrix} 25 \\ 10 \\ 100 \\ 150 \end{bmatrix}$$

On June 2, 2015 a member of the US Senate would like to have this info in US dollars. This can be readily calculated using the exchange rate \$/€. The prevailing exchange rate on that date is[5]: 1.111155. The conversion of the € investment vector into the corresponding US \$ vector $\mathbf{y}$ is simple:

$$\mathbf{y} = \begin{bmatrix} 25 \times 1.111155 \\ 10 \times 1.111155 \\ 100 \times 1.111155 \\ 150 \times 1.111155 \end{bmatrix} = \begin{bmatrix} 27.779 \\ 11.112 \\ 111.12 \\ 166.67 \end{bmatrix}$$

the idea of multiplying the vector $\mathbf{x}$ by the scalar[6] 1.111155 appears quite natural. All of us would write spontaneously:

$$\mathbf{y} = 1.111155\mathbf{x}$$

Now let us formalize this idea:

Definition 1.2.2 Given a vector $\mathbf{x} \in \mathbb{R}^n$ and a scalar $\alpha \in \mathbb{R}$, we call[7] *(scalar) product of α and $\mathbf{x}$ the n-vector:*

---

[5]Concretely, this means that, in order to purchase 1€, you have to pay 1.111155 US\$.

[6]In Linear Algebra it turns out to be frequent the combined use of (poor) numbers with (noble) arrays like vectors and matrices. In this context it is customary to stress the (poor) nature of numbers saying that they are *scalar* quantities.

[7]As before: "Scalar multiplication" is the operation and "product" is the result.

$$\alpha\mathbf{x} = \mathbf{x}\alpha = \begin{bmatrix} \alpha x_1 \\ \alpha x_2 \\ \vdots \\ \alpha x_n \end{bmatrix}$$

Another way to give the same definition would be: if $\mathbf{x} = [x_r]$ with $r = 1, 2, \ldots, n$ and α is a real number, then $\alpha\mathbf{x} = [\alpha x_r]$. Sometimes we will say that $\alpha\mathbf{x}$ is the *multiple* of $\mathbf{x}$ *according to (the coefficient)* α.

This new operation, which combines scalars and vectors, has 5 notable properties:

- (**Closure of** $\mathbb{R}^n$ **w.r.t. scalar multiplication**): the product of an n-vector with a scalar is in $\mathbb{R}^n$.
- (**Distributivity w.r.t. scalar addition**): $(\alpha + \beta)\mathbf{x} = \alpha\mathbf{x} + \beta\mathbf{x}$, with $\alpha, \beta \in \mathbb{R}$ and $\mathbf{x} \in \mathbb{R}^n$.
- (**Distributivity w.r.t. vector addition**): $\alpha(\mathbf{x} + \mathbf{y}) = \alpha\mathbf{x} + \alpha\mathbf{y}$, with $\alpha \in \mathbb{R}$ and $\mathbf{x}, \mathbf{y} \in \mathbb{R}^n$.
- (**Transgender**[8] **associativity**): given two scalars $\alpha, \beta \in \mathbb{R}$ and $\mathbf{x} \in \mathbb{R}^n$ then $\alpha(\beta\mathbf{x}) = (\alpha\beta)\mathbf{x}$.
- (**The neutral element for scalars is still in place**): The neutral element for the scalar multiplication (the number 1) is also neutral w.r.t. this new operation:

$$1 \cdot \mathbf{x} = \mathbf{x} \text{ for every } \mathbf{x} \in \mathbb{R}^n$$

We suggest our readers to control the validity of these properties with appropriate numerical examples.

The properties are exciting, checking them is less exciting, but very useful.

1.2.4   Linear Combination of Vectors

If we were asked, what is the most important notion in Linear Algebra, we would have no doubt: it is the linear combination of vectors.

Example 1.2.3 (See Example 1.2.2, on p. 7). Three European countries, not necessarily belonging to the € area, are under scrutiny by some US Senate member. The public expenditure vectors in the 3 countries are:

$$\mathbf{x} = \begin{bmatrix} 25 \\ 10 \\ 100 \\ 150 \end{bmatrix} ; \quad \mathbf{y} = \begin{bmatrix} 20 \\ 15 \\ 90 \\ 140 \end{bmatrix} ; \quad \mathbf{z} = \begin{bmatrix} 200 \\ 150 \\ 1100 \\ 1000 \end{bmatrix}$$

---

[8]"Transgender" because, while on the lhs we have two scalar multiplications (University level), on the rhs we have two different multiplications: the first one $(\alpha\beta)$ which is a primary school stuff, while the second one is a (University level) multiplication.

but they are denominated in the corresponding country's currency. The exchange rate of the three countries w.r.t. U\$ are, respectively, $1.11, 1.13, 0.6$. If our Senate member wants to evaluate the investments made in the four sectors by the three countries in U\$, he has to compute:

$$1.11\mathbf{x} + 1.13\mathbf{y} + 0.6\mathbf{z} =$$

$$= 1.11 \begin{bmatrix} 25 \\ 10 \\ 100 \\ 150 \end{bmatrix} + 1.13 \begin{bmatrix} 20 \\ 15 \\ 90 \\ 140 \end{bmatrix} + 0.6 \begin{bmatrix} 200 \\ 150 \\ 1100 \\ 1000 \end{bmatrix} =$$

$$= \begin{bmatrix} 170.35 \\ 118.05 \\ 872.70 \\ 924.70 \end{bmatrix}$$

This vector is obtained by combining the three ones with "weights" the corresponding exchange rates.

Formalizing:

Definition 1.2.3 Given k vectors $\mathbf{x}^1, \mathbf{x}^2, \ldots, \mathbf{x}^k \in \mathbb{R}^n$, we call a *linear combination* of these vectors *with coefficients (weights)* $\alpha_1, \alpha_2, \ldots, \alpha_k \in \mathbb{R}$ the vector:

$$\mathbf{x} = \alpha_1 \mathbf{x}^1 + \alpha_2 \mathbf{x}^2 + \cdots + \alpha_k \mathbf{x}^k$$

An interesting interpretation of this notion is contained in the following

Example 1.2.4 Two countries, with the same currency, have expenditure vectors:

$$\mathbf{x} = \begin{bmatrix} 25 \\ 10 \\ 100 \\ 150 \end{bmatrix}; \qquad \mathbf{y} = \begin{bmatrix} 20 \\ 15 \\ 90 \\ 140 \end{bmatrix}$$

we would like to calculate the expenditure *per capita* over the two countries. In millions the two population sizes are 1.3 and 1.5. The item *per capita* expenditure for the two countries together is the following linear combination of the two vectors, with weights the reciprocals of the population sizes:

$$\frac{1}{1.3} \begin{bmatrix} 25 \\ 10 \\ 100 \\ 150 \end{bmatrix} + \frac{1}{1.5} \begin{bmatrix} 20 \\ 15 \\ 90 \\ 140 \end{bmatrix} = \begin{bmatrix} 32.564 \\ 17.692 \\ 136.920 \\ 208.720 \end{bmatrix}$$

1.2.5   Linear Dependence/Independence of Vectors

You can combine two vectors in $\mathbb{R}^2$ and you would like to generate any of the vectors in $\mathbb{R}^2$. Why?

Example 1.2.5 A vector in $\mathbb{R}^2$ could represent a contract some Government could stipulate in handling the public debt. Think of being in a simplified world where only two dates are relevant: today (0) and tomorrow (1). If you can borrow (or lend) 100 today and you will repay (or collect) 120 tomorrow, the contract is well described by the vector:

$$\mathbf{u} = \begin{bmatrix} -100 \\ 120 \end{bmatrix}$$

You can activate this contract at various levels α. If $\alpha > 0$ you lend 100α € at 0 and you will be repaid with 120α € at 1. If $\alpha < 0$ you borrow $|100\alpha|$ € at 0 and you will repay $|120\alpha|$ € at 1. The activation at the level $\alpha = 0$ is simply non-activation.

Now imagine that another contract is available:

$$\mathbf{v} = \begin{bmatrix} -100 \\ 115 \end{bmatrix}$$

If you combine the two contracts at appropriate levels α, β, you may be able to construct a contract portfolio with very interesting characteristics: for instance you never pay and you only collect. Call $\mathbf{w}$ the portfolio paying you 1000 at each of the two maturities[9]:

$$\mathbf{w} = \begin{bmatrix} 1000 \\ 1000 \end{bmatrix}$$

Let us look at the magic values α^*, β^* of α, β, able to generate $\mathbf{w}$. The equation:

$$\alpha\mathbf{u} + \beta\mathbf{v} = \mathbf{w}$$

is equivalent to:

$$\begin{bmatrix} -100\,(\alpha + \beta) \\ 120\alpha + 115\beta \end{bmatrix} = \begin{bmatrix} 1000 \\ 1000 \end{bmatrix} \quad \text{or} \quad \begin{cases} -100\alpha - 100\beta = 1000 \\ 120\alpha + 115\beta = 1000 \end{cases}$$

hence, with elementary computations, we get: $\alpha^* = 430$, $\beta^* = -440$.

Let us now check for correctness[10]:

---

[9]Indeed very nice!

[10]In order to make the reader flexible with some popular notations, we will use indifferently both "·" and "×" for multiplication. We will adhere to the general convention that the sequence of two factors — say — : ab means $a \cdot b$ or $a \times b$. This convention will be respected also when we will generalize multiplication.

$$430 \times \begin{bmatrix} -100 \\ 120 \end{bmatrix} - 440 \times \begin{bmatrix} -100 \\ 115 \end{bmatrix} = \begin{bmatrix} -43\,000 \\ 51\,600 \end{bmatrix} - \begin{bmatrix} -44\,000 \\ 50\,600 \end{bmatrix} = \begin{bmatrix} 1000 \\ 1000 \end{bmatrix}$$

Remark 1.2.1 The preceding example could appear surprising for its political implications: constructing appropriately a portfolio one has the possibility to create the well known Land of Cockaigne, which of course is impossible in reality. Therefore a natural question should be: "Why did we succeed in constructing the vector $\begin{bmatrix} 1000 \\ 1000 \end{bmatrix}$?". The answer is geometrically evident, if we look at the two vectors **u** and **v** in the Cartesian plane or in $\mathbb{R}^2$. The Fig. 1.4 shows that, they are not aligned.

Fig. 1.4 Not aligned vectors

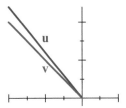

From a financial point of view, the miracle happens because the Government borrows money at 15% through the contract **v** and invests such money at 20%. As money in this market has two different prices, it is possible to make what is technically called an *arbitrage*[11]. Assume, now that the available contracts are:

$$\mathbf{u} = \begin{bmatrix} -100 \\ 120 \end{bmatrix} \text{ and } \mathbf{z} = \begin{bmatrix} 50 \\ -60 \end{bmatrix}$$

These two vectors are aligned as the Fig. 1.5 shows.

Fig. 1.5 Aligned vectors

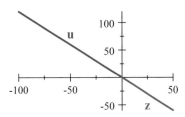

We show that with them it is impossible to construct the Land of Cockaigne. We should have:

---

[11]For Social Sciences students, it seems important to stress that an arbitrage is nothing but the opportunity to exploit market imperfections. Making arbitrages, i.e.; exploiting incoherent market positions is non-necessarily evil, but simply an opportunity.

$$\alpha \begin{bmatrix} -100 \\ 120 \end{bmatrix} + \beta \begin{bmatrix} 50 \\ -60 \end{bmatrix} = \begin{bmatrix} 1000 \\ 1000 \end{bmatrix}$$

or:

$$\begin{cases} -100\alpha + 50\beta = 1000 \\ 120\alpha - 60\beta = 1000 \end{cases}$$

Solving the first equation for α we get:

$$\alpha = \frac{1000 - 50\beta}{-100} = \frac{\beta - 20}{2}$$

Substituting in the second equation, we get:

$$120\frac{\beta - 20}{2} - 60\beta = 1000 \Leftrightarrow -1200 = 1000$$

which is impossible. This fact is financially intuitive as in both contracts the price of money is always 20%. It is interesting to show that in this case any of the two vectors can be seen as a multiple of the other:

$$\mathbf{z} = -2\mathbf{u} \text{ or, more explicitly: } \begin{bmatrix} -100 \\ 120 \end{bmatrix} = -2\begin{bmatrix} 50 \\ -60 \end{bmatrix}$$

or

$$\mathbf{u} = -\frac{1}{2}\mathbf{z} \text{ or, more explicitly: } \begin{bmatrix} 50 \\ -60 \end{bmatrix} = -\frac{1}{2}\begin{bmatrix} -100 \\ 120 \end{bmatrix}$$

The situation "aligned vectors" versus. "non-aligned" vectors can be generalized through new notions we are going to introduce:

Definition 1.2.4 Given k vectors $\mathbf{x}^1, \mathbf{x}^2, \ldots, \mathbf{x}^k \in \mathbb{R}^n$, we say that they are *linearly independent* if none of them can be represented as a linear combination of the others.[12] When these vectors are not linearly independent, we say that they are *linearly dependent*.

An important result follows:

Proposition 1.2.1 *Given n vectors* $\mathbf{x}^1, \mathbf{x}^2, \ldots, \mathbf{x}^n \in \mathbb{R}^n$*, which are linearly independent, it is possible to represent uniquely any vector* $\mathbf{v} \in \mathbb{R}^n$ *as a linear combination of the vectors* $\mathbf{x}^1, \mathbf{x}^2, \ldots, \mathbf{x}^n$:

$$\mathbf{v} = \alpha_1\mathbf{x}^1 + \alpha_2\mathbf{x}^2 + \cdots + \alpha_n\mathbf{x}^n$$

The coefficients $\alpha_1, \alpha_2, \ldots, \alpha_n$, *are the "coordinates" of* $\mathbf{v}$ *w.r.t. the given n vectors* $\mathbf{x}^1, \mathbf{x}^2, \ldots, \mathbf{x}^n$, *which (later) will be named a* basis *for* $\mathbb{R}^n$.

---

[12]In the case $k = 2$, we have informally explored above, this means that $\mathbf{u}, \mathbf{v}$ are linearly independent if and only if none of them is a multiple of the other.

Example 1.2.6 In the previous Example 1.2.5 on p. 10, the two vectors:

$$\begin{bmatrix} -100 \\ 120 \end{bmatrix} \text{ and } \begin{bmatrix} -100 \\ 115 \end{bmatrix}$$

did constitute a basis for $\mathbb{R}^2$. More generally, any two non aligned vectors in $\mathbb{R}^2$ do constitute a basis for it.

Consider these n vectors in $\mathbb{R}^n$:

$$\mathbf{e}^1 = \begin{bmatrix} 1 \\ 0 \\ 0 \\ \cdots \\ 0 \end{bmatrix} ; \mathbf{e}^2 = \begin{bmatrix} 0 \\ 1 \\ 0 \\ \cdots \\ 0 \end{bmatrix} ; \mathbf{e}^3 = \begin{bmatrix} 0 \\ 0 \\ 1 \\ \cdots \\ 0 \end{bmatrix} ; \ldots, \mathbf{e}^n = \begin{bmatrix} 0 \\ 0 \\ 0 \\ \cdots \\ 1 \end{bmatrix} \tag{1.2.1}$$

They constitute a basis for $\mathbb{R}^n$ (usually called *fundamental basis*):

$$\mathbf{x} = \begin{bmatrix} x_1 \\ x_2 \\ x_3 \\ \cdots \\ x_n \end{bmatrix} = x_1\mathbf{e}^1 + x_2\mathbf{e}^2 + x_3\mathbf{e}^3 + \cdots + x_n\mathbf{e}^n$$

We must understand that this is not the only way to generate all the n-vectors via linear combination. Let us think of $n = 2$, to make things easy. According to the formula above we have:

$$\begin{bmatrix} x_1 \\ x_2 \end{bmatrix} = x_1 \begin{bmatrix} 1 \\ 0 \end{bmatrix} + x_2 \begin{bmatrix} 0 \\ 1 \end{bmatrix}$$

The components of $\begin{bmatrix} x_1 \\ x_2 \end{bmatrix}$ are the coordinates of such vector with respect to the basis $(\mathbf{e}^1, \mathbf{e}^2)$, e.g.:

$$\begin{bmatrix} 3 \\ 4 \end{bmatrix} = 3 \begin{bmatrix} 1 \\ 0 \end{bmatrix} + 4 \begin{bmatrix} 0 \\ 1 \end{bmatrix}$$

The Fig. 1.6 illustrates it geometrically.

Fig. 1.6 Fundamental basis

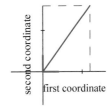

second coordinate

first coordinate

Take now any two non-aligned 2-vectors, e.g.:

$$\mathbf{u}^1 = \begin{bmatrix} 1 \\ 1 \end{bmatrix} \text{ and } \mathbf{u}^2 = \begin{bmatrix} -1 \\ 2 \end{bmatrix} \tag{1.2.2}$$

Let us try to represent $\begin{bmatrix} 3 \\ 4 \end{bmatrix}$ as a linear combination of $\mathbf{u}^1, \mathbf{u}^2$. Call α, β the two coefficients to be found, such that:

$$\alpha \begin{bmatrix} 1 \\ 1 \end{bmatrix} + \beta \begin{bmatrix} -1 \\ 2 \end{bmatrix} = \begin{bmatrix} 3 \\ 4 \end{bmatrix}$$

They must satisfy the two equations:

$$\begin{cases} \alpha - \beta = 3 \\ \alpha + 2\beta = 4 \end{cases} \tag{1.2.3}$$

Elementary computations allow us to find: $\alpha = 10/3$ and $\beta = 1/3$. Therefore, the coordinates of $\begin{bmatrix} 3 \\ 4 \end{bmatrix}$ w.r.t. the basis (1.2.2) are $\left(\dfrac{10}{3}, \dfrac{1}{3} \right)$. This simply means that:

$$\begin{bmatrix} 3 \\ 4 \end{bmatrix} = \frac{10}{3} \begin{bmatrix} 1 \\ 1 \end{bmatrix} + \frac{1}{3} \begin{bmatrix} -1 \\ 2 \end{bmatrix}$$

The geometrical interpretation could be useful. The choice of the basis (1.2.3) is nothing but the substitution of our beloved Cartesian axes of abscissae and ordinates, which respectively collect all the multiples of $\mathbf{e}^1$ and of $\mathbf{e}^2$, with a new couple of axes respectively collecting all the multiples of $\mathbf{u}^1$ and of $\mathbf{u}^2$. Look at the diagram 1.7 where $(3, 4)$ stands for $\begin{bmatrix} 3 \\ 4 \end{bmatrix}$.

Fig. 1.7 Generic basis

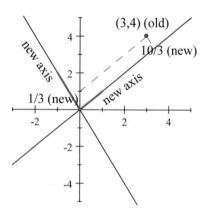

What we have done has been only to replace the old (x, y) axes with a new couple of axes, determined by $(\mathbf{u}^1, \mathbf{u}^2)$. These new axes are not orthogonal and the new coordinates of the vector are determined by the intersection points of the new axes with parallel lines to the new axes passing through the point of interest. The new coordinates:

$$\left(\frac{10}{3}, \frac{1}{3}\right)$$

are indicated in the diagram.

It should be clear that no matter how we choose new basis vectors $\mathbf{u}^1, \mathbf{u}^2$, they will allow us to span the whole plane if and only if they do not belong to the same straight line.

Similar considerations could be made for 3-vectors. The Cartesian axes, "naturally" generated by:

$$\mathbf{e}^1 = \begin{bmatrix} 1 \\ 0 \\ 0 \end{bmatrix} ; \mathbf{e}^2 = \begin{bmatrix} 0 \\ 1 \\ 0 \end{bmatrix} ; \mathbf{e}^3 = \begin{bmatrix} 0 \\ 0 \\ 1 \end{bmatrix}$$

could be replaced by any reference system generated by 3 linearly independent vectors $\mathbf{u}^1, \mathbf{u}^2, \mathbf{u}^3$. These vectors are linearly independent iff[13] they do not belong to the same plane, or, loosely speaking, if they are scrawly.

This important remark suggests that we can generalize these ideas.

Definition 1.2.5 A set of n-vectors $X = \{\mathbf{x}^1, \mathbf{x}^2, \ldots, \mathbf{x}^k\}$ is said to be a *basis* for $\mathbb{R}^n$ if:
(1) — every n-vector $\mathbf{x}$ is a linear combination of the vectors of X,
(2) — such linear combination is unique in the sense that it is impossible to represent any given vector with different coefficients.

These unique coefficients are the *coordinates of* $\mathbf{x}$ w.r.t. X.

Proposition 1.2.2 *Every basis of $\mathbb{R}^n$ is a set X of n linearly independent vectors. Every set X of linearly independent vectors of $\mathbb{R}^n$ constitutes a basis for $\mathbb{R}^n$.*

Example 1.2.7 Think of $\mathbb{R}^2$, take these two vectors, obviously linearly independent:

$$\mathbf{u} = \begin{bmatrix} 1 \\ 1 \end{bmatrix} \text{ and } \mathbf{v} = \begin{bmatrix} 1 \\ 0 \end{bmatrix}$$

It is easy to show that every vector $\mathbf{x} \in \mathbb{R}^2$ can be uniquely represented as a linear combination of $\mathbf{u}, \mathbf{v}$ with coefficients α, β. From:

$$\alpha \mathbf{u} + \beta \mathbf{v} = \mathbf{x}$$

---

[13]"iff" is beyond English. Mathematicians contract into "iff" the sequence of words "if and only if". We will use it sometimes.

we get the pair of equations (in the unknowns α, β):

$$\begin{cases} \alpha + \beta = x_1 \\ \alpha = x_2 \end{cases}$$

substituting α from the second into the first, we get univocally:

$$\begin{cases} \beta = x_1 - x_2 \\ \alpha = x_2 \end{cases}$$

1.3   Matrices

We start with:

Example 1.3.1 Take two 3−column vectors:

$$\begin{bmatrix} 1 \\ 2 \\ 3 \end{bmatrix} \text{ and } \begin{bmatrix} -1 \\ 0 \\ 4 \end{bmatrix}$$

and put them close to each other:

$$\begin{bmatrix} 1 & -1 \\ 2 & 0 \\ 3 & 4 \end{bmatrix}$$

As an alternative, take three row vectors $\begin{bmatrix} 1 & -1 \end{bmatrix}, \begin{bmatrix} 2 & 0 \end{bmatrix}, \begin{bmatrix} 3 & 4 \end{bmatrix}$ and stack them. What you get is the same:

$$\begin{bmatrix} 1 & -1 \\ 2 & 0 \\ 3 & 4 \end{bmatrix}$$

a *matrix*, with 3 rows and 2 columns.

1.3.1   Types of Matrices

The general way we will use to describe a matrix is rather easy. We organize the information:

- **Name of a matrix**: we will always use a capital letter, e.g., A:

$$A = \begin{bmatrix} 1 & -2 & 0.4 & -1 \\ 0 & 0 & 1/2 & 4 \\ 1 & 1 & 3 & -3 \end{bmatrix}$$

- **Structure of a matrix**: if it has m rows and n columns we will say that it is of type (m, n). The matrix above is of the type $(3, 4)$. This explains the symbol we will use:

$$\underset{(m,n)}{A}$$

for a matrix of type (m, n). A general way to indicate the *elements*, or *entries*, of a matrix uses the following notation:

$$\underset{(m,n)}{A} = \begin{bmatrix} a_{11} & a_{12} & \cdots & a_{1n} \\ a_{21} & a_{22} & \cdots & a_{2n} \\ \cdots & \cdots & \cdots & \cdots \\ a_{m1} & a_{m2} & \cdots & a_{mn} \end{bmatrix}$$

For each entry, the first index indicates the row and the second one the column.[14] If the number of rows and that of columns are equal (i.e.; if $m = n$) the matrix is called *square*, otherwise *rectangular*. The matrix A is rectangular while this:

$$\begin{bmatrix} 1 & -2 & -1 \\ 0 & 0 & 4 \\ 1 & 1 & -3 \end{bmatrix}$$

is square. For square matrices, the common number of rows and columns ($m = n$) is called the *order* of the matrix. For instance the matrix here above is of order 3, because it has 3 rows and 3 columns.

- **Special matrices**: If $m = 1$ the matrix can also be seen as a row vector, if $n = 1$ as a column vector, if $m = n = 1$ as a number or, better, as a *scalar*.

If all the entries of a matrix are 0, the matrix is called a *null* matrix (of type (m, n)) and we will denote it with O:

$$\underset{(m,n)}{O} = \begin{bmatrix} 0 & 0 & \cdots & 0 \\ 0 & 0 & \cdots & 0 \\ \cdots & \cdots & \cdots & \cdots \\ 0 & 0 & \cdots & 0 \end{bmatrix}$$

If we construct a square matrix using the vectors of the fundamental basis (see (1.2.1) on p. 13) we get a square matrix of order n:

---

[14]This double indexation reflects the ubiquitous nature of a matrix that can be seen as a package of stacked row vectors or a set of column vectors, placed side by side. Think of the starting Example 1.3.1. Had we to deal with the element a_{123} of some matrix A, we would like to understand whether it is the third entry in the row # 12 or the 23rd in the row # 1. In the case that this ambiguity turns out to be relevant, the two (boring, but more refined) notation items are advisable: $a_{12,3}$ or $a_{1,23}$.

$$\begin{bmatrix} 1 & 0 & \cdots & 0 \\ 0 & 1 & \cdots & 0 \\ \cdots & \cdots & \cdots & \cdots \\ 0 & 0 & \cdots & 1 \end{bmatrix}$$

which is called *identity* matrix (of order n) and usually[15] denoted with I_n or simply I, when n is implied by the context. We will soon see that these two types of matrices, O and I, will play a role similar to the one of 0 and 1 in ordinary arithmetic: this explains the choice of symbols.

Therefore a matrix $\underset{(m,n)}{A}$ can "sliced" according to rows or columns.

Call $\mathbf{a}^1, \mathbf{a}^2, \ldots, \mathbf{a}^n$ its columns, which are column m-vectors:

$$A = \begin{bmatrix} \mathbf{a}^1 | \mathbf{a}^2 | \ldots | \mathbf{a}^n \end{bmatrix}$$

If you call $\alpha^1, \alpha^2, \ldots, \alpha^m$ its rows, which are row n-vectors:

$$A = \begin{bmatrix} \alpha^1 \\ \alpha^2 \\ \cdots \\ \alpha^m \end{bmatrix}$$

For a numerical example see above at p. 16.

1.3.2 Operations on Matrices: An Inventory

Because of their evident relevance for Social Sciences, we are going to introduce some operations involving matrices:

- transposition;
- addition;
- multiplication with a scalar;
- multiplication of two conformable matrices;
- inversion of a matrix.

We will devote a subsection to each of them.

---

[15]This convention does not hold for Germans, they do prefer E, the initial letter of "*Einheit*" = unity.

1.3.3 Transposition

A matrix is nothing but the essence of a double entry table. For instance the double State\Party Table 1.2 at p. 2 contains the number of votes earned by three parties in three States.

Obviously the same info could be obtained via the Party\State Table 1.3:

Table 1.3 Party/State election

Party\State	Seaside	Nice hills	Mountains	Islands
Democratic	10000	5000	500	4000
Republican	12000	4000	500	3000
Ultraliberal	400	300	3000	220

What has happened is a simple interchange between rows and columns. That's transposition!

Definition 1.3.1 Given a matrix:

$$\underset{(m,n)}{A} = \begin{bmatrix} a_{11} & a_{12} & \cdots & a_{1n} \\ a_{21} & a_{22} & \cdots & a_{2n} \\ \cdots & \cdots & \cdots & \cdots \\ a_{m1} & a_{m2} & \cdots & a_{mn} \end{bmatrix}$$

we call the *transpose* of $\underset{(m,n)}{A}$ the matrix:

$$\underset{(n,m)}{B} = A^{\mathrm{T}} = \begin{bmatrix} a_{11} & a_{21} & \cdots & a_{m1} \\ a_{12} & a_{22} & \cdots & a_{m2} \\ \cdots & \cdots & \cdots & \cdots \\ a_{1n} & a_{2n} & \cdots & a_{mn} \end{bmatrix}$$

in other words, if $r = 1, \ldots, m$ and $s = 1, \ldots, n$:

$$b_{sr} = a_{rs}$$

Example 1.3.2 Let:

$$A = \begin{bmatrix} 1 & 2 & -3 \\ -2 & -1 & 0 \end{bmatrix}$$

then:

$$A^{\mathrm{T}} = \begin{bmatrix} 1 & -2 \\ 2 & -1 \\ -3 & 0 \end{bmatrix}$$

It is absolutely obvious that the transpose of the transpose of a matrix A is nothing but A itself. In formulas:

$$\left(A^{\mathrm{T}}\right)^{\mathrm{T}} = A$$

and that the transpose of a square matrix of order n is again square of order n.

1.3.4 Matrix Addition

If we had to assign the "LESS" prize[16] to a subsection, we would give it to this subsection and possibly to the next one. The motivation for the prize is that addition of matrices (this subsection) and multiplication of a matrix by a scalar (next subsection) simply mimic the analogous operations introduced above for vectors. This explains why we do not motivate obvious definitions.

Definition 1.3.2 Given two matrices: $\underset{(m,n)}{A}$ and $\underset{(m,n)}{B}$, of the same type, we call the *sum* of them the matrix $\underset{(m,n)}{C}$ (of the same type), obtained summing elements the same positions:

$$
\underset{(m,n)}{C} = \underset{(m,n)}{A} + \underset{(m,n)}{B} =
$$

$$
= \begin{bmatrix} a_{11} & a_{12} & \cdots & a_{1n} \\ a_{21} & a_{22} & \cdots & a_{2n} \\ \cdots & \cdots & \cdots & \cdots \\ a_{m1} & a_{m2} & \cdots & a_{mn} \end{bmatrix} + \begin{bmatrix} b_{11} & b_{12} & \cdots & b_{1n} \\ b_{21} & b_{22} & \cdots & b_{2n} \\ \cdots & \cdots & \cdots & \cdots \\ b_{m1} & b_{m2} & \cdots & b_{mn} \end{bmatrix} =
$$

$$
= \begin{bmatrix} a_{11}+b_{11} & a_{12}+b_{12} & \cdots & a_{1n}+b_{1n} \\ a_{21}+b_{21} & a_{22}+b_{22} & \cdots & a_{2n}+b_{2n} \\ \cdots & \cdots & \cdots & \cdots \\ a_{m1}+b_{m1} & a_{m2}+b_{m2} & \cdots & a_{mn}+b_{mn} \end{bmatrix}
$$

Example 1.3.3 Let:

$$
A = \begin{bmatrix} 1 & 2 & -3 \\ -2 & -1 & 0 \end{bmatrix}; B = \begin{bmatrix} 0 & 4 & 3 \\ -1 & -2 & 1 \end{bmatrix}
$$

we have:

$$
C = A + B = \begin{bmatrix} 1 & 2 & -3 \\ -2 & -1 & 0 \end{bmatrix} + \begin{bmatrix} 0 & 4 & 3 \\ -1 & -2 & 1 \end{bmatrix} = \begin{bmatrix} 1 & 6 & 0 \\ -3 & -3 & 1 \end{bmatrix}
$$

---

[16]"LESS" stands for "Least Exciting Sub Section".

The addition between matrices enjoys of the same five formal properties, we have seen for vectors. See at p. 6. The null vector **o**, there, is replaced here by an adequate null matrix O.

1.3.5 Multiplication of a Scalar by a Matrix

Once again nothing substantially new:

$$\alpha \underset{(m,n)}{A} = \alpha \begin{bmatrix} a_{11} & a_{12} & \cdots & a_{1n} \\ a_{21} & a_{22} & \cdots & a_{2n} \\ \cdots & \cdots & \cdots & \cdots \\ a_{m1} & a_{m2} & \cdots & a_{mn} \end{bmatrix} = \begin{bmatrix} \alpha a_{11} & \alpha a_{12} & \cdots & \alpha a_{1n} \\ \alpha a_{21} & \alpha a_{22} & \cdots & \alpha a_{2n} \\ \cdots & \cdots & \cdots & \cdots \\ \alpha a_{m1} & \alpha a_{m2} & \cdots & \alpha a_{mn} \end{bmatrix}$$

and no problem in multiplying from the right a matrix by a scalar since $\alpha A = A\alpha$. In other words: if you want to multiply some matrix A with a scalar α, you have simply to multiply each element of A by α. It is nothing but the extension of the operation of multiplication of a scalar by a vector, which, now, is applied to all the vectors (columns or rows) which constitute the matrix.

Example 1.3.4 The matrix:

$$-5 \begin{bmatrix} 1 & 2 & -3 \\ -2 & -1 & 0 \end{bmatrix}$$

turns out to be:

$$\begin{bmatrix} -5 & -10 & 15 \\ 10 & 5 & 0 \end{bmatrix}$$

On p. 8 we listed 5 properties of that multiplication. Their extension to matrices is straightforward:

- (**Closure of the space of** (m, n)-**matrices**): the product of an (m, n)-matrix with a scalar is again an (m, n)-matrix.
- (**Distributivity w.r.t. scalar addition**): $(\alpha + \beta) A = \alpha A + \beta A$, with $\alpha, \beta \in \mathbb{R}$ and A an (m, n)-matrix.
- (**Distributivity w.r.t. matrix addition**): $\alpha (A + B) = \alpha A + \alpha B$, with $\alpha \in \mathbb{R}$ and A, B (m, n)-matrices.
- (**Transgender[17] associativity**): given two scalars $\alpha, \beta \in \mathbb{R}$ and $\mathbf{x} \in \mathbb{R}^n$ then $\alpha (\beta \mathbf{x}) = (\alpha \beta) \mathbf{x}$.
- (**The neutral element for scalars is still in place**): The neutral element for the scalar multiplication (the number 1) is also neutral w.r.t. this new operation:

---

[17]"Transgender" because, while on the lhs we have two scalar multiplications (University stuff), on the rhs we have two different multiplications: the first one $(\alpha\beta)$ which is a primary school stuff, while the second one is a scalar multiplication.

$$1 \cdot A = A \text{ for every } (m, n) \text{-matrix } A$$

In this case too, we suggest that the reader checks the validity of these properties with numerical examples.

1.3.6 Multiplication of Matrices

This multiplication deserves attention because;

- its definition is far from being trivial;
- it's a true novelty w.r.t. the usual Arithmetic(s) and has a large number of applications, also in the field of Social Sciences.

Before introducing it, in its full generality, we start from special cases that should suggest the flavor of what we are doing.

Example 1.3.5 You are the policy maker in a State. You have the possibility to spend money in some 4 different sectors. You have a budget which you are given for each sector and are collected in a **row vector** with 4 components (in millions €):

$$\mathbf{a} = \begin{bmatrix} 50 & 30 & 70 & 10 \end{bmatrix}$$

You have decided that you will not spend the whole budget, because you want to save money for further interventions, once you have seen the effects of the first expenditure decision. You decide which fraction of each budget amount will be spent in each sector and you collect these percentages in a **column vector**:

$$\mathbf{b} = \begin{bmatrix} 25\% \\ 35\% \\ 40\% \\ 20\% \end{bmatrix}$$

You want to compute the total amount of the expenditure you have decided upon. In the trivial case of only 1 expenditure sector, you would compute:

$$\text{budget amount } \times \% \text{ you decide to spend}$$

For instance, in the case you have only the first sector, your computation would turn out to be:

$$50 \times 25\% = 50 \times 0.25 = 12.5$$

In the case of several sectors the basic idea is to multiply the vector of budget amounts by the vectors of % expenditure:

$$\mathbf{ab} = \begin{bmatrix} 50 & 30 & 70 & 10 \end{bmatrix} \begin{bmatrix} 25\% \\ 35\% \\ 40\% \\ 20\% \end{bmatrix} =$$

$$= 50 \times 0.25 + 30 \times 0.35 + 70 \times 0.4 + 10 \times 0.2 = 53$$

Example 1.3.6 Let us now consider a slightly more complex case. Instead of having only one vector of budget amounts, you are given two possibilities. This means that instead of having only a row vector you have two. It's natural to think of this pair of vectors as a (2, 4) matrix, see p. 16 above:

$$A = \begin{bmatrix} 50 & 30 & 70 & 10 \\ 60 & 20 & 70 & 20 \end{bmatrix}$$

You want to evaluate the total expenditure for each of the budget policies, under the same expenditure strategy (vector **b**). It is natural to multiply the matrix A by the vector **b** according to this scheme of calculations:

$$Ab = \begin{bmatrix} 50 & 30 & 70 & 10 \\ 60 & 20 & 70 & 20 \end{bmatrix} \begin{bmatrix} 25\% \\ 35\% \\ 40\% \\ 20\% \end{bmatrix} =$$

$$= \begin{bmatrix} 50 \times 0.25 + 30 \times 0.35 + 70 \times 0.4 + 10 \times 0.2 \\ 60 \times 0.25 + 20 \times 0.35 + 70 \times 0.4 + 20 \times 0.2 \end{bmatrix} = \begin{bmatrix} 53 \\ 54 \end{bmatrix}$$

which provides us with a (2, 1) matrix: the number of rows is determined by the number of budget policies, while the number of columns is 1, because you are considering singularly the two expenditure policies.

Example 1.3.7 In the preceding example we have made the budget policies side more complex, allowing for 2 possibilities. We will now do the same with three expenditure policy opportunities. This means that instead of having only one expenditure vector **b**, we will consider — say — 3 different expenditure policies, which naturally generate a (4, 3) matrix of expenditure policies: each column represents an expenditure policy. For instance:

$$B = \begin{bmatrix} 25\% & 50\% & 60\% \\ 35\% & 50\% & 50\% \\ 40\% & 50\% & 40\% \\ 20\% & 50\% & 30\% \end{bmatrix}$$

The product AB will be a (2, 3) matrix (# of rows = # of budget policies; # of columns = # of expenditure policies):

$$AB = \begin{bmatrix} 50 & 30 & 70 & 10 \\ 60 & 20 & 70 & 20 \end{bmatrix} \begin{bmatrix} 0.25 & 0.50 & 0.60 \\ 0.35 & 0.50 & 0.50 \\ 0.40 & 0.50 & 0.40 \\ 0.20 & 0.50 & 0.30 \end{bmatrix} = \begin{bmatrix} 53 & 80 & 76 \\ 54 & \mathbf{85} & 80 \end{bmatrix}$$

The elements of this matrix have been obtained coupling each of the possible rows of the first matrix with each of the possible columns of the second one. From a political viewpoint we have considered all the possible combinations of two budget policies with three expenditure policies and the product matrix AB provides us with the total expenditure determined by each combination (budget, expenditure). For instance, the bold **85** which appears at the position $(2, 2)$ in the AB matrix comes from the combination of the budget policy #2 (second row of the matrix A):

$$\begin{bmatrix} 60 & 20 & 70 & 20 \end{bmatrix}$$

with the expenditure policy #2 (second column of the matrix B):

$$\begin{bmatrix} 0.50 \\ 0.50 \\ 0.50 \\ 0.50 \end{bmatrix}$$

In fact:

$$\begin{bmatrix} 60 & 20 & 70 & 20 \end{bmatrix} \begin{bmatrix} 0.50 \\ 0.50 \\ 0.50 \\ 0.50 \end{bmatrix} = [85] = 85$$

From the previous examples it should be clear that in order to be able to multiply two matrices A, B it is necessary that the number of columns of A (= # of components in each of its row vector) coincides with the number of rows of B (= # of components in each of its columns). In the examples this common number was that of the intervention sectors. We will say that A, B are *conformable* in that order if A an (m, n) matrix and B an (n, p) matrix. The product of an (m, n) matrix with an (n, p) matrix is an (m, p) matrix. In the last example, we have multiplied a $(2, 4)$ matrix by a $(4, 3)$ matrix, getting a $(2, 3)$ matrix. The (mathematical) conformability condition is obviously equivalent to the (common sense) condition that the number of sectors you have to support and the number of expense percentage you handle... must be equal.

After this (rather long) preparation, we are ready to introduce the notion of the product of two conformable matrices A, B.

Definition 1.3.3 Given two matrices $\underset{(m,n)}{A}$, $\underset{(n,p)}{B}$:

$$
A = \begin{bmatrix} a_{11} & a_{12} & \cdots & a_{1n} \\ a_{21} & a_{22} & \cdots & a_{2n} \\ \cdots & \cdots & \cdots & \cdots \\ a_{m1} & a_{m2} & \cdots & a_{mn} \end{bmatrix}, \quad B = \begin{bmatrix} b_{11} & b_{12} & \cdots & b_{1p} \\ b_{21} & b_{22} & \cdots & b_{2p} \\ \cdots & \cdots & \cdots & \cdots \\ b_{n1} & b_{n2} & \cdots & b_{np} \end{bmatrix}
$$

we call *product* of A and B in that order the (m, p) -matrix C, whose generic element $c_{r,s}$ is the product of the rth row of A with the sth column of B ($r = 1, 2, \ldots, m$, $s = 1, 2, \ldots, p$):

$$
c_{rs} = \begin{bmatrix} a_{r1} & a_{r2} & \cdots & a_{rn} \end{bmatrix} \begin{bmatrix} b_{1s} \\ b_{2s} \\ \cdots \\ b_{ns} \end{bmatrix} = a_{r1}b_{1s} + a_{r2}b_{2s} + \cdots + a_{rn}b_{ns}
$$

Remark 1.3.1 If we multiply two conformable matrices — say — a $(2, 4)$-matrix with a $(4, 3)$-matrix, what you get is a $(2, 3)$-matrix: the number of rows of the product coincides with the number of rows of the first matrix (in our Example 1.3.6 they are 2), equal to the number of budget policies. In the case there are 3 use policies what you obtain is the result of any possible (budget, expenditure) decision.

We handle the multiplication of two matrices. If we change the order of the factors, some problems can arise. For instance a $(2, 3)$-matrix can be multiplied with a $(3, 3)$-matrix, getting a $(2, 3)$-matrix as a product, but in the reverse order, we would have to multiply a $(3, 3)$-matrix with a $(2, 3)$-matrix, which is an impossible task. A $(3, 2)$-matrix can be multiplied by a $(2, 3)$-matrix, getting a $(3, 3)$-matrix. In the reversed order we would obtain a $(2, 2)$-matrix, certainly different from the previous product. If the two factors are square matrices (for instance of order 2), in both the orders we get a square matrix (of order 2). Well, in general, even in this case, the order of factors could be relevant. This simple case should be convincing. Let:

$$
A = \begin{bmatrix} 1 & 2 \\ 3 & 4 \end{bmatrix} \text{ and } B = \begin{bmatrix} 1 & 0 \\ 0 & -1 \end{bmatrix}
$$

We have[18]:

$$
AB = \begin{bmatrix} 1 & 2 \\ 3 & 4 \end{bmatrix} \begin{bmatrix} 1 & 0 \\ 0 & -1 \end{bmatrix} = \begin{bmatrix} 1 & -2 \\ 3 & -4 \end{bmatrix}
$$

---

[18]Commutativity is a piece of what we learned when we were kids:

- $a + b = b + a$;
- $a \cdot b = b \cdot a$.

When we experienced some clothing, or even before, we became rationally conscious of non-commutativity:

$$
\begin{cases} \text{underwear} * \text{gown} & \neq \quad \text{gown} * \text{underwear} * \\ \text{underwear} * \text{trousers} & \neq \quad \text{trousers} * \text{underwear} * \end{cases}
$$

and

$$BA = \begin{bmatrix} 1 & 0 \\ 0 & -1 \end{bmatrix} \begin{bmatrix} 1 & 2 \\ 3 & 4 \end{bmatrix} = \begin{bmatrix} 1 & 2 \\ -3 & -4 \end{bmatrix}$$

Therefore

$$AB \neq BA$$

It can happen that for special couples of matrices A, B we find that $AB = BA$. These matrices are said *to commute*, but in general this does not happen. Special cases of commuting matrices will be met soon, for instance, if:

$$A = \begin{bmatrix} 1 & 2 \\ 3 & 4 \end{bmatrix} \text{ and } B = \begin{bmatrix} 1 & 0 \\ 0 & 1 \end{bmatrix}$$

we have:

$$\begin{bmatrix} 1 & 2 \\ 3 & 4 \end{bmatrix} \begin{bmatrix} 1 & 0 \\ 0 & 1 \end{bmatrix} = \begin{bmatrix} 1 & 2 \\ 3 & 4 \end{bmatrix} \text{ and } \begin{bmatrix} 1 & 0 \\ 0 & 1 \end{bmatrix} \begin{bmatrix} 1 & 2 \\ 3 & 4 \end{bmatrix} = \begin{bmatrix} 1 & 2 \\ 3 & 4 \end{bmatrix}$$

therefore, for instance:

$$A = \begin{bmatrix} 1 & 2 \\ 3 & 4 \end{bmatrix} \text{ and } B = \begin{bmatrix} 1 & 0 \\ 0 & 1 \end{bmatrix} \text{ are commuting matrices}$$

The reader should perceive that the matrix $\begin{bmatrix} 1 & 0 \\ 0 & 1 \end{bmatrix}$ not only commutes with A, and this would turn out to be true for any conformable A, but that the product turns out to be A. Think of numbers: if we multiply 5 by 1 we get:

$$5 \times 1 = 5$$

and 1 is the only number for which this occurs. Mathematicians say that 1 is the neutral element w.r.t. the multiplication between numbers. Well: matrices like $\begin{bmatrix} 1 & 0 \\ 0 & 1 \end{bmatrix}$ play the same role in this new multiplication. They are square matrices of some order k which have unitary elements in the *principal diagonal* (the entries a_{rr}) and null entries elsewhere:

$$\begin{bmatrix} 1 & 0 & \cdots & 0 \\ 0 & 1 & \cdots & 0 \\ \cdots & \cdots & \cdots & \cdots \\ 0 & 0 & \cdots & 1 \end{bmatrix}$$

Such square matrices are called *identity matrices*. The symbol for them is I. If we wish to indicate their order k, we can write I_k. For instance:

$$I_4 = \begin{bmatrix} 1 & 0 & 0 & 0 \\ 0 & 1 & 0 & 0 \\ 0 & 0 & 1 & 0 \\ 0 & 0 & 0 & 1 \end{bmatrix}$$

We have seen that the multiplication of a scalar by a vector or by a matrix enjoys some properties, natural for people accustomed to the multiplication of numbers. We have listed them on p. 8 for vectors and later, on p. 21 for matrices. Concerning the multiplication of matrices, please, forget the nice fact that the multiplication with a scalar was working (almost) exactly as the multiplication between numbers.

Multiplication between matrices inherits some properties from ordinary multiplication, but not all of them.

However, what is inherited is sufficient to construct a very useful theory.

In the following proposition, we bring together what remains of our old (beloved) multiplication between numbers for the case of matrices.

Proposition 1.3.1 *Assuming conformability:*
(1) — The multiplication is associative:

$$(AB)C = A(BC)$$

which authorizes us to write unambiguously ABC.
(2) — The multiplication is distributive, both on the right and on the left:

$$A(B+C) = AB + AC \text{ and } (B+C)A = BA + CA$$

(3) — Identity matrices (see above on p. 26) act as neutral elements: if A is an (m,n)-matrix then

$$I_m A = A I_n = A$$

(4) — If a factor in a multiplication is a null matrix, then the product is a null matrix:

$$OA = O \text{ and } AO = O$$

but not vice versa.[19]
(5) — a scalar factor can be put everywhere: α is a scalar, A, B are conformable matrices. We have:

$$\alpha(AB) = (\alpha A)B = A(\alpha B)$$

---

[19]Between numbers x, y, their product xy is 0 if and only if at least one of the factors is 0. For matrices the "only" part of the statement does not hold. E.g.:

$$\begin{bmatrix} 1 & 1 \\ 1 & 1 \end{bmatrix} \begin{bmatrix} 1 & 1 \\ -1 & -1 \end{bmatrix} = \begin{bmatrix} 0 & 0 \\ 0 & 0 \end{bmatrix}$$

even if no factor is a null matrix.

so, we can unambiguously write αAB.

(6) — For the product between scalars and matrices the law of null product holds like for numbers:

$$\alpha A = O \implies \begin{cases} \bullet\ \alpha = 0 \\ \quad or \\ \bullet\ A = O \\ \quad or \\ \bullet\ both \end{cases}$$

We suggest our readers examine the special cases inserted in the next example, in order to check the correctness of the assertions above.

Example 1.3.8 The numbers of the items correspond to the ones of the assertions in the Proposition 1.3.1.

(1) — Let:

$$A = \begin{bmatrix} 1 & -2 & 3 & -4 \\ 5 & -6 & 7 & -8 \end{bmatrix}, B = \begin{bmatrix} 0 & 1 & 0 \\ 0 & 0 & 1 \\ 1 & 0 & 0 \\ 0 & -1 & 0 \end{bmatrix}, C = \begin{bmatrix} 0 & -2 \\ 1 & 1 \\ 0 & -1 \end{bmatrix}$$

We have:

$$AB = \begin{bmatrix} 1 & -2 & 3 & -4 \\ 5 & -6 & 7 & -8 \end{bmatrix} \begin{bmatrix} 0 & 1 & 0 \\ 0 & 0 & 1 \\ 1 & 0 & 0 \\ 0 & -1 & 0 \end{bmatrix} = \begin{bmatrix} 3 & 5 & -2 \\ 7 & 13 & -6 \end{bmatrix}$$

hence:

$$(AB)\,C = \begin{bmatrix} 3 & 5 & -2 \\ 7 & 13 & -6 \end{bmatrix} \begin{bmatrix} 0 & -2 \\ 1 & 1 \\ 0 & -1 \end{bmatrix} = \begin{bmatrix} 5 & 1 \\ 13 & 5 \end{bmatrix}$$

In turn:

$$BC = \begin{bmatrix} 0 & 1 & 0 \\ 0 & 0 & 1 \\ 1 & 0 & 0 \\ 0 & -1 & 0 \end{bmatrix} \begin{bmatrix} 0 & -2 \\ 1 & 1 \\ 0 & -1 \end{bmatrix} = \begin{bmatrix} 1 & 1 \\ 0 & -1 \\ 0 & -2 \\ -1 & -1 \end{bmatrix}$$

and hence:

$$A\,(BC) = \begin{bmatrix} 1 & -2 & 3 & -4 \\ 5 & -6 & 7 & -8 \end{bmatrix} \begin{bmatrix} 1 & 1 \\ 0 & -1 \\ 0 & -2 \\ -1 & -1 \end{bmatrix} = \begin{bmatrix} 5 & 1 \\ 13 & 5 \end{bmatrix}$$

Therefore:

$$ABC = \begin{bmatrix} 5 & 1 \\ 13 & 5 \end{bmatrix}$$

(2) — Consider first the product:

$$\begin{bmatrix} 5 & 1 \\ 13 & 5 \end{bmatrix} \left(\begin{bmatrix} 1 \\ 2 \end{bmatrix} + \begin{bmatrix} -1 \\ 2 \end{bmatrix} \right) = \begin{bmatrix} 4 \\ 20 \end{bmatrix}$$

The result we got is the same we could obtain as follows:

$$\begin{bmatrix} 5 & 1 \\ 13 & 5 \end{bmatrix} \begin{bmatrix} 1 \\ 2 \end{bmatrix} + \begin{bmatrix} 5 & 1 \\ 13 & 5 \end{bmatrix} \begin{bmatrix} -1 \\ 2 \end{bmatrix} = \begin{bmatrix} 4 \\ 20 \end{bmatrix}$$

We could exemplify the situation also in the case of distributivity on the left.
(3) — Let:

$$A = \begin{bmatrix} 1 & -2 & 3 & -4 \\ 5 & -6 & 7 & -8 \end{bmatrix}$$

we have:

$$I_2 A = \begin{bmatrix} 1 & 0 \\ 0 & 1 \end{bmatrix} \begin{bmatrix} 1 & -2 & 3 & -4 \\ 5 & -6 & 7 & -8 \end{bmatrix} = \begin{bmatrix} 1 & -2 & 3 & -4 \\ 5 & -6 & 7 & -8 \end{bmatrix}$$

and:

$$A I_4 = \begin{bmatrix} 1 & -2 & 3 & -4 \\ 5 & -6 & 7 & -8 \end{bmatrix} \begin{bmatrix} 1 & 0 & 0 & 0 \\ 0 & 1 & 0 & 0 \\ 0 & 0 & 1 & 0 \\ 0 & 0 & 0 & 1 \end{bmatrix} = \begin{bmatrix} 1 & -2 & 3 & -4 \\ 5 & -6 & 7 & -8 \end{bmatrix}$$

(4) — Consider this (rather) trivial case:

$$\begin{bmatrix} 1 & -2 & 3 & -4 \\ 5 & -6 & 7 & -8 \end{bmatrix} \begin{bmatrix} 0 & 0 & 0 & 0 \\ 0 & 0 & 0 & 0 \\ 0 & 0 & 0 & 0 \\ 0 & 0 & 0 & 0 \end{bmatrix} = \begin{bmatrix} 0 & 0 & 0 & 0 \\ 0 & 0 & 0 & 0 \end{bmatrix}$$

(5) — Take, for instance:

$$2AB = 2 \left(\begin{bmatrix} 1 & -2 & 3 & -4 \\ 5 & -6 & 7 & -8 \end{bmatrix} \begin{bmatrix} 0 & 1 & 0 \\ 0 & 0 & 1 \\ 1 & 0 & 0 \\ 0 & -1 & 0 \end{bmatrix} \right) =$$

$$= 2 \begin{bmatrix} 3 & 5 & -2 \\ 7 & 13 & -6 \end{bmatrix} = \begin{bmatrix} 6 & 10 & -4 \\ 14 & 26 & -12 \end{bmatrix}$$

Consider this possible alternative:

$$(2A)\,B = \left(2\begin{bmatrix} 1 & -2 & 3 & -4 \\ 5 & -6 & 7 & -8 \end{bmatrix}\right)\begin{bmatrix} 0 & 1 & 0 \\ 0 & 0 & 1 \\ 1 & 0 & 0 \\ 0 & -1 & 0 \end{bmatrix} =$$

$$= \begin{bmatrix} 2 & -4 & 6 & -8 \\ 10 & -12 & 14 & -16 \end{bmatrix}\begin{bmatrix} 0 & 1 & 0 \\ 0 & 0 & 1 \\ 1 & 0 & 0 \\ 0 & -1 & 0 \end{bmatrix} =$$

$$= \begin{bmatrix} 6 & 10 & -4 \\ 14 & 26 & -12 \end{bmatrix}$$

if you are interested in extreme experiences, you could try:

$$A\,(2B) = \begin{bmatrix} 1 & -2 & 3 & -4 \\ 5 & -6 & 7 & -8 \end{bmatrix}\left(2\begin{bmatrix} 0 & 1 & 0 \\ 0 & 0 & 1 \\ 1 & 0 & 0 \\ 0 & -1 & 0 \end{bmatrix}\right) =$$

$$= \begin{bmatrix} 1 & -2 & 3 & -4 \\ 5 & -6 & 7 & -8 \end{bmatrix}\begin{bmatrix} 0 & 2 & 0 \\ 0 & 0 & 2 \\ 2 & 0 & 0 \\ 0 & -2 & 0 \end{bmatrix} = \begin{bmatrix} 6 & 10 & -4 \\ 14 & 26 & -12 \end{bmatrix}$$

with the same result.

(6) — Trivially, if:

$$\alpha\begin{bmatrix} a \\ b \end{bmatrix} = \begin{bmatrix} 0 \\ 0 \end{bmatrix}$$

we get two equations:

$$\begin{cases} \alpha a = 0 \\ \alpha b = 0 \end{cases}$$

They hold if $\alpha = 0$. If $\alpha \neq 0$, they can jointly[20] hold only if $a = b = 0$.

Later we will use a rather simple fact. Consider the product of an (m, n)-matrix A with an n-vector $\mathbf{b}$:

$$A\mathbf{b}$$

Reconsider it, thinking of A as sequence of n columns and of $\mathbf{b}$ as a a column of numbers:

---

[20]In fact, if $\alpha \neq 0$ and — say — $a \neq 0$, it should sound strange that $\alpha a = 0$. The same argument holds for b.

$$Ab = \begin{bmatrix} \mathbf{a}^1 | \mathbf{a}^2 | \cdots | \mathbf{a}^n \end{bmatrix} \begin{bmatrix} b_1 \\ b_2 \\ \cdots \\ b_n \end{bmatrix}$$

If you ignore, for a while, that the $\mathbf{a}^s$ are vectors and think of them as numbers and compute the product of these vectors. You would get:

$$b_1 \mathbf{a}^1 + b_2 \mathbf{a}^2 + \cdots + b_n \mathbf{a}^n$$

Well, rather incredibly, the computation would be right: the product of an (m, n)-matrix with an n-vector is the m-vector, linear combination of the columns of A with coefficients the corresponding components of $\mathbf{b}$. This example should be convincing:

$$\begin{bmatrix} 1 & 3 & -5 \\ 2 & -4 & 6 \end{bmatrix} \begin{bmatrix} \alpha \\ \beta \\ \gamma \end{bmatrix} = \begin{bmatrix} \alpha + 3\beta - 5\gamma \\ 2\alpha - 4\beta + 6\gamma \end{bmatrix}$$

This result can be obtained linearly combining the columns of A, with coefficients the components of $\mathbf{b}$:

$$\alpha \begin{bmatrix} 1 \\ 2 \end{bmatrix} + \beta \begin{bmatrix} 3 \\ -4 \end{bmatrix} + \gamma \begin{bmatrix} -5 \\ 6 \end{bmatrix} = \begin{bmatrix} \alpha + 3\beta - 5\gamma \\ 2\alpha - 4\beta + 6\gamma \end{bmatrix}$$

This conclusion can be generalized. Think of the product of two matrices $\underset{(m,n)}{A}$, $\underset{(n,p)}{B}$.

Think of both of them as a juxtaposition of, respectively, n and p column vectors:

$$AB = \begin{bmatrix} \mathbf{a}^1 | \mathbf{a}^2 | \cdots | \mathbf{a}^n \end{bmatrix} \begin{bmatrix} \mathbf{b}^1 | \mathbf{b}^2 | \cdots | \mathbf{b}^p \end{bmatrix}$$

the product matrix:

$$C = AB$$

can be thought of as a juxtaposition of p columns, which are m-vectors:

$$\begin{bmatrix} \mathbf{c}^1 | \mathbf{c}^2 | \cdots | \mathbf{c}^p \end{bmatrix}$$

These columns are nothing but linear combinations of the columns of A with coefficients the corresponding elements in the columns of B. To be clear:

$$\mathbf{c}^1 = A\mathbf{b}^1, \mathbf{c}^2 = A\mathbf{b}^2, \ldots, \mathbf{c}^p = A\mathbf{b}^p$$

An alternative way to tell the same story is:

$$A \begin{bmatrix} \mathbf{b}^1 | \mathbf{b}^2 | \cdots | \mathbf{b}^p \end{bmatrix} = \begin{bmatrix} A\mathbf{b}^1 | A\mathbf{b}^2 | \cdots | A\mathbf{b}^p \end{bmatrix}$$

The following numerical example illustrates our point.

Example 1.3.9 Let:

$$A = \begin{bmatrix} 1 & 3 & -5 \\ 2 & -4 & 6 \end{bmatrix}, B = \begin{bmatrix} 1 & 3 \\ 2 & 2 \\ 3 & 1 \end{bmatrix}$$

We find easily:

$$C = AB = \begin{bmatrix} 1 & 3 & -5 \\ 2 & -4 & 6 \end{bmatrix} \begin{bmatrix} 1 & 3 \\ 2 & 2 \\ 3 & 1 \end{bmatrix} = \begin{bmatrix} -8 & 4 \\ 12 & 4 \end{bmatrix}$$

Consider the first column of the product: $\begin{bmatrix} -8 \\ 12 \end{bmatrix}$. We are maintaining that it is the linear combination of the three columns of A with coefficients $1, 2, 3$ (the three components of the first column of B):

$$1 \begin{bmatrix} 1 \\ 2 \end{bmatrix} + 2 \begin{bmatrix} 3 \\ -4 \end{bmatrix} + 3 \begin{bmatrix} -5 \\ 6 \end{bmatrix} = \begin{bmatrix} -8 \\ 12 \end{bmatrix}$$

We also maintain that if we calculate the analogous linear combination of the columns of A, but using the second column of B, what we get is the second column of the product:

$$3 \begin{bmatrix} 1 \\ 2 \end{bmatrix} + 2 \begin{bmatrix} 3 \\ -4 \end{bmatrix} + 1 \begin{bmatrix} -5 \\ 6 \end{bmatrix} = \begin{bmatrix} 4 \\ 4 \end{bmatrix}$$

as previously announced.

Soon we will use these facts.

A natural question arises: "Why should a student in a Social Sciences program be interested in such an awkward attempt to compel tables (matrices) to behave like numbers?". We will try to show the interest of this calculus for political scientists.

Let us look first at this example:

Example 1.3.10 An earthquake occurs. Temporary accommodation is necessary. Two architects are asked to make proposals for temporary accommodation, on the basis of information about the population to be hosted. There are three types of temporary houses: small (S), medium (M), large (L). Each architect provides a project, suggesting a mix of S-M-L houses. We can stack their proposal in (2, 3)-matrix. Call it H. Rows correspond to architects, while columns correspond to the three types of houses

$$H = \begin{matrix} & S & M & L & \\ \begin{bmatrix} 30 & 40 & 10 \\ 25 & 45 & 10 \end{bmatrix} & \begin{matrix} \text{architect 1} \\ \text{architect 2} \end{matrix} \end{matrix}$$

Four construction firms are consulted and asked for a price list for the three types of houses. We collect these price lists in a $(3, 4)$-matrix P. Each of the four columns (a 3-vector) contains the prices, in € K, for each type of house by the 4 construction firms (F1, F2, F3, F4).

$$P := \begin{array}{c} \begin{array}{cccc} F1 & F2 & F3 & F4 \end{array} \\ \begin{bmatrix} 20 & 22 & 18 & 21 \\ 30 & 28 & 27 & 30 \\ 45 & 46 & 50 & 46 \end{bmatrix} \begin{array}{l} \text{small} \\ \text{medium} \\ \text{large} \end{array} \end{array}$$

The product HP is a $(2, 4)$ matrix collecting the cost of implementation of each of the 2 projects if assigned to each of the four construction firms. The rows correspond to the architects, while the columns refer to the construction firms

$$HP = \begin{bmatrix} 30 & 40 & 10 \\ 25 & 45 & 10 \end{bmatrix} \begin{bmatrix} 20 & 22 & 18 & 21 \\ 30 & 28 & 27 & 30 \\ 45 & 46 & 50 & 46 \end{bmatrix}$$

$$= \begin{bmatrix} 2250 & 2240 & 2120 & 2290 \\ 2300 & 2270 & 2165 & 2335 \end{bmatrix}$$

It's up to a politician to make a choice, taking into account both the plan and the costs.

We present a model of an economy, proposed by Wassily Leontief (Nobel Prize for Economics in 1973, exactly for this "input-output" model): don't worry, today the Leontief model is currently estimated by various statistical institutes.

Example 1.3.11 (*Leontief Economy*) The political problem is that of fostering growth in a multi-sector economic system. Should the Government invest in manufacturing or in banks or in utilities, or in construction? If some economic sector is isolated, the benefits of a public investment in that sector would be substantially confined to it. The connections between sectors become relevant in this setting. Leontief's model is simple, but deeply rooted and it has provided a lot of policy makers with an appropriate tool to make decisions.

An economy has n sectors, numbered with $s = 1, 2, \ldots, n$. Each of them produces a single product (the annual value of the product for the generic sector is x_s). It should be natural for any reader to think of the product quantities of the various sectors as a vector:

$$\mathbf{x} = \begin{bmatrix} x_1 \\ x_2 \\ \ldots \\ x_n \end{bmatrix}$$

Now we switch to the connections between sectors. The production process of each sector potentially needs products from other sectors. For instance, in the construction sector, when using (as usual) reinforced concrete, you need iron bars produced by the manufacturing sector. Leontief's model systematizes this aspect, assuming that each product unit of the sector s could need as an input the amount a_{rs} of the product of the sector r. Such uses of the sector products are called *intermediate* uses. It turns out to be natural to construct a *(Leontief) matrix*:

$$A = \begin{bmatrix} a_{11} & a_{12} & \cdots & a_{1n} \\ a_{21} & a_{22} & \cdots & a_{2n} \\ \cdots & \cdots & \cdots & \cdots \\ a_{n1} & a_{n2} & \cdots & a_{nn} \end{bmatrix}$$

that has been concretely estimated for several economic systems. It is usually called *"Leontief matrix of the technical coefficients"*. The production of a sector must cover two requests:

(1) — intermediate uses;

(2) — final uses (typically consumption):

$$\mathbf{c} = \begin{bmatrix} c_1 \\ c_2 \\ \cdots \\ c_n \end{bmatrix}$$

We could write, for the generic sector (number r), a balance equation:

$$x_r = \underbrace{a_{r1}x_1 + a_{r2}x_2 + \cdots + a_{rn}x_n}_{\text{intermediate uses}} + \underbrace{c_r}_{\text{final uses}}$$

These conditions can be synthetically reduced to an equation, that uses exactly the language of matrix multiplication:

$$\mathbf{x} = A\mathbf{x} + \mathbf{c}$$

hence:

$$I\mathbf{x} = A\mathbf{x} + \mathbf{c}$$

or:

$$(I - A)\mathbf{x} = \mathbf{c} \tag{1.3.1}$$

This equation suggests an interesting political problem. A planner has to decide how much of the product of each sector should be suitable for final uses. Our planner chooses $\mathbf{c}$. The (serious) problem consists in deciding which is the feasible production vector $\mathbf{x}$ which satisfies the condition above. A further reasonable question concerns the existence of a production vector $\mathbf{x}$, allowing for final uses $\mathbf{c}$. What the reader will discover is that there are viable economies, allowing for any choice of $\mathbf{c}$ and non-viable economies for which no reasonable $\mathbf{c}$ is admitted. These two economy groups do exhaust the economy populations set. We will get from Maths a discriminating criterion between viable and non-viable economies. Inversion of a matrix.

The division between matrices:

$$A/B$$

would have an important place in Linear Algebra.

The question is far from being trivial. For numbers there is no problem: the division of — say — 10 by 5 produces:

$$\frac{10}{5} = 2$$

The division by 5 can be freely replaced by the multiplication by:

$$\frac{1}{5} = 5^{-1}$$

What is the "inverse" of a number, say 5? First remember that mathematicians are a pain in the ass, so that they would replace "inverse" with the word "reciprocal", but this is a mere matter of words.

The reciprocal of 5 is a number x such that[21]:

$$x \cdot 5 = 5 \cdot x = \underset{\text{neutral el. w.r.t} \times}{1}$$

Such a number turns out to be:

$$x = \frac{1}{5} = 5^{-1}$$

Let us try to extend this idea to (m, n)-matrices. We have an (m, n)-matrix at hand and we are looking for some matrix B, such that:

$$AB = BA = I \qquad (1.3.2)$$

It should be sufficiently clear that A must be square, but we will see below, how this requirement is compelling. Follow us: in order for AB to make sense it is *necessary* that B is of the type $(n, \text{something})$, in order for BA to make sense it is necessary that the type of B is $(\text{something}, m)$. The two products:

$$AB \text{ and } BA$$

would produce respectively identity matrices of order n and m. Therefore, according to (1.3.2):

$$I_m = I_n \Rightarrow m = n$$

---

[21] Some people could think: "You're silly because $x \times 5 = 5 \times x$", but as we're working with matrices, where commutativity is not guaranteed, the precaution turns out to be justified.

Necessarily, if A is a square matrix of order n, its self-profiled inverse would be some square matrix B of order n, satisfying (1.3.2). It would be better for us if "squareness" would be sufficient, but unfortunately this is not true.

Take, for instance:

$$A = \begin{bmatrix} 0 & 0 \\ 0 & 0 \end{bmatrix}$$

It's square, but there is no B such that $AB = BA = I$. An obvious objection would be "You've chosen the 0 in that world, so, it is not surprising that it has no inverse!".

Well, let us take a "no-zero" matrix like:

$$A = \begin{bmatrix} 1 & 1 \\ 1 & 1 \end{bmatrix} \tag{1.3.3}$$

and let us try to find B such that (1.3.2) holds. Also let B be:

$$B = \begin{bmatrix} \alpha & \beta \\ \gamma & \delta \end{bmatrix}$$

The requirements from

$$AB = I$$

would be:

$$\begin{cases} \alpha + \gamma = 1 \text{ element } (1, 1) \\ \beta + \delta = 0 \text{ element } (1, 2) \\ \alpha + \gamma = 0 \text{ element } (2, 1) \\ \text{no need to continue} \end{cases}$$

The first and the third response are incompatible, so, even if A is square and has no null entry, it cannot have an inverse.

Which matrices have an inverse? It is easy to convince ourselves that if some matrix A admits an inverse B:

$$BA = AB = I$$

then B is unique. Let $C \neq B$ be another (phony) inverse:

$$CA = AC = I$$

Take:

$$BA = I$$

and multiply both sides by C on the right:

$$(BA)\,C = IC \Rightarrow B\,(AC) = C \Rightarrow BI = C \Rightarrow B = C \text{ false!}$$

Working first in the order 2 case, we can grasp the core of the question and characterize the matrices admitting an inverse.

Let:

$$AB = I$$

If

$$A = \left[\mathbf{a}^1 | \mathbf{a}^2\right], \ B = \left[\mathbf{b}^1 | \mathbf{b}^2\right]$$

and B is the inverse, then:

$$\left[\mathbf{a}^1 | \mathbf{a}^2\right]\left[\mathbf{b}^1 | \mathbf{b}^2\right] = \left[\mathbf{e}^1 | \mathbf{e}^2\right] \tag{1.3.4}$$

where:

$$\mathbf{e}^1 = \begin{bmatrix} 1 \\ 0 \end{bmatrix}; \ \mathbf{e}^2 = \begin{bmatrix} 0 \\ 1 \end{bmatrix}$$

The Eq. (1.3.4) tells us that you can generate the fundamental basis of $\mathbb{R}^2$ combining the columns of A. This means that they are a basis for $\mathbb{R}^2$ (see the proposition 1.2.2 on p. 15) and therefore that they are not aligned. In $\mathbb{R}^n$ the same argument continues to hold. Interchanging factors it is possible to state that linear independence of the columns is equivalent to the linear independence of the rows.

We can summarize these findings in the following:

Proposition 1.3.2 *Let A be a square matrix of order n. It admits the inverse matrix A^{-1} if and only if its columns (rows) are linearly independent.*

Definition 1.3.4 A square matrix, which does not admit, an inverse, because there is some linear relationship among its lines and its columns, is said to be *singular*.

Remark 1.3.2 Go back to (1.3.3). We have shown that this matrix cannot have an inverse. In fact its rows (columns) are not linearly independent.

Example 1.3.12 Let us look at a numerical example. The matrix:

$$A = \begin{bmatrix} 1 & 1 \\ 0 & 1 \end{bmatrix}$$

has linearly independent columns (rows). Let us check that its inverse is:

$$A^{-1} = \begin{bmatrix} 1 & 1 \\ 0 & 1 \end{bmatrix}^{-1} = \begin{bmatrix} 1 & -1 \\ 0 & 1 \end{bmatrix}$$

In fact:

$$\begin{bmatrix} 1 & -1 \\ 0 & 1 \end{bmatrix}\begin{bmatrix} 1 & 1 \\ 0 & 1 \end{bmatrix} = \begin{bmatrix} 1 & 1 \\ 0 & 1 \end{bmatrix}\begin{bmatrix} 1 & -1 \\ 0 & 1 \end{bmatrix} = \begin{bmatrix} 1 & 0 \\ 0 & 1 \end{bmatrix}$$

At this point, a natural question is: "Given an invertible matrix A, how can we compute A^{-1}?"

The best answer is: "Use a computer and some good software (for instance Mathcad®)!".

We will see immediately an algorithm, which is the one a computer would use: for small matrices, the computations can be made by (patient) human beings. In the next section another elegant method will be found, but, once again, we can only do it by hand for small matrices. It is (almost) a pleasure to say that, with this method computers can also have problems.

Inversion algorithm — We need the introduction of three notions, called *elementary row operations* on a matrix.

- Multiply a row of a matrix with a non-null scalar;
- Add to a row a multiple of another one;
- Exchange two rows (or columns) with each other.

Here is an

Example 1.3.13 Let:

$$A = \begin{bmatrix} 1 & 2 \\ 3 & 4 \end{bmatrix}$$

An example of the first operation is, for instance, the multiplication of the second row by -2. Performing this operation, A becomes:

$$\begin{bmatrix} 1 & 2 \\ -2 \times 3 & -2 \times 4 \end{bmatrix} = \begin{bmatrix} 1 & 2 \\ -6 & -8 \end{bmatrix}$$

An example of the second operation consists in adding the double of the second row to the first row. The matrix we get is:

$$\begin{bmatrix} 1 + 2 \times 3 & 2 + 2 \times 4 \\ 3 & 4 \end{bmatrix} = \begin{bmatrix} 7 & 10 \\ 3 & 4 \end{bmatrix}$$

Now the method:

- write the matrix A, you want to invert and put an identity matrix of the same order close to it:

$$A | I$$

- using elementary row operations, transform A into I and, parallely, apply the same transformations to the matrix on the right: when A will become I, the original identity matrix I will be A^{-1}.

Before using it in an example, let us try to grasp why it works. Consider the trivial case of order 1, in which the matrix collapses into a number:

$$A = [5]$$

The algorithm suggests that we should construct the scheme $A|I$:

$$[5] \,|\, [1]$$

In order to transform A into an identity matrix of the first order the (row) elementary operation to be used is obviously the multiplication of the unique row by the scalar $\frac{1}{5}$. We do that on both sides of the scheme:

$$\frac{1}{5}[5] \,\bigg|\, \frac{1}{5}[1]$$

hence:

$$[1] \,\bigg|\, \begin{bmatrix} \frac{1}{5} \end{bmatrix}$$

This is great: we have found the number 5^{-1} such that:

$$5^{-1} \times 5 = 5 \times 5^{-1} = 1$$

Now a non-trivial case.

Example 1.3.14 We are given $A = \begin{bmatrix} 1 & 1 \\ 0 & 1 \end{bmatrix}$. We want to find its inverse using the algorithm described above. The starting point is:

$$\begin{bmatrix} 1 & 1 \\ 0 & 1 \end{bmatrix} \begin{bmatrix} 1 & 0 \\ 0 & 1 \end{bmatrix}$$

Luckily the first column of A is already the first column we want. To adjust the second column, we need to have 0 at the position $(1, 2)$. We can reach this result simply by adding to the first row the second one multiplied by (-1). It is an elementary operation of the second type. We do that and we apply the same transformation to the matrix on the right:

$$\begin{bmatrix} 1 + (-1) \times 0 & 1 + (-1) \times 1 \\ 0 & 1 \end{bmatrix} \begin{bmatrix} 1 + (-1) \times 0 & 0 + (-1) \times 1 \\ 0 & 1 \end{bmatrix}$$

hence:

$$\begin{bmatrix} 1 & 0 \\ 0 & 1 \end{bmatrix} \begin{bmatrix} 1 & -1 \\ 0 & 1 \end{bmatrix}$$

which is simply wonderful as the matrix on the right is exactly the one we have checked above on p. 37. Now we know how it can be found.

Example 1.3.15 (*Leontief economy (again)*) Let us continue to dream. We have a matrix M and we would like to have it to solve a political problem. Back to Example 1.3.11 on p. 33, where we obtained the Eq. (1.3.1):

$$(I - A)\,\mathbf{x} = \mathbf{c} \text{ or } M\mathbf{x} = \mathbf{c}$$

where $M = I - A$. If we have M^{-1}, then we would be able to compute the production vector $\mathbf{x}$ needed to allow for a consumption vector $\mathbf{c}$:

$$\mathbf{x} = M^{-1}\mathbf{c} = (I - A)^{-1}\,\mathbf{c} \qquad (1.3.5)$$

Let us implement this path in a simple case:

$$A = \begin{bmatrix} 0.2 & 2 \\ 0.2 & 0.4 \end{bmatrix}$$

The interpretation is straightforward: a unit of product of each sector requires the product of the sector itself (20% for the first one, 40% for the second one). Each unit of the product of the second sector requires 2 units of product of the first one, while only 20% of a unit of the first one are needed for a unit of the second one. A natural question we will answer later is that of sustainability of such a scheme. The main concern is well depicted in the following Table 1.4:

Table 1.4 Leontief economy

Needs per unit of sector\Producing sector	1 (%)	2
1	20	2
2	20	40%

Reasonable, but tricky for a planner. Does a production plan exist which satisfies all of these constraints and is able to provide the desired consumption vector $\mathbf{c} = \begin{bmatrix} 100 \\ 100 \end{bmatrix}$? This is not only a political problem, but also a technical one. We derive the production plan using (1.3.5):

$$\mathbf{x} = \left(\begin{bmatrix} 1 & 0 \\ 0 & 1 \end{bmatrix} - \begin{bmatrix} 0.2 & 2 \\ 0.2 & 0.4 \end{bmatrix} \right)^{-1} \begin{bmatrix} 100 \\ 100 \end{bmatrix} = \begin{bmatrix} 3250 \\ 1250 \end{bmatrix}$$

which turns out to be very interesting: it tells us that the production volumes of a sector can be significantly higher than the direct consumption demand, because of intermediate uses.

Such intermediate uses could seriously affect the viability of the economic system. Let us assume that Mr. X, a new politician, declares, of course, in breaking news that, thanks to innovation, a new process has been introduced in the second sector. Good news for firms in the first sector, because, instead of only 2 units of the product of the sector 1, in the new setting, 3 units will be required: this means new workplaces, growth,...This candidate, Mr. X, could also announce that this change of technology has reduced the use of the product of sector 2 in the sector itself by 10%.

Stupidly, a politician like Mr. X could distill from this message several positive considerations:

$$\begin{cases} \text{more work for sector 1} \\ \text{smaller dependence of sector 2 on itself: new inspiring directions} \end{cases}$$

But let us keep to figures. In the new setting we would find:

$$\mathbf{x} = \left(\begin{bmatrix} 1 & 0 \\ 0 & 1 \end{bmatrix} - \begin{bmatrix} 0.2 & \mathbf{3} \\ 0.2 & \mathbf{0.3} \end{bmatrix} \right)^{-1} \begin{bmatrix} 100 \\ 100 \end{bmatrix} = \begin{bmatrix} -9250.0 \\ -2500.0 \end{bmatrix}$$

which is completely void of sense, because the system cannot carry on with this modification (you find in bold the new values).

We will see later how to detect such cases, looking at the Leontief matrix.

1.4 Determinants

1.4.1 Notion and Computation

What's the determinant of a square matrix A? It's a (wild) number depending on A

$$\det A$$

Another way to denote the determinant of a square matrix A is:

$$\|A\|$$

These two symbols are equivalent:

$$\det \begin{bmatrix} 1 & 2 \\ 3 & 4 \end{bmatrix} \quad \text{and} \quad \begin{Vmatrix} 1 & 2 \\ 3 & 4 \end{Vmatrix}$$

We do not directly reveal the general formula for this number[22], but we indicate how to compute it.

Let's start from $n = 1$. Given:

$$A = [a_{11}]$$

---

[22]It's possible, but there is no need for the reader to learn it.

its determinant is simply the value of its only element:

$$\det A = \det [a_{11}] = a_{11}$$

So, for instance:

$$\det [-5] = -5$$

If we think of the Cartesian representation (on an axis) of the number a_{11} it is the *length with sign* of an arrow starting at 0 and ending exactly at a_{11}. Note that length is the measure of objects living in a straight line (the axis).

Please, note also that the determinant turns out to be 0 if and only if $a_{11} = 0$.

Now let $n = 2$. Take this matrix:

$$A = \begin{bmatrix} \mathbf{1} & 2 \\ 3 & \mathbf{4} \end{bmatrix}$$

The two bold elements constitute the *principal diagonal* (see above on p. 26), while the others constitute the *secondary one*. Well, the determinant of A is the difference between the products of the elements of the principal diagonal and that of the secondary one:

$$\det A = 1 \times 4 - 2 \times 3 = -2$$

How to interpret this number? A nice idea is to think of a square matrix of order 2 as a couple of vectors (row/column vectors on your choice). We choose column vectors and we think of A as:

$$\left[\begin{bmatrix} 1 \\ 3 \end{bmatrix} \begin{bmatrix} 2 \\ 4 \end{bmatrix} \right]$$

We represent them in a Cartesian plane, together with the two equivalent possibilities to construct:

$$\begin{bmatrix} 1 \\ 3 \end{bmatrix} + \begin{bmatrix} 2 \\ 4 \end{bmatrix} = \begin{bmatrix} 2 \\ 4 \end{bmatrix} + \begin{bmatrix} 1 \\ 3 \end{bmatrix} = \begin{bmatrix} 3 \\ 7 \end{bmatrix}$$

We have carried out this exercise (see Fig. 1.3 at p. 7) and we have shown that these two vectors do create a parallelogram.

With some tedious computations we could prove that det A is nothing but the *area*[23] of this parallelogram as shown in Fig. 1.8.

---

[23]Please pay attention to language. Notions like "piece of surface" and its extension (area) are frequently interchanged. Sentences like "In the San Diego area there is massive illegal immigration of people from Mexico to the USA" are accepted, even if instead of "area" a mathematician would use another word.

Fig. 1.8 Determinant as an area

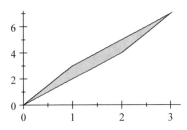

Please note that, when we are on a plane, the natural extension measure of any 2-dimensional object is its area.[24] Also note, please, that if the two vectors are aligned, the area of the generated parallelogram is 0.

In order to strengthen the reader conviction that this interpretation is correct, look at the following:

Example 1.4.1 We have a square matrix of order 2:

$$A = \begin{bmatrix} 2 & 0 \\ 0 & 3 \end{bmatrix} \text{ with } \det A = 6$$

The two (column) vector it consists of are:

$$\begin{bmatrix} 2 \\ 0 \end{bmatrix} \text{ and } \begin{bmatrix} 0 \\ 3 \end{bmatrix}$$

The parallelogram they generate is a rectangle (Fig. 1.9) whose area is trivially 6.

Fig. 1.9 Rectangle

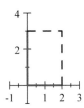

---

[24]If we move along the boundaries of this parallelogram, starting from the first side (the one from $\begin{bmatrix} 0 \\ 0 \end{bmatrix}$ to $\begin{bmatrix} 1 \\ 3 \end{bmatrix}$ and so on, the parallelogram always stays on our right: this explains the "−" sign of the area. Interchanging the columns of A we would move along the perimeter, with the parallelogram on our left: this would bring us to a positive area:

$$\det \begin{bmatrix} 2 & 1 \\ 4 & 3 \end{bmatrix} = 2$$

Working with row columns, we would obtain an analogous rectangle with sides of length 3, 2, instead of 2, 3, but with the same area. Note that starting from the side from $\mathbf{o}$ to $\begin{bmatrix} 0 \\ 2 \end{bmatrix}$, the rectangle is on our left, so that its area turns out to be positive.

Now $n = 3$. No length, no area, but volume.

We are given a square matrix of the third order:

$$A = \begin{bmatrix} a_{11} & a_{12} & a_{13} \\ a_{21} & a_{22} & a_{23} \\ a_{31} & a_{32} & a_{33} \end{bmatrix}$$

The determinant of A, det A or $\|A\|$ can be computed using a rule, thanks to Pierre Frédéric Sarrus. It consists first in expanding A copying on its right its first two columns and inserting some arrows according to the following scheme:

$$
\begin{array}{ccccc}
a_{11} & a_{12} & a_{13} & a_{11} & a_{12} \\
& \searrow & \searrow & \searrow & \\
a_{21} & a_{22} & a_{23} & a_{21} & a_{22} \\
& & \searrow & \searrow & \searrow \\
a_{31} & a_{32} & a_{33} & a_{31} & a_{32} \\
& & + & + & +
\end{array}
$$

This scheme suggests the following expression:

$$a_{11}a_{22}a_{33} + a_{12}a_{23}a_{31} + a_{13}a_{21}a_{32}$$

A similar scheme can be constructed:

$$
\begin{array}{ccccc}
a_{11} & a_{12} & a_{13} & a_{11} & a_{12} \\
& & \swarrow & \swarrow & \swarrow \\
a_{21} & a_{22} & a_{23} & a_{21} & a_{22} \\
& \swarrow & \swarrow & \swarrow & \\
a_{31} & a_{32} & a_{33} & a_{31} & a_{32} \\
- & - & - & &
\end{array}
$$

that suggests the following expression:

$$-a_{13}a_{22}a_{31} - a_{11}a_{23}a_{32} - a_{12}a_{21}a_{33}$$

Well:

$$\det A = \begin{Vmatrix} a_{11} & a_{12} & a_{13} \\ a_{21} & a_{22} & a_{23} \\ a_{31} & a_{32} & a_{33} \end{Vmatrix} =$$

$$= a_{11}a_{22}a_{33} + a_{12}a_{23}a_{31} + a_{13}a_{21}a_{32} - a_{13}a_{22}a_{31} - a_{11}a_{23}a_{32} - a_{12}a_{21}a_{33}$$

At first sight it could seem a bit difficult, but look at this:

Example 1.4.2 Let:

$$A = \begin{bmatrix} 1 & 2 & 3 \\ 4 & 5 & 6 \\ 7 & 8 & 9 \end{bmatrix}$$

We get:

$$\det A = 1 \times 5 \times 9 + 2 \times 6 \times 7 + 3 \times 4 \times 8 - 3 \times 5 \times 7 - 1 \times 6 \times 8 - 2 \times 4 \times 9 = 0$$

The geometrical interpretation of the determinant of a matrix of order 3 is the volume of a solid. We consider a very simple example. Let:

$$A = \begin{bmatrix} 1 & 0 & 0 \\ 0 & 2 & 0 \\ 0 & 0 & 3 \end{bmatrix}$$

Its determinant is $1 \times 2 \times 3 = 6$. The three (column) vectors constituting it are aligned with the three coordinate axes and determine a parallelepiped (Fig. 1.10) whose volume is $1 \times 2 \times 3 = 6$.

Fig. 1.10 Parallelepided

For matrices of order > 3 there is no simple rule, but we can count on a general rule allowing us to compute determinants of order n in terms of determinants of order $n - 1$. It is clear at this point that, in principle, we can compute determinants of any order.[25]

This rule is due to P.S. de Laplace. It is not difficult, but it requires the introduction of some vocabulary, which however will turn out to be useful later on.

Definition 1.4.1 Given any-(m, n) matrix A we call *submatrix* of A any matrix which is obtained eliminating rows from A (maximum $m - 1$, minimum 0) and/or columns, (maximum $n - 1$, minimum 0). Let B be any square submatrix of order

---

[25]One of the Authors (LP), when young and silly, tried to compute a determinant of order 7 by hand, with no success. But, at that time, no computer was available.

$k, k = 1, \ldots, n$, extracted from A, the determinant of B is called *minor* of order k of the matrix A (corresponding to the square submatrix B).

Note that every matrix A is a submatrix of itself and that any submatrix must contain at least 1 row and 1 column.

Remark 1.4.1 Any element a_{rs} of A may be considered as one of its square submatrices having order 1. Therefore every a_{rs} of A is one of its minor of order 1.

Note that if A is a square matrix of order n and the same number d of rows and columns are eliminated, the submatrix is still square and its order is $n - d$. Moreover in a square matrix A it is possible to construct square submatrices keeping the intersection of the first row and the first column. In fact, by eliminating the *last* d rows and d columns, the resulting square submatrix is nothing but the intersection between the first $k = n - d$ rows and the first $k = n - d$ columns. Such special submatrices are said of *North-West* because of their position in A

$$
\begin{array}{ccccc}
\text{NW} & & \text{N} & & \\
\nwarrow & & \uparrow & & \\
& \boxed{\begin{array}{cc} a_{11} & a_{12} \end{array}} & \cdots & a_{n1} & \\
\text{W} \quad \longleftarrow & \begin{array}{cc} a_{21} & a_{22} \end{array} & \cdots & \cdots & \longrightarrow \quad \text{E} \\
& \cdots \quad \cdots & \cdots & \cdots & \\
& \downarrow & & & \\
& \text{S} & & &
\end{array}
$$

We summarize in the following

Definition 1.4.2 Let A be a square matrix of order n.

(i) The square submatrices A_k resulting from the intersection of the first k rows and the first k columms of A, $k = 1, \ldots, n$, are called *principal North-West (NW) submatrices*

$$
A_1 = [a_{11}]; \ A_2 = \begin{bmatrix} a_{11} & a_{12} \\ a_{21} & a_{22} \end{bmatrix}; \ A_3 = \begin{bmatrix} a_{11} & a_{12} & a_{13} \\ a_{21} & a_{22} & a_{23} \\ a_{31} & a_{32} & a_{33} \end{bmatrix}; \cdots; A_n = A
$$

(ii) The determinants M_k, $k = 1, \ldots, n$, of the principal NW submatrices are called *principal NW minors of* A.

$$
M_1 = \det A_1 = a_{11}; \ M_2 = \det A_2; \cdots; M_n = \det A_n = \det A
$$

For square matrices we may add further notions

Definition 1.4.3 Let A be a square matrix of order n. For everyone of its elements a_{rs}, $r, s = 1, \ldots, n$, we call:

(i) *complementary matrix* of a_{rs} the square submatrix of order $n-1$ you get by erasing the row and the column intersecting exactly at a_{rs} (i.e; the row r and the column s);

(ii) *complementary minor* of a_{rs} the determinant M_{rs} of its complementary matrix;

(iii) *cofactor* or *algebraic complement* a_{rs}^* of a_{rs} the complementary minor multiplied by $(-1)^{r+s} = (-)^{r+s}$:

$$a_{rs}^* = (-)^{r+s} M_{rs} = \begin{cases} M_{rs} & \text{if } r+s \text{ is even} \\ -M_{rs} & \text{if } r+s \text{ is odd} \end{cases}$$

As we will soon see, the cofactors play a crucial role in computing determinants of order higher than 3.

Example 1.4.3 Let us now see these notions in practice. First:

$$A = \begin{bmatrix} 1 & 0 & 2 & -3 & -4 \\ 0 & 1 & -1 & 2 & 2 \\ -1 & 1 & -3 & -4 & -3 \\ 1 & 2 & 3 & 4 & -6 \end{bmatrix}$$

A submatrix B of A, can be obtained eliminating its second row and its first two columns:

$$B = \begin{bmatrix} 2 & -3 & -4 \\ -3 & -4 & -3 \\ 3 & 4 & -6 \end{bmatrix}$$

Now let A be a square matrix of order 4:

$$A = \begin{bmatrix} 1 & 0 & 2 & -3 \\ 0 & 1 & -1 & 2 \\ -1 & 1 & -3 & -4 \\ 1 & 2 & 3 & 4 \end{bmatrix}$$

Its principal NW submatrices are:

$$A_1 = [1]; \quad A_2 = \begin{bmatrix} 1 & 0 \\ 0 & 1 \end{bmatrix}; \quad A_3 = \begin{bmatrix} 1 & 0 & 2 \\ 0 & 1 & -1 \\ -1 & 1 & -3 \end{bmatrix}; \quad A_4 = \begin{bmatrix} 1 & 0 & 2 & -3 \\ 0 & 1 & -1 & 2 \\ -1 & 1 & -3 & -4 \\ 1 & 2 & 3 & 4 \end{bmatrix}$$

its principal NW minors turn out to be:

$$M_1 = \det\,[1] = 1$$

$$M_2 = \det \begin{bmatrix} 1 & 0 \\ 0 & 1 \end{bmatrix} = 1$$

$$M_3 = \det \begin{bmatrix} 1 & 0 & 2 \\ 0 & 1 & -1 \\ -1 & 1 & -3 \end{bmatrix} = 0$$

$$M_4 = \det \begin{bmatrix} 1 & 0 & 2 & -3 \\ 0 & 1 & -1 & 2 \\ -1 & 1 & -3 & -4 \\ 1 & 2 & 3 & 4 \end{bmatrix} = 27$$

Let us now take the square matrix again:

$$A = \begin{bmatrix} 1 & 0 & 2 & -3 \\ 0 & 1 & -1 & 2 \\ -1 & 1 & -3 & -4 \\ 1 & 2 & 3 & 4 \end{bmatrix}$$

Let us focus on the (bold) -1 at the position $(2, 3)$. Let us cancel the row and the column it belongs to:

$$\begin{bmatrix} 1 & 0 & & -3 \\ & & & \\ -1 & 1 & & -4 \\ 1 & 2 & & 4 \end{bmatrix}$$

Its complementary submatrix is:

$$\begin{bmatrix} 1 & 0 & -3 \\ -1 & 1 & -4 \\ 1 & 2 & 4 \end{bmatrix}$$

Its complementary minor is:

$$M_{23} = \begin{Vmatrix} 1 & 0 & -3 \\ -1 & 1 & -4 \\ 1 & 2 & 4 \end{Vmatrix} = \det \begin{bmatrix} 1 & 0 & -3 \\ -1 & 1 & -4 \\ 1 & 2 & 4 \end{bmatrix} = 21$$

The corresponding cofactor or algebraic complement is:

$$a_{23}^* = (-)^{2+3}\, 21 = -21$$

At this point, we are able to indicate how to reduce the computation of a determinant of order n to the computation of determinants of order $n - 1$.

Rule (P.S. de Laplace) — You are given a square matrix A of order n:

$$A = \begin{bmatrix} a_{11} & a_{12} & \cdots & a_{1n} \\ a_{21} & a_{22} & \cdots & a_{2n} \\ \cdots & \cdots & \cdots & \cdots \\ a_{n1} & a_{n2} & \cdots & a_{nn} \end{bmatrix}$$

Choose any line,[26] for instance a row, the rth one:

$$\mathbf{a}^r = \begin{bmatrix} a_{r1} & a_{r2} & \cdots & a_{rn} \end{bmatrix}$$

Construct the vector collecting the cofactors of each element in that row:

$$\alpha^r = \begin{bmatrix} a_{r1}^* & a_{r2}^* & \cdots & a_{rn}^* \end{bmatrix}$$

and transpose it:

$$(\alpha^r)^{\mathrm{T}} = \begin{bmatrix} a_{r1}^* \\ a_{r2}^* \\ \cdots \\ a_{rn}^* \end{bmatrix}$$

Then multiply $\mathbf{a}^r$ with $(\alpha^r)^{\mathrm{T}}$ or, more elementarily, compute the sum of the products of the elements of the chosen line (the row $\# r$) with their cofactors.

Take the following matrix:

$$\begin{bmatrix} 1 & 0 & -3 \\ -1 & 1 & -4 \\ 1 & 2 & 4 \end{bmatrix}$$

We know very well that its determinant is:

$$\det \begin{bmatrix} 1 & 0 & -3 \\ -1 & 1 & -4 \\ 1 & 2 & 4 \end{bmatrix} = 21$$

Well, we choose its second column:

$$\begin{bmatrix} 0 \\ 1 \\ 2 \end{bmatrix}$$

the relevant complementary submatrices are:

---

[26]"Line" means row or column, as you prefer.

$$A_{12} = \begin{bmatrix} -1 & -4 \\ 1 & 4 \end{bmatrix}, \ A_{22} = \begin{bmatrix} 1 & -3 \\ 1 & 4 \end{bmatrix}, \ A_{32} = \begin{bmatrix} 1 & -3 \\ -1 & -4 \end{bmatrix}$$

Hence the cofactors of each of its elements are:

$$\begin{cases} a_{12}^* = (-1)^{1+2} \left\| \begin{matrix} -1 & -4 \\ 1 & 4 \end{matrix} \right\| = 0 \\[2em] a_{22}^* = (-1)^{2+2} \left\| \begin{matrix} 1 & -3 \\ 1 & 4 \end{matrix} \right\| = 7 \\[2em] a_{32}^* = (-)^{2+3} \left\| \begin{matrix} 1 & -3 \\ -1 & -4 \end{matrix} \right\| = 7 \end{cases}$$

Therefore:

$$\det A = 0 \times a_{12}^* + 1 \times a_{22}^* + 2 \times a_{32}^* = 0 \times 0 + 1 \times 7 + 2 \times 7 = 21$$

1.4.2   How to Invert a Matrix Using Determinants

We are sure that this section on determinants is not very sexy. In order to make it more attractive, first we show how a matrix can be inverted, using determinants (a very easy procedure, for instance, in the (2, 2) case).

Definition 1.4.4 Given a square matrix A, we call *adjoint* of A and we denote it by A^* the square matrix obtained replacing each element with its cofactor.

Proposition 1.4.1 *Given an invertible matrix A, its inverse can be computed as follows:*

$$A^{-1} = \frac{1}{\det A} A^{*\mathrm{T}}$$

In words: the inverse matrix is the transpose of the adjoint matrix, divided by[27] the determinant of A.

Example 1.4.4 Take:

$$A = \begin{bmatrix} 1 & 2 \\ 3 & 4 \end{bmatrix}$$

We want to invert it using determinants. First of all we compute its determinant:

$$\det \begin{bmatrix} 1 & 2 \\ 3 & 4 \end{bmatrix} = -2$$

---

[27] In a rigorous textbook, instead of "divided by…" you would read "multiplied by the reciprocal of…", but this textbook is deliberately non-rigorous.

We then construct the adjoint matrix:

$$A^* = \begin{bmatrix} a_{11}^* & a_{12}^* \\ a_{21}^* & a_{22}^* \end{bmatrix} = \begin{bmatrix} 4 & -3 \\ -2 & 1 \end{bmatrix}$$

Then we transpose it:

$$A^{*T} = \begin{bmatrix} 4 & -2 \\ -3 & 1 \end{bmatrix}$$

and we multiply it by the scalar:

$$\frac{1}{\det A} = \frac{1}{-2} = -\frac{1}{2}$$

getting:

$$A^{-1} = \frac{1}{\det A} A^{*T} = -\frac{1}{2} \begin{bmatrix} 4 & -2 \\ -3 & 1 \end{bmatrix} = \begin{bmatrix} -2 & 1 \\ 3/2 & -1/2 \end{bmatrix}$$

We can easily check for correctness multiplying the matrix we started from with the one we have found:

$$\begin{bmatrix} -2 & 1 \\ 3/2 & -1/2 \end{bmatrix} \begin{bmatrix} 1 & 2 \\ 3 & 4 \end{bmatrix} = \begin{bmatrix} 1 & 0 \\ 0 & 1 \end{bmatrix}$$

1.4.3 Determinants and "Viability" of a Leontief System

This is another interesting application of determinants. It concerns the Leontief model we already met twice in the Examples 1.3.11 on p. 33 and 1.3.15 on p. 40. In the second example we have discovered the possibility that a Leontief system has too high coefficients of intermediate uses which make it non-viable. A bright use of determinants allows us to detect viable systems.

Proposition 1.4.2 (Hawkins-Simon) *A Leontief system with coefficient matrix A is viable if and only if the principal NW minors of the matrix I − A are positive.*

Example 1.4.5 Let us take the case we already examined in the Example 1.3.15 on p. 40, where we worked with two Leontief matrices:

$$\begin{bmatrix} 0.2 & 2 \\ 0.2 & 0.4 \end{bmatrix} \text{ and } \begin{bmatrix} 0.2 & 3 \\ 0.2 & 0.3 \end{bmatrix}$$

The first one was that of a viable system, while the second one that of a non-viable system. Let us use the Hawkins-Simon condition to make a diagnosis. With the first matrix, the matrix I − A turns out to be:

$$\begin{bmatrix} 1 & 0 \\ 0 & 1 \end{bmatrix} - \begin{bmatrix} 0.2 & 2 \\ 0.2 & 0.4 \end{bmatrix} = \begin{bmatrix} 0.8 & -2 \\ -0.2 & 0.6 \end{bmatrix}$$

Its two NW principal minors are:

$$0.8 \quad \text{and} \quad 0.8 \times 0.6 - (-2) \times 0.2 = 0.88$$

both positive.

We repeat the ceremony with the second coefficient matrix. The matrix $I - A$ now is:

$$\begin{bmatrix} 1 & 0 \\ 0 & 1 \end{bmatrix} - \begin{bmatrix} 0.2 & 3 \\ 0.2 & 0.3 \end{bmatrix} = \begin{bmatrix} 0.8 & -3 \\ -0.2 & 0.7 \end{bmatrix}$$

and its NW principal minors are:

$$0.8 \text{ and } 0.8 \times 0.7 - (-3)(-0.2) = -0.04$$

Thanks to the Hawkins-Simon conditions, the negativity of the second NW minor reveals quickly the non-viability of the system.

1.5 Rank of a Matrix

We know that an (m, n)-matrix can be seen as a set of n (column) vectors $\mathbf{a}^s \in \mathbb{R}^m$ (with $s = 1, 2, \ldots, n$):

$$A = \left[\mathbf{a}^1 | \mathbf{a}^2 | \cdots | \mathbf{a}^n \right]$$

or as a pile of m (row) vectors $\mathbf{b}^r \in \mathbb{R}^n$ (with $r = 1, 2, \ldots, m$):

$$A = \begin{bmatrix} \mathbf{b}^1 \\ \mathbf{b}^2 \\ \cdots \\ \mathbf{b}^m \end{bmatrix}$$

A pair of questions will turn out to be interesting[28]:

- Out of the n vectors $\mathbf{a}^s \in \mathbb{R}^m$, how many of them are linearly independent, the others being linear combinations of them?
- Out of the m vectors $\mathbf{b}^r \in \mathbb{R}^n$, how many of them are linearly independent, the others being linear combinations of them?

---

[28] See, for instance, the Example 1.2.5 on p. 10.

At first sight it seems that the two answers should be somehow independent, but this is not true as the two numbers are necessarily equal and their common value is called *rank* of the matrix, which will turn out to play an important role in the sequel.

We try to let our readers understand why, in this respect, columns and rows are strictly connected. Look at the following:

Example 1.5.1 Let us construct a square matrix A of order 2 with a clear linear dependence between *rows*. In order to do that a strategy consists in taking as second row of A a multiple of the first. E.g.:

$$A = \begin{bmatrix} \alpha & \beta \\ k\alpha & k\beta \end{bmatrix} \tag{1.5.1}$$

It is easy to check the (rather) sensational fact that *having introduced linear dependence between the rows of A, we have introduced linear dependence between the columns of A*: in fact, both of them are multiples of the column vector:

$$\begin{bmatrix} 1 \\ k \end{bmatrix}$$

according to α and β, respectively.

We are ready to introduce the announced notion of rank.

Definition 1.5.1 Given a matrix A of type (m, n), we call *rank* of A, noted $\rho(A)$, the maximum number of linearly independent rows (columns) that can be extracted from A.

For instance, in the case of (1.5.1):

$$\begin{cases} \text{if } \alpha = \beta = 0 & \text{then } \rho(A) = 0 \\ \text{if not} & \text{then } \rho(A) = 1 \end{cases}$$

A reasonable question is:

• Given some (m, n)-matrix A, which is its rank $\rho(A)$?

The answer is surprisingly articulated as it can be raised in two different contexts:

1. We are given A and we want to find its rank $\rho(A)$.
2. We are given a (square) matrix A and we associate with it a family of matrices depending on some parameter λ, so that A generates a family of matrices $B(\lambda)$ depending on A and λ. Studying the Leontief model we have stumbled upon the matrix $(I - A)$, which is a very special and very interesting case of $B(\lambda) = \lambda I - A$. Important applications in Economics and Statistics require to find the values of λ such that $\rho[B(\lambda)] < n$.

We will try to be more precise:

- **Rank of an empirical matrix** — You are given some (m, n)-matrix A, whose
  entries a_{rs} are determined by wise empirical (= statistical) evaluations With probability 1, the rank of such a matrix is:

$$\rho(A) = \min(m, n)$$

For instance in a $(2, 5)$ empirical matrix we are almost sure that its rank is 2, because
$2 < 5$. For a $(4, 2)$ empirical matrix, its rank is almost certainly 2, because $4 > 2$.
- In an empirical (m, n)-matrix, $\rho(A) = \min(m, n)$,
- In an empirical square matrix A of order n, the rank is $\rho(A) = n$.
- Maybe you are interested in finding values of a parameter λ such that the rank of
  a square matrix $B(\lambda)$:

$$B(\lambda) = \lambda I - A$$

is smaller than n. For such a problem, look below at p. 87.

When dealing with non-empirical matrices,[29] the simple comparison between the
numbers of rows and of columns is generally insufficient.

Therefore, we need to construct strategies to determine the rank of a matrix A.

The possible strategies are :

$$\left\{ \begin{array}{l} \text{1. Use some software on } A \\[1em] \text{2. Use the reduction method on } A \\[1em] \text{3. Use determinants (Kronecker method)} \end{array} \right.$$

Let us explore these possibilities.

1. At first sight the first possibility seems to be the best, but you need (necessarily)
   a computer and an appropriate software.

   With Mathcad, for instance, if you define a matrix A :

$$A := \begin{bmatrix} 1 & 2 & 3 \\ 3 & 4 & 7 \\ 4 & 6 & 10 \end{bmatrix}$$

and you ask for the rank of A, you get:

$$\text{rank}(A) = 2$$

That's fine, but the information you receive is partial because, even if you know
the rank, i.e.; the maximum number of linearly independent columns (rows), can

---

[29]This happens frequently when constructing simple models to illustrate pieces of some Social
Science.

be extracted from A, but you do not know which columns (rows) are linearly independent.

2. The *elimination/reduction method*[30] is the one we have already encountered before for inverting a matrix. You take A and using the elementary row operations (see p. 38), construct the largest identity matrix you are able in the NW corner of the matrix. The results you can get are immediately summarized (the arrow $\rightarrow$ means what you get from A (on the left) via elementary operations (on the right):

$$A \rightarrow \begin{cases} \begin{bmatrix} I_k & B \\ O & O \end{bmatrix} & \text{in this case } k < m, n \text{ and } \rho(A) < m, n \\[2ex] \begin{bmatrix} I_n \\ O \end{bmatrix} & \text{in this case } m > n \text{ and } \rho(A) = n \\[2ex] \begin{bmatrix} I_m & O \end{bmatrix} & \text{in this case } m < n \text{ and } \rho(A) = m \\[2ex] I_n & \text{in this case } m = n \text{ and } \rho(A) = m = n \end{cases} \qquad (1.5.2)$$

In the last three case, A is said to have *full rank*, as its rank is the maximum physically allowed by its type. (If a $(3, 4)$ -matrix or a $(4, 3)$ — matrix have rank 3, they are full rank matrices).

3. The Kronecker method is useful for treating (small) cases at hand. We are working on an (m, n)-matrix. The idea behind this algorithm is simple: first of all, you can freely interchange rows with other rows and columns with other columns. You look for a non-null minor M of order k (being $k < m, n$). Such a minor M is determined by the intersection of k rows and k columns, that we can think of as the first k (rows and columns). We can say that $k \le \rho(A)$. We can try to increase the order of M bordering it with the appropriate elements from the remaining $(m - k)$ rows and $(n - k)$ columns. If a bordering leads to a non-null new minor M' of order, then $k + 1 \le \rho(A)$ and we can try again. If no admissible bordering is successful, then $\rho(A) = k$.

We try to illustrate the methods 2 and 3 in the next examples.

Example 1.5.2 (*Reduction method*) We're given:

$$A = \begin{bmatrix} 0 & 0 & 0 \\ 1 & 2 & 3 \\ 4 & 5 & 6 \\ 5 & 7 & 9 \end{bmatrix}$$

Its rank $\rho(A)$ cannot exceed 3. First of all we can move the first row to last place:

---

[30]It is the algorithm currently used by the most efficient software packages.

$$\begin{bmatrix} 1 & 2 & 3 \\ 4 & 5 & 6 \\ 5 & 7 & 9 \\ 0 & 0 & 0 \end{bmatrix}$$

Adding to the second row, the first multiplied by -4, we get:

$$\begin{bmatrix} 1 & 2 & 3 \\ 4+(-4)\,1 & 5+(-4)\,2 & 6+(-4)\,3 \\ 5 & 7 & 9 \\ 0 & 0 & 0 \end{bmatrix} = \begin{bmatrix} 1 & 2 & 3 \\ 0 & -3 & -6 \\ 5 & 7 & 9 \\ 0 & 0 & 0 \end{bmatrix}$$

Analogously, in the new matrix, adding to the third row the first one, multiplied by (-5), we get

$$\begin{bmatrix} 1 & 2 & 3 \\ 0 & -3 & -6 \\ 5+(-5)\,1 & 7+(-5)\,2 & 9+(-5)\,3 \\ 0 & 0 & 0 \end{bmatrix} = \begin{bmatrix} 1 & 2 & 3 \\ 0 & -3 & -6 \\ 0 & -3 & -6 \\ 0 & 0 & 0 \end{bmatrix}$$

If we continue, we can reduce the matrix to the following one:

$$\begin{bmatrix} 1 & 2 & 3 \\ 0 & -3 & -6 \\ 0 & -3 & -6 \\ 0 & 0 & 0 \end{bmatrix} \rightarrow \begin{bmatrix} 1 & 0 & a' & b' \\ 0 & 1 & a'' & b'' \\ 0 & 0 & 0 \\ 0 & 0 & 0 \end{bmatrix}$$

hence:

$$\rho(A) = 2$$

Example 1.5.3 (*Kronecker method*) The first move mimics the first one in the previous example:

$$A = \begin{bmatrix} 0 & 0 & 0 \\ 1 & 2 & 3 \\ 4 & 5 & 6 \\ 5 & 7 & 9 \end{bmatrix} \rightarrow \begin{bmatrix} 1 & 2 & 3 \\ 4 & 5 & 6 \\ 5 & 7 & 9 \\ 0 & 0 & 0 \end{bmatrix}$$

The NW minor M of order 2 is:

$$\det \begin{bmatrix} 1 & 2 \\ 4 & 5 \end{bmatrix} = -3 \neq 0$$

The rank $\rho(A) \geq 2$. If we border this minor with the last row (made of zeroes), we cannot prove that the rank is > 2 because a null line provides a null determinant (Laplace rule), therefore, we need to explore only one potentially relevant bordering:

$$\det \begin{bmatrix} 1 & 2 & 3 \\ 4 & 5 & 6 \\ 5 & 7 & 9 \end{bmatrix} = 0$$

We can conclude that $\rho(A) = 2$, as above.

Remark 1.5.1 In both the preceding examples, where the rank has turned out to be 2, a pair of independent column vectors is made by the two first ones, while a pair of independent row vectors is made by the second and the third ones. Let us check for linear independence. The linear combination of the first two columns with coefficients α, β is:

$$\alpha \begin{bmatrix} 1 \\ 4 \\ 5 \\ 0 \end{bmatrix} + \beta \begin{bmatrix} 2 \\ 5 \\ 7 \\ 0 \end{bmatrix} = \begin{bmatrix} \alpha + 2\beta \\ 4\alpha + 5\beta \\ 5\alpha + 7\beta \\ 0 \end{bmatrix}$$

We look for α, β such that it is null. We get the system of equations:

$$\begin{cases} \alpha + 2\beta = 0 \\ 4\alpha + 5\beta = 0 \\ 5\alpha + 7\beta \end{cases}$$

Substituting $\alpha\,(= -2\beta)$ from the first into the second, we get:

$$4(-2\beta) + 2\beta = 0 \Longrightarrow \beta = \alpha = 0$$

values forced by the first two equations, that also turn out to satisfy the third. Let us move to the row control. It is analogous. From:

$$\alpha \begin{bmatrix} 1 & 2 & 3 \end{bmatrix} + \beta \begin{bmatrix} 4 & 5 & 6 \end{bmatrix} = \begin{bmatrix} 0 & 0 & 0 \end{bmatrix}$$

we get:

$$\begin{cases} \alpha + 4\beta = 0 \\ 2\alpha + 5\beta = 0 \\ 3\alpha + 6\beta = 0 \end{cases}$$

hence:

$$\alpha = \beta = 0$$

What is relevant for us about the rank of a generic (m, n)-matrix A, whose rank is k, is that you can always transform it into this format:

$$\begin{bmatrix} B & C \\ O_1 & O_2 \end{bmatrix}$$

where B is a square non-singular matrix of order k, C is a matrix of type $(k, n - k)$, O_1 is a null matrix of type $(m - k, k)$ and O_2 a null matrix of type $(m - k, n - k)$. If some of the dimensions of these matrices turns out to be 0, the corresponding matrix does not exist. A complete picture of what can happen is provided by (1.5.2) above.

1.6   Statistical Applications of Linear Algebra

In Social Sciences some simple linear algebra skills are very frequently used.
    They can be interpreted in two symmetric ways, we label Statistics and Probability.

1. **Statistics** — A *discrete statistical variable* $\mathbf{X}$ is some quantity that can take a list of n *possible values*:

$$x_1 < x_2 < \cdots < x_n$$

and we have data about the percentages of the times these possible values have been observed. These percentages are respectively:

$$p_1, p_2, \ldots, p_n \geq 0$$

Of course, their sum must be equal to 1:

$$\sum_{s=1}^{n} p_s = 1$$

2. **Probability** — A *discrete random variable*[31] $\mathbf{X}$ is some quantity, determined by a random experiment, that can take a list[32] of n *possible values*[33]:

$$x_1 < x_2 < \cdots < x_n$$

To each of them the probability it occurs is assigned:

---

[31] The AA. would call it *random number*, but the prevailing nomenclature is adopted in this textbook.

[32] The order requirement is not necessary, but helpful in managing several applications.

[33] In principle, the possible values could be an infinity:

$$x_1, x_2, \ldots, x_n, \ldots \qquad (1.6.1)$$

To handle this case, we should introduce appropriate mathematical tools, like the sum of an infinity of addenda, called *series*. We have decided to omit this theme, thinking that models of the type (1.6.1) can be substituted for our readers with models:

$$x_1, x_2, \ldots, x_n$$

with n large.

$$p_1, p_2, \ldots, p_n$$

Of course, as above, their sum must be equal to 1:

$$\sum_{s=1}^{n} p_s = 1$$

In both cases, we will adopt the notation:

$$\mathbf{X} \sim \begin{cases} x_1 & x_2 & \cdots & x_n & \leftarrow \text{ possible values} \\ \\ p_1 & p_2 & \cdots & p_n & \leftarrow \text{ frequency/probability} \end{cases} \tag{1.6.2}$$

The function which associates to each possible value x_s its frequency or its probability is usually called *frequency function* or, respectively, *probability function*.

The difference between Stats and Prob can be seen by our readers as follows:

$$\begin{cases} \text{Statistics looks back and collects frequencies} \\ \\ \text{Probability looks ahead and works on assessed probabilities} \end{cases}$$

An example should help to understand the two perspectives.

Example 1.6.1 (*Films watching*) We investigate the habits of some population as far as the number of films typically watched per day. The possible numbers of watched films per day are:

$$\mathbf{x} = \begin{bmatrix} 0 & 1 & 2 & 3 \end{bmatrix}$$

(wow! A (row) vector.) The corresponding observed percentages are:

$$\mathbf{p} = \begin{bmatrix} 0.4 \\ 0.3 \\ 0.2 \\ 0.1 \end{bmatrix}$$

(a (column)[34] vector). Therefore, a compact version of (1.6.2) would be:

$$\mathbf{X} \sim \begin{cases} \mathbf{x} & \leftarrow \text{ possible values} \\ \\ \mathbf{p}^T & \leftarrow \text{ frequency} \end{cases}$$

or, in detail, the frequency distribution is:

---

[34]The subtle reason for finding first a row vector and then a column vector, will turn out to be clear quickly.

$$\mathbf{X} \sim \begin{cases} \begin{array}{cccc} 0 & 1 & 2 & 3 \end{array} \quad \leftarrow \text{ possible values} \\ \\ \begin{array}{cccc} 0.4 & 0.3 & 0.2 & 0.1 \end{array} \quad \leftarrow \text{ frequency} \end{cases}$$

Now imagine we extract a person in this population and ascertain his/her number of films watched per day. Such a number is a random variable. In the case every member of the population has the same chances to be selected, it is natural[35] to interpret frequencies like probabilities and the (random) number of films per day is a random variable with probability distribution:

$$\mathbf{X} \sim \begin{cases} \begin{array}{cccc} 0 & 1 & 2 & 3 \end{array} \quad \leftarrow \text{ possible values} \\ \\ \begin{array}{cccc} 0.4 & 0.3 & 0.2 & 0.1 \end{array} \quad \leftarrow \text{ probability} \end{cases}$$

Mean - Expected Values

We will use systematically the notation:

$$\mathbf{x} = \begin{bmatrix} x_1 & x_2 & \cdots & x_n \end{bmatrix}$$

and:

$$\mathbf{p}^{\mathsf{T}} = \begin{bmatrix} p_1 & p_2 & \cdots & p_n \end{bmatrix}$$

introduced in perfect pirate style in the preceding Example.

$$\mathbf{X} \sim \begin{cases} \mathbf{x} \quad \leftarrow \text{ possible values} \\ \\ \mathbf{p}^{\mathsf{T}} \quad \leftarrow \text{ frequency/probability} \end{cases}$$

Both in Statistics and in Probability the quantity:

$$m = \mathbf{x}\mathbf{p} = \sum_{s=1}^{n} x_s p_s$$

turns out to be crucial:

- M $(\mathbf{X})$, called *mean value* of the statistical variable $\mathbf{X}$ is the most famous *location measure*;
- E $(\mathbf{X})$, called *expected value* of the random variable $\mathbf{X}$ is the long-range average value of values taken[36] by $\mathbf{X}$.

---

[35]In reality the problem of transforming frequency into probability is very delicate, but it is not central for this textbook.

[36]Imagine you observe several times random quantities $\mathbf{X}_1, \mathbf{X}_2, \ldots, \mathbf{X}_n$ with probability distribution:

Consider now the difference:

$$\mathbf{x} - m\mathbf{1}$$

where:

- $\mathbf{x}$ is a vector of possible values of a statistical/random values;
- $\mathbf{1} \in \mathbb{R}^n$ is a (row) vector of ones[37]:

$$\mathbf{1} = \begin{bmatrix} 1 & 1 & \cdots & 1 \end{bmatrix}$$

Explicitly:

$$\mathbf{x} - m\mathbf{1} = \begin{bmatrix} x_1 & x_2 & \cdots & x_n \end{bmatrix} - m\begin{bmatrix} 1 & 1 & \cdots & 1 \end{bmatrix} =$$
$$= \begin{bmatrix} x_1 - m & x_2 - m & \cdots & x_n - m \end{bmatrix}$$

Sometimes, the vector:

$$\mathbf{m} = m\mathbf{1} = \begin{bmatrix} m & m & \cdots & m \end{bmatrix}$$

will turn out to be useful:

$$\mathbf{x} - m\mathbf{1} = \mathbf{x} - \mathbf{m} = \begin{bmatrix} x_1 - m & x_2 - m & \cdots & x_n - m \end{bmatrix}$$

The components of this vector are called *deviations* from the mean/expected value. The mean/expected value of deviations is non-informative because it is always 0:

$$(\mathbf{x} - \mathbf{m})\,\mathbf{p} = (\mathbf{x} - m\mathbf{1})\,\mathbf{p} = \mathbf{xp} - m\,\underbrace{\mathbf{1p}}_{1} = \mathbf{xp} - m = m - m = 0$$

Example 1.6.2 Back to the Example 1.6.1. We have:

$$m = \mathbf{xp} = \begin{bmatrix} 0 & 1 & 2 & 3 \end{bmatrix} \begin{bmatrix} 0.4 \\ 0.3 \\ 0.2 \\ 0.1 \end{bmatrix} = 1$$

Therefore:

$$\mathbf{xp} - m = 1 - 1 = 0$$

$$\begin{cases} \mathbf{x} & \leftarrow \text{ possible values} \\ \\ \mathbf{p}^{\mathsf{T}} & \leftarrow \text{ frequency} \end{cases}$$

Under broad conditions the (random) average of the observations:

$$\mathbf{Y}_n = \frac{\sum_{s=0}^{n} \mathbf{X}_s}{n}$$

approaches indefinitely E $(\mathbf{X})$.

[37] Usually called 'fat one'.

When trying to measure the variability/randomness of a statistical/probability distribution the most commonly used parameter is *variance*, generally denoted with $\sigma^2 [\mathbf{X}]$.

Its expression is:

$$\sigma^2 [\mathbf{X}] = \sum_{s=1}^{n} (x_s - m)^2 \, p_s$$

This parameter can be represented with the language of linear algebra.

Take the vector $\mathbf{x} - \mathbf{m}$ and the matrix P, collecting the frequency/probability distribution in its principal diagonal[38]:

$$P = \begin{bmatrix} p_1 & 0 & & 0 \\ 0 & p_2 & & 0 \\ & & \ddots & \\ 0 & 0 & & p_n \end{bmatrix}$$

The variance of the distribution turns out to be:

$$\sigma^2 [\mathbf{X}] = (\mathbf{x} - \mathbf{m}) \, P \, (\mathbf{x} - \mathbf{m})^{\mathrm{T}}$$

Example 1.6.3 Once again, back to the film watching habits (1). In this case:

$$\mathbf{x} - \mathbf{m} = \begin{bmatrix} 0 & 1 & 2 & 3 \end{bmatrix} - \begin{bmatrix} 1 & 1 & 1 & 1 \end{bmatrix} = \begin{bmatrix} -1 & 0 & 1 & 2 \end{bmatrix}$$

and:

$$P = \begin{bmatrix} 0.4 & 0 & 0 & 0 \\ 0 & 0.3 & 0 & 0 \\ 0 & 0 & 0.2 & 0 \\ 0 & 0 & 0 & 0.1 \end{bmatrix}$$

The variance obtains:

$$\begin{bmatrix} -1 & 0 & 1 & 2 \end{bmatrix} \begin{bmatrix} 0.4 & 0 & 0 & 0 \\ 0 & 0.3 & 0 & 0 \\ 0 & 0 & 0.2 & 0 \\ 0 & 0 & 0 & 0.1 \end{bmatrix} \begin{bmatrix} -1 \\ 0 \\ 1 \\ 2 \end{bmatrix} = 1$$

---

[38] Such a matrix (with all 0's outside the principal diagonal) is called *diagonal matrix*. we think it is generated by the n-vector $\mathbf{p}$, putting its components in the principal diagonal of a square matrix of order n. Frequent symbols employed for this transformation are:

$$P = \mathrm{diag}\,(\mathbf{p}) = \widehat{\mathbf{p}}$$

The logical dimension of the variance of a statistical/random variable is the square of the dimension of **X**.

If **X** is measured in meters (m), its variance is in square meters (m²). It's OK.

But, if **X** is measured in €, its variance is measured in €², a notion which should seem to be a bit exotic also for the courageous category of financial mathematicians. This problem is usually solved using the square root of variance:

$$\sigma [\mathbf{X}] = \sqrt{\sigma^2 [\mathbf{X}]}$$

called *standard deviation*:

$$\sigma [\mathbf{X}] = \sqrt{(\mathbf{x} - \mathbf{m}) \, P \, (\mathbf{x} - \mathbf{m})^{\mathrm{T}}}$$

In the case of the preceding example, as the variance is unitary, the standard deviation is unitary too, but in general, this coincidence does not occur. Think of the case $\sigma^2 = 4$, which entails $\sigma = 2$.

Another important application of linear algebra concerns the case of *bivariate distributions*.

We start from:

Example 1.6.4 A population is studied under a double profile: income-consumption. We are given the statistic:

Table 1.5 Statistic

Income \Consumption	80	110	180
100	0.4	0.1	0
200	0.1	0.1	0.3

For instance the 0.4 in the NW cell tells us that 40% of the population has income 100 and consumption level 80, analogously for the other cells. We can study income and consumption separately. It suffices to work on the *marginal distributions* obtained through the row/columns totals:

Table 1.6 Row totals

Income	Frequency
100	0.5
200	0.5

or:

Table 1.7 Column totals

Consumption	80	110	180
Frequency	0.5	0.2	0.3

We can compute both mean values and standard deviations for both distributions. But what is relevant for Social Sciences is to try to investigate how relevant is the connection between the two marginal aspects summarized in the Tables 1.6 and 1.7. Concretely 'How much the income level turns out to determine the consumption level?'. Of course, such a question is crucial not only for Economics, but for other Social Sciences (Sociology, Political Science and Demography, at least). Here are the data we have are summarized in the frequency matrix:

$$
P = \begin{bmatrix} 0.4 & 0.1 & 0 \\ 0.1 & 0.1 & 0.3 \end{bmatrix}
$$

A crucial statistical problem is to quantify how strongly the two dimensions (income and consumption) are related.

The first step consists in a generalization of variance, a key notion we met above at p. 62.

In general, the frequency/probability key data are concentrated in an (m, n) matrix:

$$
P = \begin{bmatrix} p_{1,1} & \cdots & p_{1,n} \\ \cdots & \cdots & \cdots \\ p_{m,1} & \cdots & p_{m,n} \end{bmatrix}
$$

The possible values of the first variable $\mathbf{X}$ (in the Example above: income) are collected in an m-(row) vector:

$$
\mathbf{x} = \begin{bmatrix} x_1 & x_2 & \cdots & x_m \end{bmatrix}
$$

while the ones of the second variable $\mathbf{Y}$ (in the Example above: consumption) are collected in an n-(column) vector:

$$
\mathbf{y} = \begin{bmatrix} y_1 \\ y_2 \\ \cdots \\ y_n \end{bmatrix}
$$

Their 'separate' distributions, the marginal ones for the two variables, are easily obtained from the matrix P. If we multiply (on the right) P by the n-(column vector of ones) $\mathbf{1}_n$ we get the m-(row) vector

$$
\xi = P\mathbf{1}_n = \begin{bmatrix} \xi_1 \\ \xi_2 \\ \cdots \\ \xi_m \end{bmatrix}^{\mathrm{T}}
$$

providing us with the marginal frequency/probability distribution for $\mathbf{X}$.

Practically, you can get this frequency/probability simply computing the m row totals:

$$\xi_r = \sum_{s=1}^{n} p_{r,s} \text{ with } r = 1, 2, \ldots, m$$

As announced above, from these distributions, we can derive the mean/expected value of $\mathbf{X}$:

$$(\text{M/E}) [\mathbf{X}] = x\xi = x P \mathbf{1}_n = m_\mathbf{X}$$

As far as the columns totals are concerned, we get:

$$\theta_s = \sum_{r=1}^{m} p_{r,s}$$

Of course the n-(column vector):

$$\theta = \mathbf{1}_m P$$

will provide the marginal frequency/probability distribution for $\mathbf{Y}$ and the mean/expected value of $\mathbf{Y}$:

$$(\text{M/E})\big[\mathbf{Y}\big] = \theta y = \mathbf{1}_m P y = m_\mathbf{Y}$$

Also the standard deviations $\sigma [\mathbf{X}]$ and $\sigma [\mathbf{X}]$ can be obtained according to what we have seen above.

The basis of the analysis of the correlation between $\mathbf{X}$ and $\mathbf{Y}$, is the covariance between them.

It can be represented as:

$$\text{cov} [\mathbf{X}, \mathbf{Y}] = \text{E} \big[(\mathbf{x} - m_\mathbf{X}) \, \text{P} \, (\mathbf{y} - m_\mathbf{Y})\big]$$

and the *linear correlation coefficient*[39] $r [\mathbf{X}, \mathbf{Y}]$ (a masterpiece of Statistical Analysis in Social Sciences) is nothing but:

$$r [\mathbf{X}, \mathbf{Y}] = \frac{\text{cov} [\mathbf{X}, \mathbf{Y}]}{\sigma [\mathbf{X}] \, \sigma [\mathbf{Y}]}$$

Let us see this in practice.

First of all, we must translate the notion of covariance in terms different from linear algebra:

---

[39]It can take values only in the interval $[-1, 1]$. When its value is ± 1, there is an exact linear affine relationship between the variables. When it takes values close to ± 1 such a relationship is only approximate. Its square $(r [\mathbf{X}, \mathbf{Y}])^2$ is called *determination coefficient* and has an exciting interpretation. Assume that $r = 0.9$ and, therefore, $r^2 = 0.81$. If the two statistical variables are income level and consumption level this means that the differences in income levels explain 81% of the variability (variance) of consumption levels.

$$\text{cov}\,[\mathbf{X},\mathbf{Y}] = \sum_{r=1}^{m}\sum_{s=1}^{n}(x_r - m_{\mathbf{X}})\,(y_s - m_{\mathbf{Y}})\,p_{r,s}$$

Now back to the Example 1.6.4.

Example 1.6.5 In this case, for income and consumption, we have:

$$\mathbf{x} = \begin{bmatrix} 100 & 200 \end{bmatrix};\; \mathbf{y} = \begin{bmatrix} 80 \\ 110 \\ 180 \end{bmatrix}$$

The respective mean values are:

$$m_{\mathbf{X}} = 150 \text{ and } m_{\mathbf{X}} = 116$$

their standard deviations are:

$$\sigma\,[\mathbf{X}] = 50\;;\; \sigma\,[\mathbf{Y}] = 44.14$$

The two deviation vectors are:

$$\mathbf{x} - 150 \cdot \mathbf{1}_2 = \begin{bmatrix} -50 & 50 \end{bmatrix}$$

and

$$\mathbf{y} - 116 \cdot \mathbf{1}_3 = \begin{bmatrix} -36 \\ -6 \\ 64 \end{bmatrix}$$

The covariance is:

$$\text{cov}\,[\mathbf{X},\mathbf{Y}] = \begin{bmatrix} -50 & 50 \end{bmatrix} \cdot \begin{bmatrix} 0.4 & 0.1 & 0 \\ 0.1 & 0.1 & 0.3 \end{bmatrix} \cdot \begin{bmatrix} -36 \\ -6 \\ 64 \end{bmatrix} = 1500$$

The linear correlation coefficient is:

$$r\,[\mathbf{X},\mathbf{Y}] = \frac{1500}{50 \cdot 44.14} = 0.68 > 0$$

According to what we have written in the footnote 39 at p. 65 an completing those news, we inform the reader that the positivity of r tells us that trendily consumption grows with the income, which is not sensational, but also that the fact that:

$$r^2 = 0.68^2 = 0.462\,4$$

only 46.24% of the variance of consumption is 'explained' by the income level opens wide horizons to research for Social Sciences.

1.7 Linear Applications

Let us start with an example of political interest,[40] we are going to use it repeatedly in this chapter.

Example 1.7.1 A policy maker would like to reach a certain number m of *objectives*. Think of GNP, defense level, education investments, net balance of foreign trade, etc. The policy maker can control a certain number n of variables called *tools* or *(policy instruments)*. These tools could be, for instance, level of the public debt, public expenditure, taxes and subsidies, etc. Potentially every decision concerning a tool can have an effect on every objectives. We can list the objectives in an m-vector:

$$\mathbf{y} = \begin{bmatrix} y_1 \\ y_2 \\ \cdots \\ y_m \end{bmatrix}$$

and the tools in an n-vector:

$$\mathbf{x} = \begin{bmatrix} x_1 \\ x_2 \\ \cdots \\ x_n \end{bmatrix}$$

At least in a first approximation, we can think of the dependence of the objectives on the tools as described by a linear system of algebraic equations:

$$\begin{cases} y_1 = a_{11}x_1 + a_{12}x_2 + \cdots + a_{1n}x_n \\ y_2 = a_{21}x_1 + a_{22}x_2 + \cdots + a_{2n}x_n \\ \cdots \\ y_m = a_{m1}x_1 + a_{m2}x_2 + \cdots + a_{mn}x_n \end{cases}$$

or, more synthetically, using the (m, n)-matrix:

$$A = \begin{bmatrix} a_{11} & a_{12} & \cdots & a_{1n} \\ a_{21} & a_{22} & \cdots & a_{2n} \\ \cdots & \cdots & \cdots & \cdots \\ a_{m1} & a_{m2} & \cdots & a_{mn} \end{bmatrix}$$

we can write:

$$\mathbf{y} = A\mathbf{x} \tag{1.7.1}$$

This formula transforms tools $\mathbf{x}$ into objectives $\mathbf{y}$.

We will learn later that this is a special case of a more general one. Here the structure of the dependence of $\mathbf{y}$ on $\mathbf{x}$ is very simple. We will talk about these schemes

---

[40]This model is due to Jan Tinbergen, 1969 Nobel Prize for Economics.

as *vector functions* and we will write symbolically:

$$\mathbf{y} = \mathbf{f}(\mathbf{x})$$

An example of such more general dependence could be:

$$\begin{cases} y_1 = 3x_1 + 2x_2^2 \\ y_2 = \ln x_1 + 3x_2 \end{cases}$$

Here we have two objectives and two tools but the way the first ones depend on the second ones is not as simple as in the (1.7.1) case: for instance there is a square (x_2^2) and a log (ln).

In the special case of the application (1.7.1), the vector function $\mathbf{f}$ enjoys two important properties:

- **Homogeneity**: if we multiply $\mathbf{x}$ by any constant α, then $\mathbf{y}$ turns out to be multiplied by α:

$$\mathbf{f}(\alpha\mathbf{x}) = A(\alpha\mathbf{x}) = \alpha(A\mathbf{x}) = \alpha\mathbf{f}(\mathbf{x})$$

- **Additivity**: the vector $\mathbf{y}$ associated with the sum of two tool vectors $\mathbf{x}^1$, $\mathbf{x}^2$ is the sum of their objective vectors:

$$\mathbf{f}\left(\mathbf{x}^1 + \mathbf{x}^2\right) = A\left(\mathbf{x}^1 + \mathbf{x}^2\right) = A\mathbf{x}^1 + A\mathbf{x}^2 = \mathbf{f}\left(\mathbf{x}^1\right) + \mathbf{f}\left(\mathbf{x}^2\right)$$

Look at the following example to check these properties in a numerical case.

Example 1.7.2 Let:

$$A = \begin{bmatrix} 1 & 2 & 3 \\ 4 & 5 & 6 \end{bmatrix}$$

Let also:

$$\mathbf{x} = \begin{bmatrix} a \\ b \\ c \end{bmatrix}$$

We have:

$$\mathbf{f}(\mathbf{x}) = \begin{bmatrix} 1 & 2 & 3 \\ 4 & 5 & 6 \end{bmatrix} \begin{bmatrix} a \\ b \\ c \end{bmatrix} = \begin{bmatrix} a + 2b + 3c \\ 4a + 5b + 6c \end{bmatrix}$$

Let us now multiply $\mathbf{x}$ by 3. We have:

$$3\mathbf{x} = 3 \begin{bmatrix} a \\ b \\ c \end{bmatrix} = \begin{bmatrix} 3a \\ 3b \\ 3c \end{bmatrix}$$

and:

$$\mathbf{f}(3\mathbf{x}) = \begin{bmatrix} 1 & 2 & 3 \\ 4 & 5 & 6 \end{bmatrix} \begin{bmatrix} 3a \\ 3b \\ 3c \end{bmatrix} = \begin{bmatrix} 3a + 6b + 9c \\ 12a + 15b + 18c \end{bmatrix} =$$

$$= 3 \begin{bmatrix} a + 2b + 3c \\ 4a + 5b + 6c \end{bmatrix} = 3\mathbf{f}(\mathbf{x})$$

As far as additivity is concerned, consider the same matrix A. Let:

$$\mathbf{x}^1 = \begin{bmatrix} a \\ b \\ c \end{bmatrix}, \mathbf{x}^2 = \begin{bmatrix} d \\ e \\ f \end{bmatrix}$$

and notice that:

$$\mathbf{f}(\mathbf{x}^1) = \begin{bmatrix} 1 & 2 & 3 \\ 4 & 5 & 6 \end{bmatrix} \begin{bmatrix} a \\ b \\ c \end{bmatrix} = \begin{bmatrix} a + 2b + 3c \\ 4a + 5b + 6c \end{bmatrix};$$

$$\mathbf{f}(\mathbf{x}^2) = \begin{bmatrix} 1 & 2 & 3 \\ 4 & 5 & 6 \end{bmatrix} \begin{bmatrix} d \\ e \\ f \end{bmatrix} = \begin{bmatrix} d + 2e + 3f \\ 4d + 5e + 6f \end{bmatrix}$$

We have:

$$\mathbf{f}(\mathbf{x}^1 + \mathbf{x}^2) = \begin{bmatrix} 1 & 2 & 3 \\ 4 & 5 & 6 \end{bmatrix} \left(\begin{bmatrix} a \\ b \\ c \end{bmatrix} + \begin{bmatrix} d \\ e \\ f \end{bmatrix} \right) =$$

$$= \begin{bmatrix} 1 & 2 & 3 \\ 4 & 5 & 6 \end{bmatrix} \begin{bmatrix} a \\ b \\ c \end{bmatrix} + \begin{bmatrix} 1 & 2 & 3 \\ 4 & 5 & 6 \end{bmatrix} \begin{bmatrix} d \\ e \\ f \end{bmatrix} =$$

$$= \begin{bmatrix} a + 2b + 3c \\ 4a + 5b + 6c \end{bmatrix} + \begin{bmatrix} d + 2e + 3f \\ 4d + 5e + 6f \end{bmatrix} = \mathbf{f}(\mathbf{x}^1) + \mathbf{f}(\mathbf{x}^2)$$

An important fact is that (1.7.1) defines all the conceivable linear applications. We could prove the following:

Proposition 1.7.1 *The formula:*

$$\mathbf{f}(\mathbf{x}) = A\mathbf{x} \qquad (1.7.2)$$

where: A is an (m, n)-matrix, $\mathbf{x} \in \mathbb{R}^n$, $\mathbf{f}(\mathbf{x}) \in \mathbb{R}^m$ defines an application[41] $\mathbf{f}$ from $\mathbb{R}^n$ to $\mathbb{R}^m$ which is homogeneous and additive. Vice versa: given any application $\mathbf{f}$, from

---

[41] In other words, a law that associates to each n-vector exactly one m-vector.

$\mathbb{R}^n$ *to* $\mathbb{R}^m$, *which is homogeneous and additive, there exists a (unique)* (m, n)-*matrix* A *such that:* $\mathbf{f}(\mathbf{x}) = A\mathbf{x}$.

Remark 1.7.1 We would like to comment this result. We do that from different perspectives:

(1) — Mathematician perspective. For a mathematician, Proposition 1.7.1 is exciting, because it shows that *only* homogeneity and additivity force the world to accept that the dependence of objectives on instruments is captured by linearity (1.7.2).

(2) — Social scientist perspective. For a political scientist the two crucial properties (homogeneity and additivity) appear to have less appeal than the one we would expect. Let us start from homogeneity and face two cases:

$$\begin{cases} \text{1st case: } \mathbf{x} \text{ is the instrument vector and we switch to } 1.05\mathbf{x} \\ \\ \text{2nd case: } \mathbf{x} \text{ is the instrument vector and we switch to } 2\mathbf{x} \end{cases}$$

In the first case, the variation "up 5%" of each instrument suggests us that each of the objectives will register an increase of 5%. This could happen, but, if we assume that instead of 1.05, we consider — say — 2, the fact that if you double the inputs, then you double the outputs is not so natural.

About additivity. If we consider a variation:

$$\mathbf{h} = \begin{bmatrix} h_1 \\ h_2 \\ \ldots \\ h_n \end{bmatrix}$$

in the tools vector, we can write:

$$\mathbf{f}(\mathbf{x} + \mathbf{h}) = A(\mathbf{x} + \mathbf{h}) = A\mathbf{x} + A\mathbf{h}$$

This means that the effect on objectives $\mathbf{y}$ is proportional to the variations we are introducing on the decision variables $\mathbf{x}$. Once again we could accept this evaluation for small $\mathbf{h}$, but for large $\mathbf{h}$, this could be disputable. It is incredible, but this question is the core one in the next chapter, devoted to Differential Calculus.

The important conclusion we are suggesting to the reader is that the (linear) model objective-instruments can be thought of as relevant in the case we are using it for small variations in the input (= instrument) variables, but that could be misleading in the case the instruments are exposed to not necessarily small variations. We suggest our readers look at the next chapter in this perspective.

1.8 Linear Algebraic Systems

We start with two concrete cases, in which the problem we want to study turns out to be relevant.

Example 1.8.1 Nothing new w.r.t. what we have seen in the Example 1.3.11, when discussing the Leontief model. We are given a vector of final uses $\mathbf{c} \in \mathbb{R}^n$. We are looking for a production vector $\mathbf{x}$, which allow them, or, such that:

$$(I - A)\,\mathbf{x} = \mathbf{c}$$

Mathematically, we are given a set of n linear algebraic equations in n unknowns and we would like to find such unknowns (the components of the vector $\mathbf{x}$).

Example 1.8.2 Once again looking back at the previous Example 1.7.1, based on the linear application $\mathbf{f} : \mathbb{R}^n \to \mathbb{R}^m$:

$$\mathbf{y} = A\mathbf{x}$$

The Government decides the objectives to be reached:

$$\mathbf{y} = \mathbf{b} \text{ given}$$

and should consider the system of equations:

$$A\mathbf{x} = \mathbf{b} \tag{1.8.1}$$

We want to find all the vectors $\mathbf{x}$ such that this equation holds. In principle, may be that there is no vector $\mathbf{x}$ satisfying the equation: this would mean that there is no policy allowing it to attain the objectives. May be there is only a single vector $\mathbf{x}^*$ satisfying Eq. (1.8.1), in which case there is exactly only one policy allowing the Government to fulfil its plan; maybe there are many[42] vectors satisfying the equation. In this case the Government has several ways to try to reach its objectives.

---

[42]It is easy to show that if there is more than one vector satisfying (1.8.1), then there is an infinity of vectors with that property. Let $\mathbf{x}^*$ and $\mathbf{x}^{**}$ be two different solutions:

$$A\mathbf{x}^* = \mathbf{b} \text{ and } A\mathbf{x}^{**} = \mathbf{b}$$

take any two numbers α and $(1 - \alpha)$ adding up to 1. The vector $\mathbf{v}(\alpha) = \alpha\mathbf{x}^* + (1 - \alpha)\mathbf{x}^{**}$ is another solution for every α:

$$A\mathbf{v}(\alpha) = A\left[\alpha\mathbf{x}^* + (1 - \alpha)\mathbf{x}^{**}\right] =$$
$$= \alpha A\mathbf{x}^* + (1 - \alpha) A\mathbf{x}^{**} =$$
$$= \alpha\mathbf{b} + (1 - \alpha)\mathbf{b} = \mathbf{b}$$

We will use a simple vocabulary in studying these systems:

- A is called *coefficient matrix*;
- **b** is called *known term*;
- **x***, a vector satisfying:

$$A\mathbf{x}^* = \mathbf{b}$$

  is said to be a *solution* of the system[43];
- a system with no solution is called *impossible* or *overdetermined*;
- a system with one or more solutions is called *possible*;
- systems with a unique solution are said to be *determined*;
- systems with an infinity of solutions are called *underdetermined*, for such systems, possibly after reordering the n unknowns, their solution vector **x** can be split into two blocks (subvectors):

$$\mathbf{x} = \left[\begin{array}{c} \mathbf{x}^1 \\ -- \\ \mathbf{x}^2 \end{array} \right]$$

The first block contains[44] ρ unknowns, the other gathers the remaining $n - \rho$ unknowns. The mechanism of this indeterminacy, we will discover later, boils down to the fact that, for such systems, we can choose arbitrarily values to be assigned to the $\mathbf{x}^2$ components, whose number is $n - \rho$. Once this choice has been made, the values of the first ρ unknowns will turn out to be determined by the choice(s) of $\mathbf{x}^2$. For this reason we usually say that the system allows for $n - \rho$ *degrees of freedom*. For determined systems $\rho = n$ and therefore the number of degrees of freedom for such systems will be 0.

According to our standard exposition style, we will first study a special case and then we will move to the general one.

1.8.1 A Special Case: Cramer's Systems

The characteristics of such systems, called[45] *Cramer's systems*, are two:

- The number of equations m and the number of unknowns n are equal: $m = n$, hence the coefficient matrix A is square.
- The coefficient matrix A is non-singular. Consequently there exists a unique inverse A^{-1} for it.

---

[43] Please, pay attention to the fact that a solution is a vector with n components. The solution is the vector. The components constitute the solution, but a single component is not a solution. Several times we met students that tried to convince us that each component of a vector $\mathbf{x}^*$ is a solution. They did not succeed.

[44] The choice of the letter "ρ" has been made with hindsight.

[45] Gabriel Cramer (Genève, July 31, 1704 – Bagnols-sur-Cèze, January 4, 1752) - Professor of Philosophy and Mathematics at Geneva University. He got a *PhD* at 18.

These systems are determined: there is exactly one solution for them:

$$\mathbf{x}^* = A^{-1}\mathbf{b}$$

Let us look at the simple:

Example 1.8.3 Consider the system $A\mathbf{x} = \mathbf{b}$:

$$\underbrace{\begin{bmatrix} 1 & 2 \\ 3 & 4 \end{bmatrix}}_{A} \underbrace{\begin{bmatrix} x_1 \\ x_2 \end{bmatrix}}_{\mathbf{x}} = \underbrace{\begin{bmatrix} 10 \\ 10 \end{bmatrix}}_{\mathbf{b}}$$

or:

$$\begin{cases} x_1 + 2x_2 = 10 \\ 3x_1 + 4x_2 = 10 \end{cases}$$

We are aware that A is non-singular. In fact:

$$\det \begin{bmatrix} 1 & 2 \\ 3 & 4 \end{bmatrix} = -2 \neq 0$$

We can compute its inverse:

$$A^{-1} = \begin{bmatrix} 1 & 2 \\ 3 & 4 \end{bmatrix}^{-1} = \begin{bmatrix} -2 & 1 \\ 3/2 & -1/2 \end{bmatrix}$$

and consequently:

$$\mathbf{x}^* = \begin{bmatrix} -2 & 1 \\ 3/2 & -1/2 \end{bmatrix} \begin{bmatrix} 10 \\ 10 \end{bmatrix} = \begin{bmatrix} -10 \\ 10 \end{bmatrix} \tag{1.8.2}$$

Remark 1.8.1 When dealing with a Leontief scheme, with a non-singular matrix $I - A$, the system of equations:

$$(I - A)\mathbf{x} = \mathbf{c}$$

is a Cramer's system, with solution:

$$\mathbf{x} = (I - A)^{-1}\mathbf{c}$$

1.8.2   The General Case

We are given a system of m equations in n unknowns:

$$\begin{cases} a_{11}x_1 + a_{12}x_2 + \cdots + a_{1n}x_n = b_1 \\ a_{21}x_1 + a_{22}x_2 + \cdots + a_{2n}x_n = b_2 \\ \cdots \\ a_{m1}x_1 + a_{m2}x_2 + \cdots + a_{mn}x_n = b_m \end{cases}$$

or:

$$\begin{bmatrix} a_{11} & a_{12} & \cdots & a_{1n} \\ a_{21} & a_{22} & \cdots & a_{2n} \\ \cdots & \cdots & \cdots & \cdots \\ a_{m1} & a_{m2} & \cdots & a_{mn} \end{bmatrix} \begin{bmatrix} x_1 \\ x_2 \\ \cdots \\ x_n \end{bmatrix} = \begin{bmatrix} b_1 \\ b_2 \\ \cdots \\ b_m \end{bmatrix}$$

or also, more synthetically:

$$\underset{(m,n)}{A} \underset{(n,1)}{\mathbf{x}} = \underset{(m,1)}{\mathbf{b}}$$

May be that $m \neq n$ or that, even if $m = n$, the matrix A is singular, i.e.; its rows (columns) are not linearly independent or, which is equivalent, $\det A = 0$. The strategy to solve these systems consists in manipulating the equations in order to simplify their structure, so that their eventual solution(s) appear.

We first illustrate the philosophy of the procedure in a very simple case, for which we know already the solution, then we will describe the general strategy.

Go back to the (Cramer's) system we studied in the preceding Example 1.8.3, at p. 73:

$$\begin{cases} x_1 + 2x_2 = 10 \\ \\ 3x_1 + 4x_2 = 10 \end{cases}$$

we try to obtain from each equation the value of one of the unknowns. To start the procedure it is important that the coefficient of the first unknown in the first equation is unitary. In this case it already is. Would it have been — say — 5, well we would have first divided the first equation by 5. We divide both sides of the second equation by 3, getting the new system:

$$\begin{cases} x_1 + 2x_2 = 10 \\ \\ x_1 + \dfrac{4}{3}x_2 = \dfrac{10}{3} \end{cases}$$

If we subtract from the second equation the first one, the unknown x_1 disappears in the second equation:

$$\begin{cases} x_1 + 2x_2 = 10 \\ 0x_1 + \left(\dfrac{4}{3} - 2\right) x_2 = \left(\dfrac{10}{3} - 10\right) \end{cases}$$

hence:

$$\begin{cases} x_1 + 2x_2 = 10 \\ -\dfrac{2}{3}x_2 = \dfrac{-20}{3} \end{cases}$$

Now, let us manipulate the second equation in order that the second unknown acquires a unitary coefficient. It is sufficient to multiply the second equation by $-\dfrac{3}{2}$:

$$\begin{cases} x_1 + 2x_2 = 10 \\ x_2 = 10 \end{cases}$$

Last step: we want to ban x_2 from the first equation. In order to do that it is sufficient that we add -2 times the second to it:

$$\mathbf{x}^* = \begin{cases} x_1^* = -10 \\ x_2^* = 10 \end{cases}$$

Even if the computations we have made turn out to be a bit tedious, they have brought us to the solution (1.8.2) we already knew.

It is possible to speed up the computations extracting from the system of the equations what really counts in order to find the solution. We repeat the previous computations using such a shortcut.

We start from the matrix collecting both the coefficients and the known terms:

$$[A|\mathbf{b}] = \begin{bmatrix} 1 & 2 & 10 \\ 3 & 4 & 10 \end{bmatrix}$$

frequently labelled as *complete matrix*, because it includes both the coefficients and the column of the known terms. We work on it using only elementary row operations. The strategy consists in trying to transform the block of the coefficients into an identity matrix, progressively "cleaning" each of its columns.

We will use the symbol "$\sim$" between two matrices to declare that we pass from one to the other using some elementary row operation.

First we divide the second row by 3:

$$\begin{bmatrix} 1 & 2 & 10 \\ 3 & 4 & 10 \end{bmatrix} \sim \begin{bmatrix} 1 & 2 & 10 \\ 1 & \dfrac{4}{3} & \dfrac{10}{3} \end{bmatrix}$$

then we subtract the first row from the second:

$$\begin{bmatrix} 1 & 2 & 10 \\ 1 & 4 & 10 \\ & 3 & 3 \end{bmatrix} \sim \begin{bmatrix} 1 & 2 & 10 \\ 0 & -\dfrac{2}{3} & -\dfrac{20}{3} \end{bmatrix}$$

getting a "clean" first column. Then we multiply the second row by $-\dfrac{3}{2}$, getting:

$$\begin{bmatrix} 1 & 2 & 10 \\ 0 & -\dfrac{2}{3} & -\dfrac{20}{3} \end{bmatrix} \sim \begin{bmatrix} 1 & 2 & 10 \\ 0 & 1 & 10 \end{bmatrix}$$

In order also to "clean" the second column, it suffices to subtract from the first row the double of the second:

$$\begin{bmatrix} 1 & 2 & 10 \\ 0 & 1 & 10 \end{bmatrix} \sim \begin{bmatrix} 1 & 0 & -10 \\ 0 & 1 & 10 \end{bmatrix} \tag{1.8.3}$$

We have carefully grown an identity matrix on the area previously occupied by the coefficients and we are awarded with the last column providing us with the solution.

The algorithm used on the complete matrix is exactly the one used by computers to solve systems.

What happens in general?

Take a complete matrix:

$$[A|\mathbf{b}]$$

and manipulate it (elementary row operations + possible re-ordering of the columns of A), trying to grow in the coefficients area, the largest identity matrix I_ρ you are able to construct:

$$[A|\mathbf{b}] \sim \begin{bmatrix} \underset{(\rho,\rho)}{I} & \underset{(\rho,n-\rho)}{C} & \underset{(\rho,1)}{\beta^1} \\ \underset{(m-\rho,\rho)}{O} & \underset{(m-\rho,n-\rho)}{O} & \underset{(m-\rho,1)}{\beta^2} \end{bmatrix} \tag{1.8.4}$$

It will turn out that ρ, the maximum order you can reach is nothing but $\rho(A)$, the rank of A.

It is easy to illustrate the structure of the matrix in its more general form. We will add later a few words about the two special cases

$$\begin{cases} \rho = n \\ \rho = m \end{cases}$$

The NW block is always the identity matrix of order equal to the rank of A. This rank is the number ρ of unknowns the system allows to compute, once the remaining $n - \rho$ "free ones" have been fixed, if the system has solution. The block C contains the coefficients of the free unknowns to be used to compute the others. The two O blocks are null matrices. The last column in general is split into two subcolumns. The upper block has ρ components and the lower one $m - \rho$ components. They are obtained from the vector $\mathbf{b}$ of known terms via the various elementary operations that have been done.

We learn how to extract from the matrix (1.8.4) all that we could need about the system.

- **Is the system possible or impossible?** — The answer is immediate. It is sufficient to look at the SE corner, or at the subvector β^2. There are two possibilities:

$$\begin{cases} (1) \; \beta^2 = \mathbf{0} \\[2mm] (2) \; \beta^2 \neq \mathbf{0} \end{cases}$$

or, in words: (1) all the components of such a vector are zero versus. (2) at least one component of such a vector is different from 0.
The diagnosis is straightforward: in the case (1) ($\beta^2 = \mathbf{0}$) the system is possible, while in the case (2) it is impossible. The reason is evident: in the case (2), legitimately manipulating the equations, we can obtain an equation of this type[46]:

$$0x_1 + 0x_2 + \cdots + 0x_n = \beta \neq 0$$

which is impossible. Another way to depict this situation is that using such a line in the $[A|\mathbf{b}]$ we could obtain that:

$$\rho\left([A|\mathbf{b}]\right) = \rho\left(A\right) + 1$$

so, the incomplete and the complete matrix would have different ranks. In the case (1) the two ranks are equal.
- **If the system is possible, is it determined or not?** — Again the answer is immediate. Compare the common value of the rank of both the incomplete and the complete matrix:

$$\rho\left([A|\mathbf{b}]\right) = \rho\left(A\right)$$

with n, the number of unknowns. The rank $\rho\left([A|\mathbf{b}]\right) = \rho\left(A\right)$ tells us how many truly independent equations we are given: if $\rho\left(A\right) = \rho\left([A|\mathbf{b}]\right) = n$, we can reduce the system to a Cramer's system, which is determined, while, if $\rho\left(A\right) = \rho\left([A|\mathbf{b}]\right) < n$ the system is underdetermined: it has $n - \rho\left(A\right)$ *degrees of freedom*.
- **If the system is possible, how can we find its solution(s)?** — First of all we have to note that the last $m - \rho$ equations of the system, corresponding to the blocks $\left[O|O|\beta^2\right]$ are identical and non-informative. They are identical because all of them tell us the same (uninteresting) story:

---

[46] The null coefficients do come from the O blocks, while $\beta \neq 0$, comes from $\beta^2 \neq \mathbf{0}$.

$$0x_1 + 0x_2 + \cdots + 0x_n = 0$$

and non-informative as they are satisfied by every vector $\mathbf{x} \in \mathbb{R}^n$ They can safely be dropped by the system that becomes:

$$\begin{cases} \left[\underset{(n,n)}{I} \quad \underset{(n,1)}{\beta^1} \right] & \text{in the case } \rho = n \\ \quad\quad\quad\quad \text{or} \\ \left[\underset{(\rho,\rho)}{I} \quad \underset{(\rho,n-\rho)}{C} \quad \underset{(n,1)}{\beta^1} \right] & \text{in the case } \rho < n \end{cases}$$

Once again no problem because, if $\rho(A) = n$, the structure of the matrix boils down to:

$$\left[\underset{(n,n)}{I} \quad \underset{(n,1)}{\beta^1} \right]$$

and, as in (1.8.3), you read directly the solution:

$$\mathbf{x}^* = \underset{(n,1)}{\beta^1}$$

while, if $\rho(A) < n$, then only ρ unknowns, collected in the ρ-vector $\mathbf{x}^{1*}$, can be determined as a function of the remaining $n - \rho$ unknowns, collected in $\underset{(n-\rho,1)}{\mathbf{x}^{2*}}$:

$$\underset{(\rho,1)}{\mathbf{x}^{1*}} = \underset{(\rho,1)}{\beta^2} - \underset{(\rho,n-\rho)}{C} \underset{(n-\rho,1)}{\mathbf{x}^{2*}}$$

- Substantially, the unknown vector $\mathbf{x}^*$ is splitted into two subvectors $\underset{(\rho,1)}{\mathbf{x}^*}$ and $\underset{(n-\rho,1)}{\mathbf{x}^{2*}}$:

$$\mathbf{x}^* = \left[\begin{array}{c} \mathbf{x}^{1*} \\ \hline \mathbf{x}^{2*} \end{array} \right]$$

Before a numerical example, we state in a different form the necessary and sufficient condition of possibility (in terms of equality of the two ranks). Such proposition is frequently mentioned as the *Rouché -Capelli Theorem*, even if it is somehow trivial.

Remark 1.8.2 Take the lhs of the equation $A\mathbf{x} = \mathbf{b}$ and rewrite it highlighting the columns of the coefficient matrix A:

$$A\mathbf{x} = \left[\mathbf{b}^1 | \mathbf{b}^2 | \cdots | \mathbf{b}^n \right] \left[\begin{array}{c} x_1 \\ x_2 \\ \cdots \\ x_n \end{array} \right] = x_1 \mathbf{b}^1 + x_2 \mathbf{b}^2 + \cdots + x_n \mathbf{b}^n$$

Such a lhs is a linear combination of a certain number $\rho(A)$ of columns of A that are linearly independent. If the system is possible and only in this case, the vector **b** can be obtained via a linear combination of that vectors. Should the system be impossible, this would imply that if we associate the vector **b** to the columns of A, the rank increases by 1 unit.

Remark 1.8.3 We have seen how to treat a system in the case its matrix $[A|\mathbf{b}]$ is reduced elementarily to the form (1.8.4). As the computations involved are huge, without appropriate software tools, we signal that there is an intermediate interesting possibility, which turns out to be very efficient in the case of small scale systems. The basic idea is that, instead of reaching the possibility configuration of (1.8.4), we could stop before at the following configuration:

$$\begin{bmatrix} \underset{(\rho,\rho)}{B} & \underset{(\rho,n-\rho)}{C} & \underset{(\rho,1)}{\beta^1} \\ \underset{(m-\rho,\rho)}{O} & \underset{(m-\rho,n-\rho)}{O} & \underset{(m-\rho,1)}{\mathbf{0}} \end{bmatrix} \tag{1.8.5}$$

where B is non-singular. In this case all the solutions of the system can be obtained from the Cramer's system w.r.t. $\mathbf{x}^1$:

$$B\mathbf{x}^1 = \beta^1 - C\mathbf{x}^2$$

hence:

$$\mathbf{x}^1 = B^{-1}\left[\beta^1 - C\mathbf{x}^2\right]$$

Some examples will follow soon.

A rapid series of practical examples could help our readers to become familiar with the treatment of some simple cases. We will avoid the use of the form (1.8.4) because of its huge computational cost, in the absence of appropriate tools, and we will see how to treat some interesting cases.

Example 1.8.4 Consider the following system (of the standard type $A\mathbf{x} = \mathbf{b}$):

$$\begin{bmatrix} 1 & 2 & 2 & 4 \\ 2 & k & 1 & 0 \\ 1 & 0 & 1 & 0 \end{bmatrix} \begin{bmatrix} x_1 \\ x_2 \\ x_3 \\ x_4 \end{bmatrix} = \begin{bmatrix} 10 \\ 10 \\ 10 \end{bmatrix}$$

$k \in \mathbb{R}$ is a parameter. The rank of the matrix is 3 for every k as the submatrix obtained canceling its second column:

$$\begin{bmatrix} 1 & 2 & 4 \\ 2 & 1 & 0 \\ 1 & 1 & 0 \end{bmatrix}$$

has determinant:

$$\det \begin{bmatrix} 1 & 2 & 4 \\ 2 & 1 & 0 \\ 1 & 1 & 0 \end{bmatrix} = 4 \neq 0$$

We can therefore rewrite fruitfully the system as follows:

$$\begin{bmatrix} 1 & 2 & 4 \\ 2 & 1 & 0 \\ 1 & 1 & 0 \end{bmatrix} \begin{bmatrix} x_1 \\ x_3 \\ x_4 \end{bmatrix} = \begin{bmatrix} 10 \\ 10 \\ 10 \end{bmatrix} - \begin{bmatrix} 2 \\ k \\ 0 \end{bmatrix} x_2$$

Once we choose an arbitrary value for x_2:

$$x_2 = t \in \mathbb{R}$$

for the remaining unknowns a unique value as a function of t is at hand:

$$\begin{bmatrix} x_1 \\ x_3 \\ x_4 \end{bmatrix} = \begin{bmatrix} 1 & 2 & 4 \\ 2 & 1 & 0 \\ 1 & 1 & 0 \end{bmatrix}^{-1} \left(\begin{bmatrix} 10 \\ 10 \\ 10 \end{bmatrix} - \begin{bmatrix} 2 \\ k \\ 0 \end{bmatrix} t \right)$$

hence:

$$\begin{cases} x_2^* = t \text{ arbitrary} \\ \begin{bmatrix} x_1^*(t) \\ x_3^*(t) \\ x_4^*(t) \end{bmatrix} = \begin{bmatrix} -kt \\ kt + 10 \\ -\dfrac{1}{2}t - \dfrac{1}{4}kt - \dfrac{5}{2} \end{bmatrix} \end{cases}$$

This system has exactly 1 degree of freedom, as exactly one of the unknowns can be chosen arbitrarily and the others follow.

Example 1.8.5 Consider the system $\mathbf{Ax} = \mathbf{b}$:

$$\begin{bmatrix} 1 & 2 & 3 & 4 \\ 0 & 1 & 2 & 3 \\ 1 & 3 & 5 & 7 \\ 1 & 1 & 1 & 1 \end{bmatrix} \begin{bmatrix} x_1 \\ x_2 \\ x_3 \\ x_4 \end{bmatrix} = \begin{bmatrix} 10 \\ 10 \\ 10 \\ 0 \end{bmatrix}$$

Pay attention to the submatrix at the intersection of the first 2 rows, with its first 2 columns:

$$\begin{bmatrix} 1 & 2 \\ 0 & 1 \end{bmatrix}$$

The corresponding minor:

$$\begin{Vmatrix} 1 & 2 \\ 0 & 1 \end{Vmatrix} = 1 \neq 0 \tag{1.8.6}$$

hence the rank $\rho(A) \geq 2$. Now we will try to border it in all possible ways (Kronecker method), combining the third and the fourth row with the third and the fourth column. Here are the results:

(third row and third column):

$$\begin{Vmatrix} 1 & 2 & 3 \\ 0 & 1 & 2 \\ 1 & 3 & 5 \end{Vmatrix} = 0$$

(third row and fourth column):

$$\begin{Vmatrix} 1 & 2 & 4 \\ 0 & 1 & 3 \\ 1 & 3 & 7 \end{Vmatrix} = 0$$

(fourth row and third column):

$$\begin{Vmatrix} 1 & 2 & 3 \\ 0 & 1 & 2 \\ 1 & 1 & 1 \end{Vmatrix} = 0$$

(fourth row and fourth column):

$$\begin{Vmatrix} 1 & 2 & 4 \\ 0 & 1 & 3 \\ 1 & 1 & 1 \end{Vmatrix} = 0$$

Kronecker tells us that the rank of A is 2:

$$\rho(A) = 2$$

If we complete A with the column $\mathbf{b} = \begin{bmatrix} 10 \\ 10 \\ 10 \\ 0 \end{bmatrix}$, we have to test two more minors in order to ascertain the rank of the complete matrix $[A|\mathbf{b}]$:

$$\begin{bmatrix} 1 & 2 & 3 & 4 & 10 \\ 0 & 1 & 2 & 3 & 10 \\ 1 & 3 & 5 & 7 & 10 \\ 1 & 1 & 1 & 1 & 0 \end{bmatrix}$$

The minors, we are interested in, involve the third and the fourth row, together with the last (new entry) fifth column. They turn out to be both null:

$$\begin{Vmatrix} 1 & 2 & 10 \\ 0 & 1 & 10 \\ 1 & 3 & 10 \end{Vmatrix} = \begin{Vmatrix} 1 & 2 & 10 \\ 0 & 1 & 10 \\ 1 & 1 & 0 \end{Vmatrix} = 0$$

hence, the system is possible ("Rouché-Capelli Theorem"). The minor (1.8.6) can be called *rank minor* as it is different from 0 and its order is equal to the rank of the matrix. Such information allows us to drop the equations outside its corresponding submatrix (the third and the fourth) and to state that solutions for the system can be obtained giving to the "outside variables" x_3, x_4 arbitrary values t, u and determining the first two as a function of these choices:

$$\begin{bmatrix} 1 & 2 & 3 & 4 \\ 0 & 1 & 2 & 3 \end{bmatrix} \begin{bmatrix} x_1 \\ x_2 \\ x_3 \\ x_4 \end{bmatrix} = \begin{bmatrix} 10 \\ 10 \end{bmatrix}$$

$$\Downarrow$$

$$\begin{bmatrix} 1 & 2 \\ 0 & 1 \end{bmatrix} \begin{bmatrix} x_1 \\ x_2 \end{bmatrix} + \begin{bmatrix} 3 & 4 \\ 2 & 3 \end{bmatrix} \begin{bmatrix} x_3 \\ x_4 \end{bmatrix} = \begin{bmatrix} 10 \\ 10 \end{bmatrix}$$

Let now:

$$\begin{cases} x_3^* = t \in \mathbb{R} \text{ arbitrary} \\[2ex] x_4^* = u \in \mathbb{R} \text{ arbitrary} \end{cases}$$

hence:

$$\begin{bmatrix} x_1^*(t,u) \\ x_2^*(t,u) \end{bmatrix} = \begin{bmatrix} 1 & 2 \\ 0 & 1 \end{bmatrix}^{-1} \left(\begin{bmatrix} 10 \\ 10 \end{bmatrix} - \begin{bmatrix} 3 & 4 \\ 2 & 3 \end{bmatrix} \begin{bmatrix} t \\ u \end{bmatrix} \right) = \begin{bmatrix} t + 2u - 10 \\ 10 - 3u - 2t \end{bmatrix}$$

This system has 2 degrees of freedom, as two unknowns can be chosen arbitrarily and the others follow:

$$\mathbf{x} = \begin{bmatrix} x_1^* \\ x_2^* \\ x_3^* \\ x_4^* \end{bmatrix} = \begin{bmatrix} t + 2u - 10 \\ 10 - 3u - 2t \\ t \\ u \end{bmatrix}$$

Example 1.8.6 Now compare these two systems $A\mathbf{x} = \mathbf{b}^1$ and $A\mathbf{x} = \mathbf{b}^2$, sharing the same coefficient matrix A, but with different known terms vectors $\mathbf{b}^1$, $\mathbf{b}^2$:

$$\begin{cases} 2x_1 + x_2 = 5 \\ 3x_1 = 10 \\ 6x_1 = 20 \end{cases} \quad \text{and} \quad \begin{cases} 2x_1 + x_2 = 5 \\ 3x_1 = 10 \\ 6x_1 = 0 \end{cases}$$

we can rewrite them in terms of coefficient matrix and known terms vectors:

$$\begin{bmatrix} 2 & 1 \\ 3 & 0 \\ 6 & 0 \end{bmatrix} \begin{bmatrix} x_1 \\ x_2 \end{bmatrix} = \begin{bmatrix} 5 \\ 10 \\ 20 \end{bmatrix} \text{ and } \begin{bmatrix} 2 & 1 \\ 3 & 0 \\ 6 & 0 \end{bmatrix} \begin{bmatrix} x_1 \\ x_2 \end{bmatrix} = \begin{bmatrix} 5 \\ 10 \\ 0 \end{bmatrix}$$

We can put both in the matrix form $[A|\mathbf{b}]$, the reader should be already familiar with:

$$\begin{bmatrix} 2 & 1 & 5 \\ 3 & 0 & 10 \\ 6 & 0 & 20 \end{bmatrix} \text{ and } \begin{bmatrix} 2 & 1 & 5 \\ 3 & 0 & 10 \\ 6 & 0 & 0 \end{bmatrix}$$

In both cases, the coefficient matrix has rank 2, in both cases, as $\det \begin{bmatrix} 2 & 1 \\ 3 & 0 \end{bmatrix} = -3 \neq$ 0. As the coefficient matrix has only two columns, its rank cannot exceed 2. Now let us look at the rank of the complete matrices. Computing their determinants we get:

$$\det \begin{bmatrix} 2 & 1 & 5 \\ 3 & 0 & 10 \\ 6 & 0 & 20 \end{bmatrix} = 0 \text{ and } \det \begin{bmatrix} 2 & 1 & 5 \\ 3 & 0 & 10 \\ 6 & 0 & 0 \end{bmatrix} = 60 \neq 0$$

While for the first system the rank of the complete matrix is equal to the one of the incomplete: $\rho(A) = \rho([A|\mathbf{b}^1]) = 2$ and hence by Rouché-Capelli proposition the system is possible, for the second system $\rho(A) = 2 \neq \rho([A|\mathbf{b}^2]) = 3$ and hence, by the same theorem, the system turns out to be impossible. For the first system, the solution can be obtained as follows[47]:

$$\begin{bmatrix} 2 & 1 \\ 3 & 0 \end{bmatrix} \begin{bmatrix} x_1 \\ x_2 \end{bmatrix} = \begin{bmatrix} 5 \\ 10 \end{bmatrix}$$

$$\Downarrow$$

$$\begin{bmatrix} x_1^* \\ x_2^* \end{bmatrix} = \begin{bmatrix} 2 & 1 \\ 3 & 0 \end{bmatrix}^{-1} \begin{bmatrix} 5 \\ 10 \end{bmatrix} =$$

$$= \begin{bmatrix} 0 & 1/3 \\ 1 & -2/3 \end{bmatrix} \begin{bmatrix} 5 \\ 10 \end{bmatrix} =$$

$$= \begin{bmatrix} 10/3 \\ -5/3 \end{bmatrix}$$

---

[47]The inverse:

$$\begin{bmatrix} 2 & 1 \\ 3 & 0 \end{bmatrix}^{-1}$$

could be obtained using determinants. Please note the recurrence of 3 in some denominators. It depends on the fact that the determinant of the (uncomplete) coefficient matrix is -3. See above the Example 1.8.3 on p. 73.

In this example the linear structure connecting the rows of the coefficient matrix, which is replicated in $\mathbf{b}^1$, but not in $\mathbf{b}^2$ is rather evident: in A, the third row is the double of the second. The power of the methods we have seen also lies in the fact that with reasonably simple computation we can formulate a diagnosis even in the presence of relationships that are not so evidence: think for instance of a case in which, in the coefficient matrix, the connection consists in the fact that the third row is equal to 47 times the first plus (-37) times the second: such a coefficient matrix would be:

$$
\begin{bmatrix}
3 & -1 \\
2 & 2 \\
47 \times 3 - 37 \times 2 & 47 \times (-1) - 37 \times 2
\end{bmatrix}
=
\begin{bmatrix}
3 & -1 \\
2 & 2 \\
67 & -121
\end{bmatrix}
$$

We know our readers are very smart, but we would be surprised that somebody is able to detect the linear relation among the rows simply looking at the matrix:

$$
\begin{bmatrix}
3 & -1 \\
2 & 2 \\
67 & -121
\end{bmatrix}
$$

In the analysis of the two systems in this Example, we have used determinants. An interesting exercise we suggest to our readers would consist in working out the analysis, using the reduction method. Good luck!

The AA. think that previous pages could be something as a "mathematical headache". A lot of computations, but nothing clearly relevant for Social Sciences.

It is a natural opinion, but we can easily see how difficult is to maintain the position. Let's look at the following:

Example 1.8.7 Back to the Tinbergen's objectives-tools models, we already met above, at p. 67. You are the Premier, you are interested into some two political objectives:

$$
\mathbf{y} =
\begin{bmatrix}
y_1 \\
y_2
\end{bmatrix}
$$

you can count on two instruments:

$$
\mathbf{x} =
\begin{bmatrix}
x_1 \\
x_2
\end{bmatrix}
$$

The way your targets $\mathbf{y}$ turn out to depend on your decisions $\mathbf{x}$ is very simple. It's a linear dependence:

$$
\mathbf{y} = A\mathbf{x}
$$

more clearly:

$$
\begin{bmatrix}
y_1 \\
y_2
\end{bmatrix}
=
\begin{bmatrix}
1 & 2 \\
3 & k
\end{bmatrix}
\begin{bmatrix}
x_1 \\
x_2
\end{bmatrix}
\quad \text{being } k \text{ a real parameter}
$$

or, more explicitly:

$$\begin{cases} y_1 = x_1 + 2x_2 \\ y_2 = 3x_1 + kx_2 \end{cases}$$

You could think that two tools under your control, should suffice to govern two objectives. Concretely, this means that for every choice of the objectives:

$$\mathbf{y} = \mathbf{b}$$

you have (at least) a tool vector $\mathbf{x}$, allowing you to attain your target. The problem boils down to explore the solutions of the linear system:

$$\begin{bmatrix} 1 & 2 \\ 3 & k \end{bmatrix} \begin{bmatrix} x_1 \\ x_2 \end{bmatrix} = \begin{bmatrix} b_1 \\ b_2 \end{bmatrix} \tag{1.8.7}$$

It is clear that if the coefficient matrix $\begin{bmatrix} 1 & 2 \\ 3 & k \end{bmatrix}$ is non-singular, then the system is a Cramer system with the unique solution:

$$\begin{bmatrix} x_1^* \\ x_2^* \end{bmatrix} = \begin{bmatrix} 1 & 2 \\ 3 & k \end{bmatrix}^{-1} \begin{bmatrix} b_1 \\ b_2 \end{bmatrix} = \begin{bmatrix} \dfrac{kb_1 - 2b_2}{k-6} \\ \dfrac{3b_2 - 9b_1}{k-6} \end{bmatrix} \quad \text{for } k \neq 6$$

describing the policy $\begin{bmatrix} x_1^* \\ x_2^* \end{bmatrix}$ you have to choose in order to attain your objectives.

Assume now that $k = 6$. If so, the coefficient matrix is singular (with rank 1). Back, immediately to the system (1.8.7):

$$\begin{bmatrix} 1 & 2 \\ 3 & k \end{bmatrix} \begin{bmatrix} x_1 \\ x_2 \end{bmatrix} = \begin{bmatrix} b_1 \\ b_2 \end{bmatrix} \quad \text{or} \quad \begin{cases} x_1 + 2x_2 = b_1 \\ 3x_1 + 6x_2 = b_2 \end{cases}$$

Dividing by 3 both sides of the second equation we get:

$$\begin{cases} x_1 + 2x_2 = b_1 \\ x_1 + 2x_2 = b_2/3 \end{cases}$$

If $b_1 \neq b_2/3$, in which case the Rouché-Capelli condition is violated, there is no policy $\mathbf{x}^*$ allowing for the success in attaining the objectives. If the Rouché-Capelli condition is satisfied, i.e.; if $b_1 = b_2/3$, the second equation is superfluous and there is an infinity of vectors $\mathbf{x}^*$ allowing for the success. From $x_1 + 2x_2 = b_1$ we get:

$$x_2 = (b_1 - x_1)/2$$

hence:

$$\mathbf{x}^* = \begin{bmatrix} \alpha \\ (b_1 - \alpha)/2 \end{bmatrix}, \text{ with } \alpha \in \mathbb{R} \text{ arbitrary}$$

A concrete conclusion suggested by this example is that if the dependence of the political objectives on the instruments is "affected" by some linear dependence, such dependence must be taken into account when establishing political targets.

1.9 Applications to Networks

An important tool, recently emerged in Social Sciences, is the *network analysis*.

Think of some population of n individuals (later called *agents*), numbered $s = 1, 2, \ldots, n$. Think also of a matrix A collecting the connections between them.

An n-square matrix could describe this basic connection state:

$$C = \begin{bmatrix} 1 & 0 & 1 \\ 1 & 1 & 0 \\ 0 & 0 & 1 \end{bmatrix}$$

Rows must be interpreted as follows:

$$c_{r,s} = \begin{cases} 1 & \text{if Mr./Ms. } r \text{ is connected to Mr./Ms. } s \\ 0 & \text{f Mr./Ms. } r \text{ is not connected to Mr./Ms. } s \end{cases}$$

It is interesting to study whether there are indirect connections.

The 0s and 1s you read in C are not "normal" numbers, but *Boolean numbers*.

Boole's numeric system is absolutely coherent with common sense if you interprete 0 as false, 1 as true. The Boolean Arithmetic is very simple:

- Multiplication works as usual;
- Addition suggests us that 1(truth) +1 (truth) = 1 ($n > 1 \rightarrow 1$).

As for 2nd order relationships, we get, in Boolean arithmetic:

$$\begin{bmatrix} 1 & 0 & 1 \\ 1 & 1 & 0 \\ 0 & 0 & 1 \end{bmatrix} \cdot \begin{bmatrix} 1 & 0 & 1 \\ 1 & 1 & 0 \\ 0 & 0 & 1 \end{bmatrix} = \begin{bmatrix} 1 & 0 & 1 \\ 1 & 1 & 1 \\ 0 & 0 & 1 \end{bmatrix}$$

As we are working with a matrix of the third order, we would like to to explore the third power[48] of C, according to Boolean arithmetic:

---

[48] What happens if the dimension is greater than n is irrelevant.

$$\begin{bmatrix} 1 & 0 & 1 \\ 1 & 1 & 1 \\ 0 & 0 & 1 \end{bmatrix} \cdot \begin{bmatrix} 1 & 0 & 1 \\ 1 & 1 & 0 \\ 0 & 0 & 1 \end{bmatrix} = \begin{bmatrix} 1 & 0 & 1 \\ 1 & 1 & 1 \\ 0 & 0 & 1 \end{bmatrix}$$

The answers we get is that the connections between people are: directly

$$\begin{cases} \text{Mr/Ms 1 are directly or non-directly connected with Mr/Ms 3} \\ \text{Mr/Ms 2 are directly or non-directly connected with all agents} \\ \text{Mr/Ms 3 are connected only with themselves} \end{cases}$$

Such connections can be easily qualified as information exchanges.

In a network an obvious interest concerns subjects which turn out to be relevant in this process of information exchange.

Some technical tools are offered to evaluate the central relevance of each agent.

Freeman index suggests us, with reference to a population of n agents, to compute the difference between the average number of connections of some certain agent and the average number of connections, to be divided by $(n-1)(n-2)$.

Example 1.9.1 The connections of the agent 1 are 1, the ones of agent 2 are equally, while the ones of agent 3 are 0. The total number of connections is $1 + 1 = 2$. The average number of connections turns out to be:

$$c_{\text{average}} = \frac{1 + 2 + 0}{3} = 1$$

The *centrality indices* of the three agents are respectively:

$$\begin{cases} \text{agent 1} & c_1 = \frac{1}{3} \\ \text{agent 2} & c_2 = \frac{2}{3} \\ \text{agent 3} & c_3 = 0 \end{cases}$$

The centrality index is:

$$\sum_{s}^{3} \frac{c_s - c_{\text{average}}}{2 \cdot 1} = \frac{\left(\frac{1}{3} - 1\right) + \left(\frac{2}{3} - 1\right) + (0 - 1)}{2 \cdot 1} = -1$$

This negative results tells us that in this population there is no "central agent".

1.10 Some Complements on Square Matrices

Research in Social Sciences often requires more advanced notions in Linear Algebra. We will use them later, when studying Dynamic Systems (see Chap. 4). Some long-run properties of Dynamic Systems turn out to depend crucially on what we are going to see here.

Let us consider an n-square matrix A:

$$A = \begin{bmatrix} a_{11} & a_{12} & \cdots & a_{1n} \\ a_{21} & a_{22} & \cdots & a_{2n} \\ \cdots & \cdots & \cdots & \cdots \\ a_{n1} & a_{n2} & \cdots & a_{nn} \end{bmatrix}$$

It characterizes a linear applications from $\mathbb{R}^n$ to $\mathbb{R}^n$, defined by equations of the type:

$$\mathbf{y} = A\mathbf{x} \tag{1.10.1}$$

being $\mathbf{x}$, $\mathbf{y}$ in $\mathbb{R}^n$.

The matrix A can be thought of as describing transformation of a system (now at $\mathbf{x}$) into the new position (at $\mathbf{y}$). In general the new position is far from the initial one: a linear transformation mixes things...

Consider:

Example 1.10.1 — Let:

$$A = \begin{bmatrix} 1 & 2 \\ 0 & 3 \end{bmatrix}$$

Take the vector $\mathbf{x} = \begin{bmatrix} 2 \\ 1 \end{bmatrix}$. Transform it through (1.10.1). You get:

$$\begin{bmatrix} 1 & 2 \\ 0 & 3 \end{bmatrix} \begin{bmatrix} 2 \\ 1 \end{bmatrix} = \begin{bmatrix} 4 \\ 3 \end{bmatrix}$$

the new vector $\begin{bmatrix} 4 \\ 3 \end{bmatrix}$ is somehow extraneous to $\mathbf{x}$. A linear transformation acts on vectors under two aspects: (1) size the vector $\begin{bmatrix} 4 \\ 3 \end{bmatrix}$ is larger in some sense than $\begin{bmatrix} 2 \\ 1 \end{bmatrix}$; (2) direction. The first aspect is easily expected, while the second is not so trivial, because it concerns the relative weight of the components. In $\mathbf{x}$ the first component is the double of the second one. In $\mathbf{y}$ the first component is only $1/3$ greater than the second one.

These phenomena depend substantially on the fact that the standard reference system in $\mathbb{R}^n$ are non-necessarily the natural reference scheme for the description of a given social situation.

We borrow from Economics an interpretation of this scheme.

Consider an economic system, consisting of n se about a historic contribution in [10].

This is Economics, but there is Politics too. A political issue could be that of allowing an economic system to grow harmonically, i.e.; with the same growth rate $\lambda - 1$ for each sector.[49]

Let $\mathbf{x}$ be the vector of the stocks of capital at some date. Let also A be the matrix transforming the current state $\mathbf{x}$ into the new one. A balanced evolution would be described by the idea that the 'new' stocks of capital evolve according to the *same growth rate* across sectors.

Easily this idea of harmonic growth can be translated into this equation:

$$A\mathbf{x} = \lambda\mathbf{x} \tag{1.10.2}$$

Its concrete interpretation is that the transformation of $\mathbf{x}$ into a new state $\mathbf{y}$ boils down to multiply the current state $\mathbf{x}$ with some constant λ, guaranteeing the same growth rate to each sector.

Example 1.10.2 Take (1.10.2) and assume that, for some non-null vector $\mathbf{x}$ we have:

$$A\mathbf{x} = 1.1\mathbf{x}$$

This concretely means that the capital stock invested in each of the n sectors grows of $10\% = 1.1 - 1$. The policy implication is that the current distribution of capital across sectors (the vector $\mathbf{x}$), modified by the structure of the economic system (the matrix A), allows for the harmonic growth (at 10%) of each sector.

Two ingredients must be mixed in focusing this interesting situation:

- some non-null[50] n-vector $\mathbf{x}$;
- some appropriate scalar λ.

Remark 1.10.1 Please, note that, if there is some non-null vector $\mathbf{x}$ satisfying (1.10.2), there is an infinity of them: $\mathbf{x}^\alpha = \alpha\mathbf{x}$. That is: all the real multiples of $\mathbf{x}$, which would be in such a family with $\alpha = 1$: $\mathbf{x}^1 = 1 \cdot \mathbf{x}$:

$$A\mathbf{x}^\alpha = A\alpha\mathbf{x} = \lambda\mathbf{x} = \lambda\mathbf{x}^\alpha$$

Look at this special case:

$$A = \begin{bmatrix} 1 & 2 \\ 0 & 3 \end{bmatrix}$$

The vector $\mathbf{x}^1 = \begin{bmatrix} 1 \\ 0 \end{bmatrix}$ satisfies (1.10.2) with $\lambda = 1$:

---

[49]Trivially: the growth rate $\lambda - 1 = 10\%$ implies the growth factor $\lambda = 1 + 10\% = 1.1$.

[50]It's trivial, but frequently ignored: if $\mathbf{x} = \mathbf{o}$, the equality $A\mathbf{x} = \lambda\mathbf{x}$ is obviously satisfied for every λ. The economic interpretation of this mathematical fact is obvious: an economy without any capital stock can grow at any rate, which will bring it to... zero-stock everywhere.

$$A\mathbf{x} = \begin{bmatrix} 1 & 2 \\ 0 & 3 \end{bmatrix} \begin{bmatrix} 1 \\ 0 \end{bmatrix} = \begin{bmatrix} 1 \\ 0 \end{bmatrix} \text{ and } 1 \cdot \begin{bmatrix} 1 \\ 0 \end{bmatrix} = \begin{bmatrix} 1 \\ 0 \end{bmatrix} \qquad (1.10.3)$$

Well, for any real α, generating $\mathbf{x}^\alpha = \alpha \begin{bmatrix} 1 \\ 0 \end{bmatrix} = \begin{bmatrix} \alpha \\ 0 \end{bmatrix}$, we reach the same conclusion as in (1.10.3):

$$A\mathbf{x}^\alpha = \begin{bmatrix} 1 & 2 \\ 0 & 3 \end{bmatrix} \begin{bmatrix} \alpha \\ 0 \end{bmatrix} = \begin{bmatrix} \alpha \\ 0 \end{bmatrix} \text{ and } 1 \cdot \begin{bmatrix} \alpha \\ 0 \end{bmatrix} = \begin{bmatrix} \alpha \\ 0 \end{bmatrix}$$

Definition 1.10.1 Given a square matrix A of order n, we say that it admits as *eigenvalue* λ if there exists a non-null vector $\mathbf{v}$, such that:

$$A\mathbf{v} = \lambda\mathbf{v} \qquad (1.10.4)$$

Such a non-null vector $\mathbf{v}$ is said to be an *eigenvector* for A, *associated with the eigenvalue* λ.

Example 1.10.3 Recall Remark 1.10.1. For the matrix $\begin{bmatrix} 1 & 2 \\ 0 & 3 \end{bmatrix}$, $\lambda = 1$ is an eigenvalue and the infinite family of vectors $\begin{bmatrix} \alpha \\ 0 \end{bmatrix}$, $\alpha \in \mathbb{R}$, $\alpha \neq 0$, collects the eigenvectors associated with the eigenvalue $\lambda = 1$.

How to find eigenvalues and eigenvectors for a given square matrix A?
Back to the fundamental equation (1.10.2), which can be trivially rewritten as:

$$[\lambda I - A]\mathbf{x} = \mathbf{o}$$

With respect to $\mathbf{x}$, it is a linear algebraic system in n unknowns (the components of $\mathbf{x}$). If it is a Cramer's system it has the unique trivial solution $\mathbf{x} = \mathbf{o}$, to which we are not interested. See p. 89.

In order that the system (1.10.4) is not a Cramer's system, it is necessary and sufficient that its coefficient matrix $\lambda I - A$ be singular. Using determinants:

$$\det [\lambda I - A] = 0$$

The l.h.s. of the above equation always is a polynomial $P(\lambda)$ of nth degree in λ:

$$\det [\lambda I - A] = \lambda^n + \alpha_{n-1}\lambda^{n-1} + \alpha_{n-2}\lambda^{n-2} + \cdots + \alpha_1\lambda + \alpha_0$$

where the coefficients α_k, $k = 0, 1, \cdots, n-1$ depend on the elements of the square matrix A.

Definition 1.10.2 Let A be a square matrix of order n. The polynomial $P(\lambda)$ of nth degree in λ:

$$P(\lambda) = \det[\lambda I - A] = \lambda^n + \alpha_{n-1}\lambda^{n-1} + \alpha_{n-2}\lambda^{n-2} + \cdots + \alpha_1\lambda + \alpha_0$$

is called *characteristic polynomial* of A. The equation

$$P(\lambda) = \lambda^n + \alpha_{n-1}\lambda^{n-1} + \alpha_{n-2}\lambda^{n-2} + \cdots + \alpha_1\lambda + \alpha_0 = 0$$

is called *characteristic equation* of A.

The *eigenvalues* of A are nothing but the solutions of its characteristic equation:

$$\lambda_k \text{ is an eigenvalue of } A \Longleftrightarrow \lambda_k \text{ solves } P(\lambda) = 0$$

Remark 1.10.2 Look at the very special case:

$$A = \begin{bmatrix} a_{11} & a_{12} \\ a_{21} & a_{22} \end{bmatrix}$$

The characteristic polynomial for A is:

$$\det\left(\lambda\begin{bmatrix} 1 & 0 \\ 0 & 1 \end{bmatrix} - \begin{bmatrix} a_{11} & a_{12} \\ a_{21} & a_{22} \end{bmatrix}\right) = \lambda^2 - \lambda(a_{11} + a_{22}) + a_{11}a_{22} - a_{12}a_{21}$$

In a square matrix A, the sum of its principal diagonal elements $(a_{11}, a_{22}, \ldots, a_{nn})$ is called *trace* of A:

$$\text{tr}(A) = a_{11} + a_{22} + \ldots + a_{nn}$$

in our case:

$$\text{tr}(A) = a_{11} + a_{22}$$

The term independent of λ in the polynomial is nothing but the determinant of A. therefore for 2-square matrices the characteristic polynomial turns out to be:

$$P(\lambda) = \lambda^2 - \text{tr}(A)\lambda + \det(A)$$

Example 1.10.4 Take:

$$A = \begin{bmatrix} 1 & 2 \\ 0 & 3 \end{bmatrix}$$

Its characteristic polynomial is[51]:

$$P(\lambda) = \det\left(\lambda\begin{bmatrix} 1 & 0 \\ 0 & 1 \end{bmatrix} - \begin{bmatrix} 1 & 2 \\ 0 & 3 \end{bmatrix}\right) = \lambda^2 - 4\lambda + 3$$

---

[51] Please note that $\text{tr}(A) = 1 + 3 = 4$ and that $\det A = 3$.

Equating to 0 the characteristic polynomial, we get two eigenvalues $\lambda_1 = 1, \lambda_2 = 3$ for the matrix. All the eigenvectors associated to λ_1 have been already found above at p. 89. They are:

$$\mathbf{v} = \begin{bmatrix} \alpha \\ 0 \end{bmatrix} = \alpha \begin{bmatrix} 1 \\ 0 \end{bmatrix}, \alpha \neq 0$$

As concerns the ones associated to the eigenvalue $\lambda_2 = 3$, we have to explore the linear system:

$$\left(3 \begin{bmatrix} 1 & 0 \\ 0 & 1 \end{bmatrix} - \begin{bmatrix} 1 & 2 \\ 0 & 3 \end{bmatrix} \right) \begin{bmatrix} v_1 \\ v_2 \end{bmatrix} = \begin{bmatrix} 0 \\ 0 \end{bmatrix}$$

or, less cryptically:

$$\begin{cases} 2v_1 - 2v_2 = 0 \\ 0v_1 + 0v_2 = 0 \end{cases}$$

The second equation is far from being informative, while the first one produces the family of eigenvectors:

$$\mathbf{v} = \begin{bmatrix} \beta \\ \beta \end{bmatrix} = \beta \begin{bmatrix} 1 \\ 1 \end{bmatrix}, \beta \neq 0$$

Concluding. The matrix:

$$A = \begin{bmatrix} 1 & 2 \\ 0 & 3 \end{bmatrix}$$

has two eigenvalues:

$$\lambda_1 = 1 \text{ and } \lambda_2 = 3$$

the associated families of eigenvectors are respectively:

$$\text{for } \lambda_1 = 1 \text{ the eigenvectors are } \begin{bmatrix} \alpha \\ 0 \end{bmatrix}$$

$$\text{for } \lambda_2 = 3 \text{ the eigenvectors are } \begin{bmatrix} \beta \\ \beta \end{bmatrix}$$

Let us look now at the political implications of these computations. If we interpret (1.10.1) as the evolution rule of some economy. Maths tells us that if we think of steady state growth possibilities, given the interactions between the two sectors 1, 2 of this economy (described by the matrix A), there are two possibilities:

(1) — Option A: capital is concentrated in sector 1. Sector 2 does not exist and sector 1 will exhibit a null growth rate 0%:

$$\lambda_1 - 1 = 1 - 1 = 0$$

(2) — Option B: capital is equally allocated to the two sectors: something is given to the sector 1 and the same to sector 2. The growth rate of the economy jumps to 200%:

$$\lambda_2 - 1 = 3 - 1 = 2$$

A natural question for a politician would be 'Well, guys, we are happy for these interesting insights, but what happens in the case the capital allocation is not one of the the extreme ones envisaged above?'. The answer is postponed to Chap. 4.

1.11 Exercises

Exercise 1.11.1 *Let A and B be two matrices defined by*

$$A = \begin{bmatrix} -1 & 0 & 2 \\ 3 & 1 & -2 \end{bmatrix} \quad B = \begin{bmatrix} 2 & -1 \\ 3 & 5 \\ 0 & -2 \end{bmatrix}$$

Compute the following products: (a) AB; (b) AA^{T} and BB^{T}; (c) BA.

Exercise 1.11.2 *Let*

$$A = \begin{bmatrix} 1 & 2 \\ 2 & 0 \\ 1 & -2 \\ 5 & -2 \end{bmatrix} \quad B = \begin{bmatrix} 1 & 0 & -1 \\ 2 & 1 & 4 \end{bmatrix} \quad C = \begin{bmatrix} 4 & -4 & 3 & 1 \\ 1 & 0 & -1 & 0 \end{bmatrix}$$

Find AB, AC and CA.

Exercise 1.11.3 *Let's consider the following linear systems*

$$(3.1) \begin{cases} 2x_1 + x_2 = 3 \\ x_1 - x_2 = 1 \end{cases} ; \quad (3.2) \begin{cases} x_1 + x_2 - x_3 = 3 \\ x_1 - x_2 + 2x_3 = 1 \end{cases}$$

$$(3.3) \begin{cases} x_1 + 2x_2 - x_3 = 2 \\ 2x_1 - x_2 + 2x_3 = 1 \end{cases}$$

(a) *Write the coefficient matrix A and the augmented matrix $[A|\mathbf{b}]$ of the linear system.*
(b) *Write, using elementary row operations, $[A|\mathbf{b}]$ in upper-triangular form[52].*
(c) *Using back substitution, find the solution of the system.*

Exercise 1.11.4 *Let's consider the following second order square matrix*

$$A = \begin{bmatrix} 1 & 2 \\ 1 & \alpha \end{bmatrix}$$

and α real number (i.e. $\alpha \in \mathbb{R}$).

---

[52] All the entries under the principal diagonal are null.

(a) *Calculate the determinant of A and discuss for which values of α, if they exist, the inverse matrix of A exists. From now on, let's continue the exercise considering $\alpha = 4$.*

(b) *Find the inverse of A and calculate the unique solution of the following linear system*

$$Ax = \begin{bmatrix} 1 \\ 1 \end{bmatrix}$$

(c) *Let's consider the equation $Ax = x + b$, where*

$$A = \begin{bmatrix} 1 & 2 \\ 1 & 4 \end{bmatrix} \quad and \quad b = \begin{bmatrix} 4 \\ 7 \end{bmatrix}$$

*Find the vector **x**.*

Exercise 1.11.5 *Let a be a real number. Consider the following second order square matrix*

$$A = \begin{bmatrix} 2a & 3 \\ a & a+4 \end{bmatrix}$$

(a) *Find the values of a for which the matrix A is not invertible.*

(b) *For all other values of a, find the expression of the inverse matrix A.*

Exercise 1.11.6 *Let's consider the following matrices*

$$A = \begin{bmatrix} 2 & 1 \\ -1 & 3 \end{bmatrix} \quad and \quad B = \begin{bmatrix} 1 & 1 \\ 0 & 1 \end{bmatrix}$$

(a) *Find the following inverse matrices*

$$A^{-1} \;\; ; \;\; B^{-1} \;\; ; \;\; (A+B)^{-1} \;\; ; \;\; (A+B)^{-2}$$

(b) *Calculate A^3 and B^3.*

(c) *Solve the equation $Ax + Bx = I_2$, where I_2 is the identity matrix with order 2.*

Exercise 1.11.7 *The government of the country Tintaly is well aware of Tinbergen's objectives-tools model. The numbers of objectives and tools are $m = n = 2$. The relationship between the tool vector $x = \begin{bmatrix} x_1 \\ x_2 \end{bmatrix}$ and the objective vector $y = \begin{bmatrix} y_1 \\ y_2 \end{bmatrix}$ is the standard one*

$$y = Ax \qquad\qquad (1.11.1)$$

where

$$A = \begin{bmatrix} 0.1 & 0.3 \\ 0.5 & 1 \end{bmatrix}$$

(a) *Determine the matrix B which allows us to find the policy tools* $\mathbf{x}$ *as a linear function of the objectives* $\mathbf{x} = B\mathbf{y}$.

(b) *Assume that the "clean" endogenous relationship (1.11.1) is disturbed by some exogenous factor, so that (1.11.1) becomes:*

$$\mathbf{y} = A\mathbf{x} + \mathbf{b}(t)$$

where $\mathbf{b}(t) \in \mathbb{R}^2$, *and t is an exogenous parameter (for instance the surplus or the deficit of the Balance of payments of Tintaly). The Government policy, consisting in choosing* $\mathbf{x}$, *turns out to depend on t, so that* $\mathbf{x}$ *becomes* $\mathbf{x}(t)$. *Find an expression of* $\mathbf{x}(t)$ *in terms of* $\mathbf{y}$, A, $\mathbf{b}(t)$.

(c) *A research institute maintains that if the type of* $\mathbf{b}(t)$ *is*

$$\mathbf{b}(t) = \alpha + \beta t \qquad \alpha, \beta \in \mathbb{R}^2$$

then the policy $\mathbf{x}(t)$, *in the presence of* $\mathbf{b}(t)$, *is of the type*

$$\mathbf{x}(t) = \mathbf{u} + \mathbf{v}t$$

What is your opinion about this?

Exercise 1.11.8 *Consider a Leontief system with two sectors. The input-output coefficient matrix is:*

$$A = \begin{bmatrix} 0.1 & \alpha \\ 5 & 0.1 \end{bmatrix}$$

(a) *Compute* $\det[I - A]$.

(b) *Mr. Bean, a member of Parliament, maintains that even if the coefficient* α *at position* $(1, 2)$ *is not known for positive values of* α, *the matrix* $[I - A]^{-1}$ *always exists. Is he right or not? Answer and justify your answer.*

(c) *Again Mr. Bean has declared his total confidence in this productive system: no matter what the true value of* α *is, the system is viable. What is your opinion based on the Hawkins-Simon conditions? For the values of* α *such that the system is viable, compute the production vector*

$$\begin{bmatrix} x_1 \\ x_2 \end{bmatrix}$$

which allows the vector of final uses $\begin{bmatrix} 500 \\ 500 \end{bmatrix}$.

Exercise 1.11.9 *Consider a Leontief economy with two sectors. The coefficient matrix is*

$$A = \begin{bmatrix} 0.1 & b \\ b & 0.2 \end{bmatrix}$$

(a) *Using Hawkins-Simon conditions, determine for which values of b (obviously non-negative) the system is viable.*

(b) *Under the viability assumption, compute the production vector x^* which allows for the final uses*

$$\mathbf{c} = \begin{bmatrix} 100 \\ 200 \end{bmatrix}$$

(c) *Compute the vector*

$$\mathbf{i} = \begin{bmatrix} i_1 \\ i_2 \end{bmatrix}$$

of the implied intermediate uses.

Exercise 1.11.10 *In a country, the economic system includes only three sectors. The input-output coefficient matrix is*

$$A = \begin{bmatrix} 0.1 & 0.4 & 0 \\ 0 & 0.1 & 0.1 \\ 0 & 0 & 0.2 \end{bmatrix}$$

(a) *Construct the matrix $I - A$ and compute $\det [I - A]$.*

(b) *Using the Hawkins-Simon conditions decide whether the economic system is viable or not.*

(c) *Consider the following vector of final uses*

$$\mathbf{c} = \begin{bmatrix} c \\ c \\ c \end{bmatrix}$$

being $c > 0$. Determine the output vector $\mathbf{x} = \begin{bmatrix} x_1 \\ x_2 \\ x_3 \end{bmatrix}$ which allows for the final uses $\mathbf{c}$.

Exercise 1.11.11 *We are in a bipolar political system. There are two parties B, R. Voters change their preferences weekly. This evolution is modeled according to a Markov model. The voter groups are B, R, U: B = Blue, R = Red and U = Uncertain. The following table shows the % expected movements from group to group*

↙	B	R	U
B	0.5	0.1	0.4
R	0.1	0.6	0.4
U	0.4	0.3	0.2

We can extract from the above the standard transition matrix

$$A = \begin{bmatrix} 0.5 & 0.1 & 0.4 \\ 0.1 & 0.6 & 0.4 \\ 0.4 & 0.3 & 0.2 \end{bmatrix}$$

The initial composition of the voter population is described by the following vector

$$\mathbf{x}^0 = \begin{bmatrix} 1100 \\ 1000 \\ 900 \end{bmatrix}$$

in which the rows correspond respectively to B, R and U.

(a) Compute the expected composition of the vector population, one week later

$$\mathbf{x}^1 = A\,\mathbf{x}^0$$

(b) Do you think that the matrix $(I - A)^{-1}$ exists or not? Why?
(c) Compute A^2 and determine the composition of the population two weeks later.

Chapter 2
Differential Calculus

2.1 What's a Function

A candidate for some election plans her/his campaign. Call x the number of voters she/he will contact and y the cost of the campaign. Such a cost depends on the number x of contacted voters. Such dependence is represented mathematically as follows:

$$y = f(x)$$

to be read "y equal to f of x". We can think of f as a black box which transforms the input x, the number of people you contact, into the output y, the number of votes you gather:

$$x \longrightarrow \boxed{\;\| \; f \; \| \;} \longrightarrow y$$

As x is a real number (an element of $\mathbb{R}$) and the same is true for y, a popular way to write, we already met while studying Linear Algebra, which illustrates the nature of f, is:

$$f : \mathbb{R} \to \mathbb{R}$$

Now assume that the cost components of the electoral campaign are:

- a fixed amount € 5000;
- a variable amount, proportional to x, the number of voters reached: € 0.5 per voter, usually called *unitary variable cost*;

The total cost of the campaign is:

$$y = 5000 + 0.5x$$

This is a very simple example of function in which the unitary variable cost is $v = 0.5$.

A slightly more complex example would be obtained in the case the unitary variable cost, instead of being constant ($= 0.5$), i.e., € 0.5, would realistically diminish

© Springer Nature Switzerland AG 2018
L. Peccati et al., *Maths for Social Sciences*, UNITEXT - La Matematica
per il 3 + 2 113, https://doi.org/10.1007/978-3-030-02336-2_2

Fig. 2.1 Cost function

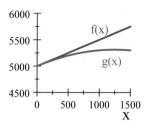

if x increases, for instance:

$$v(x) = 0.5 - 0.2 \cdot \frac{x}{1000}$$

A new cost function g would be defined by:

$$g(x) = 5000 + xv(x) = 5000 + x\left(0.5 - 0.2 \cdot \frac{x}{1000}\right) = 5000 + 0.5x - 0.0002x^2$$

Functions f, g are usually represented through their *graph*, the set of the points in a Cartesian plane with abscissas x and corresponding ordinates $f(x)$ or $g(x)$, as Fig. 2.1 shows. We will learn how to construct such graphs also in the case of "difficult" functions, which turn out to be useful in socio-political applications.

2.1.1 Intervals

In handling functions, we will use the notion of *interval*. There are four types of intervals:

1. $[a, b]$, the set of numbers x such that $a \le x \le b$, they are called *closed* as they include both their endpoints a, b;
2. (a, b), the set of numbers x such that $a < x < b$, they are called *open* as they do not include their endpoints a, b;
3. $[a, b)$, the set of numbers x such that $a \le x < b$, they are called *semi-open to the right* as they do not include their upper endpoint b;
4. $(a, b]$, the set of numbers x such that $a < x \le b$, they are called *semi-open to the left* as they do not include their lower endpoint a.

It is customary to include in the appropriate type of semi-open intervals:

- $[a, +\infty)$, the set of numbers x such that $x \ge a$;
- $(-\infty, b]$, the set of numbers x such that $x \le b$;

and, among the open intervals:

- $(a, +\infty)$, the set of numbers x such that $x > a$;

- $(-\infty, b)$, the set of numbers x such that $x < b$;
- $(-\infty, +\infty) = \mathbb{R}$.

2.1.2 Easy Functions

Functions of the type :

$$y = mx + q$$

with parameters $m, q \in \mathbb{R}$ can be graphed easily. Their technical name is *linear* (*affine*) functions because their graph is a (non-vertical) straight line. The parameter m is called *slope* of the straight line. Positive values of m characterize straight lines for which the ordinate y increases when the abscissa x increases, moving from the left to the right; negative values of m characterize decreasing ordinate, while $m = 0$ (for instance $y = 0x + 5$ or $y = 5$) characterizes horizontal straight lines. The parameter q is called *intercept* and it is nothing but the ordinate of the point where the straight line crosses the vertical axis. For instance the easy function:

$$y = 3x - 5$$

is represented by a straight line with slope 3 and intercept -5 as shown in Fig. 2.2:

Fig. 2.2 Linear affine

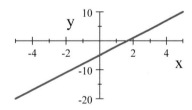

We will call these functions *easy*, because the construction of their graph is easy. By joke, we will call *difficult* the functions whose graph is not immediate. For instance, a difficult function is:

$$y = x - x^3$$

In this chapter, among other things, we will learn how to draw the graph of several difficult functions.

2.1.3 Elementary Functions

Many models, of interest for Social Sciences, are based on functions, that even "difficult", in the fun sense above, turn out to be rather elementary. It is important that the interested readers become familiar with these functions, both for counting on them as basic examples and for using them to approach modeling of social facts.

2.1.3.1   Quadratic Functions

These functions are of the type:

$$y = ax^2 + bx + c \text{ with } a \neq 0.$$

Their graph is a parabola with vertical axis. The quantity:

$$\Delta = b^2 - 4ac$$

is called *discriminant* of the polynomial $ax^2 + bx + c$.

The parabola can be concave (of the type "hat"), if a is negative, or convex (of the type "cup"), if a is positive.

Here are two examples.

Example 2.1.1 Let's consider the following quadratic functions:

$$y_1 = 2x^2 - 4x \quad \text{and} \quad y_2 = 1 - x^2$$

The coefficient a of y_1 is positive (i.e.; $a = 2$), therefore y_1 is convex. Meanwhile the coefficient a of y_2 is negative (i.e.; $a = -1$), therefore y_2 is concave. In general, the *vertex* of the parabola has coordinates:

$$\begin{bmatrix} x_V \\ y_V \end{bmatrix} = \begin{bmatrix} -\dfrac{b}{2a} \\ -\dfrac{\Delta}{4a} \end{bmatrix}$$

If Δ is negative, the parabola does not cross the abscissas axis, if Δ is 0, it touches it at exactly one point, if Δ is positive it crosses the abscissas axis at the two points:

$$\left.\begin{matrix} x_1 \\ x_2 \end{matrix}\right\} = \frac{-b \pm \sqrt{\Delta}}{2a}$$

We have already seen above (at p. 100), for instance, the relevance of such functions in evaluating an electoral campaign.

Example 2.1.2 Let's consider the quadratic function defined by

$$f(x) = x^2 - 4x - 5$$

Fig. 2.3 Parabolas

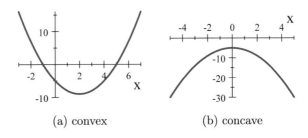

(a) convex   (b) concave

First of all, we observe that $a = 1 > 0$ then f is convex. The discriminant of $x^2 - 4x - 5$ is $\Delta = 36 > 0$. The parabola representing the function on the plane has vertex $V(2, -9)$ (the turning point on the parabola) and from the formulas

$$x_1 = \frac{-4 - \sqrt{36}}{2} = -5 \quad \text{and} \quad x_2 = \frac{-4 + \sqrt{36}}{2} = 1$$

we obtain that it crosses the x-axes at $(-1, 0)$ and $(5, 0)$. The Fig. 2.3a shows it.

The quadratic function defined by

$$f(x) = -5 - x^2$$

corresponds to a parabola with vertex $V(0, -5)$ and from $a = -1 < 0$ we deduce that f is concave. The discriminant associated to $-5 - x^2$ is $\Delta = -20 < 0$. Therefore the parabola does not intersect the x-axis (see Fig. 2.3b).

2.1.3.2   Power Functions

We call *power function* a function of the type:

$$y = x^\alpha, \text{ defined for } x \geq 0, \text{ with positive exponent } \alpha$$

The behavior of this function strictly depends on the value of the parameter α. In particular three cases must be distinguished:

$$\begin{cases} \alpha > 1 \\ \alpha = 1 \\ 0 < \alpha < 1 \end{cases}$$

The Fig. 2.4 collects the graphs corresponding to $x^2, x, x^{1/2} = \sqrt{x}$.

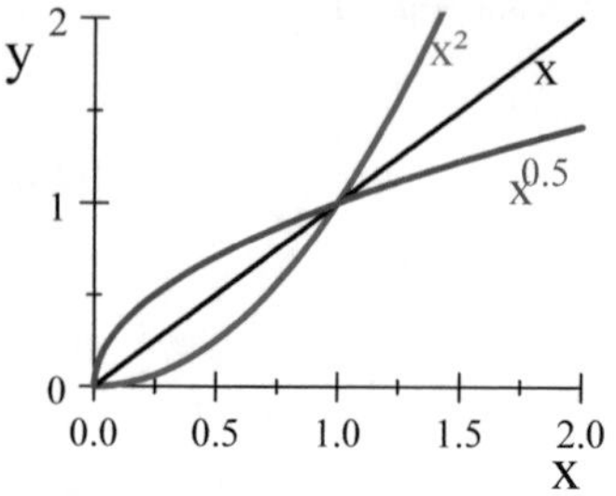

Fig. 2.4 Powers 2, 1 and 0.5

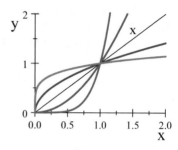

Fig. 2.5 Power onion

The next Fig. 2.5 is enriched with the graphs of x^3, $x^{1/3} = \sqrt[3]{x}$ and provides us with the nice picture of the (Tropea[1, 2]-like) "power onion".

If x increases, y too does. With exponents $\alpha < 1$, the initial increase is very quick, but when x increases, the growth rate of y declines. Exactly the reverse if $\alpha > 1$.

In some countries, like in Italy, such functions have been used by the Government in constructing income tax rates schedules.

2.1.3.3 Exponential Functions

The notion of exponential function is absolutely relevant:

- in itself, and we will see later why it provides us with the simplest way to model a very natural variation model: the one in which the variation of some quantity (income, population size, electoral supporters) is proportional to the current size of that quantity[3];

---

[1] The AA. are Italian. When looking at this graph they cannot ignore the celebrated Tropea onion.

[2] If it's the case, look for Tropea onions on Wikipedia.

[3] At the time of the publication of this book, there has been a "serious" debate in the Authors country about the observed number of deaths. Nobody has taken into account nor the population size nor its age composition.

It's sufficient to have (a friend with) a dog or more.

You have two dogs: one of them decides to leave you for the great prairies. For you it's an important loss.

You have some twenty dogs: one of them makes the same decision.

- it is usually (and dangerously) confused with the power function.

Let's dispose immediately of the second problem.
We consider two examples.

Example 2.1.3 You want to have a square[4] bedroom, and you are considering the size of your (square) bedroom. Let x be the side length. The area you can count on is:

$$p(x) = x^2$$

For instance, a square room with side 3m provides you with $p(3) = 9\text{m}^2$. If you decide the room is too small, you can look for a larger room, for instance, with side 4m. The new area, you will enjoy, would be $p(4) = 16\text{m}^2$.

Example 2.1.4 Your bedroom is assaulted by small, nice, white, but intolerable rats. At time 0, the initial size of this rat population $r(0)$ is almost negligible. Say, it stays at minimal physiological size dimension[5] $r(0) = 1$. They grow rather rapidly. Each of them produces 3 new rats per year. After 1 year, the number of (Ms.) rats[6] will be:

$$r(1) = r(0) + r(0)\frac{3}{2} = 1 + \frac{3}{2} = 2.5$$

After 2 years they will be:

$$r(2) = r(0)\left(1 + \frac{3}{2}\right)^2 = 6.25$$

After x years, the size of the rat population will be:

$$r(x) = \left(1 + \frac{3}{2}\right)^x = 2.5^x$$

This is a nice example of exponential function. Its name is appropriate because it recalls us that what does vary is the exponent.

Here is general scheme. All of us are acquainted with powers:

$$c = a^b, \text{ for instance: } 2^3 = 2 \times 2 \times 2 = 8$$

In both cases you loose one dog, but the situation is slightly different.

[4]The fact that the room is square is logically unnecessary, but it helps a lot in explaining the difference between power-exponential we're worrying about. Anyway, there should be no problem for people wanting a square bedroom.

[5]All of us are conscious that we should start from $r(0) = 2$. But, for counting, it's sufficient we consider only females, so $r(0) = 1$ suffices. The total nibbling population can be reconstructed through some appropriate adjustment factor. We will see significant examples in Chap. 4.

[6]Not all new rats turn out to be perfect (=female).

Basically, the *power structure* can generate two types of functions, according to what is thought of as fixed and what is thought of as variable.

In the case of your bedroom the exponent was fixed: 2. At least in the absence of drugs.

If we think of a power, with *fixed exponent*, we deal with *power functions*:

$$f(x) = x^\alpha, \text{ restricted to } x \geq 0$$

the ones we have met in the previous subsection.

If we think of a power, with *fixed basis*, we deal with an *exponential functions*:

$$f(x) = a^x, \text{ being the basis } a \text{ some positive number.}$$

In order to avoid confusion, the following scheme should help:

$a^b \rightarrow$	x^b	the exponent is fixed	power function	example: x^2
	a^x	the basis is fixed	exponential function	example: 2^x

Let us now focus our attention on exponential functions:

$$y = a^x$$

where a is a positive parameter. This fact implies that *an exponential function takes everywhere positive values.*

Their behavior depends on the position of a w.r.t. 1.

- If $a = 1$ the value of a^x is equal to 1 for every x.
- If $a > 1$ the quantity a^x increases more and more rapidly with x.
- if $a < 1$ the value of a^x decreases more and more slowly when x increases.

The Fig. 2.6 illustrates the three cases.[7]

If we consider the family of exponential curves ($y = a^x$) with $a > 1$, and if we look at their behavior at the origin $x = 0$, we can notice that:

- all of the corresponding ordinates is 1 (a trivial consequence of the fact that $a^0 = 1$);
- their slopes at 0 are different: high values of a provide curves with slopes greater than 1, while curves with values of a sufficiently close to 1, are... a bit lazy in going up and their slope at 0 is positive but rather small.

We could ask ourselves: "For which value of a, the slope at the origin of the exponential function is not too high, nor too small, for instance the same of the

---

[7]Politicians need media. Media tell us stories. We are not worrying about fake news, but about standard news. Several journalists love to describe exploding dynamics of hot phenomena, as "exponential" phenomena, probably ignoring that if $0 < a < 1$ what would happen is exactly the reverse they think.

Fig. 2.6 Plots of a^x

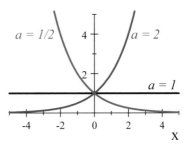

bisector of the first and third quadrant, with equation $y = x$ and corresponding slope 1?".

The answer is far from being evident, as the value to be given to a is the number $e = 2.718\ldots$, which appears regularly in several sectors of mathematics.

To celebrate its centrality (in mathematics), at Sorbonne University (Paris, France) there is a commemorative plaque, remembering all of us that:

$$e^{i\pi} + 1 = 0$$

Such a number puts together in a straightforward manner five leading numbers in Mathematics:

- π: no need to present it as it has been one of our childhood nightmares:

$$\pi = \frac{\text{length of the circle}}{\text{diameter}}$$

- 0: the neutral element for addition, or $x + 0 = x$ for every x;
- 1: the neutral element for multiplication, or $x \cdot 1 = x$ for every x;
- i: the imaginary unit ($i^2 = -1$): crucial in solving several classic mathematical problems
- e: possibly a new entry for some readers.

Remark 2.1.1 A possible alternative notation for exponential functions is:

$$y = a^x = \underbrace{\exp_a x}_{\text{new notation}}$$

In the case the basis is e:

$$y = e^x = \underbrace{\exp x}_{\text{new notation}}$$

We will see later why it can be useful in some cases.

2.1.3.4 Logarithmic Functions

These functions stem from the inversion of exponential functions.

For instance, take the exponential function, with basis 10:

$$y = 10^x$$

The following assertions about it can be rephrased using the notion of logarithm.

- If $x = 0$ then $y = 1$, because $10^0 = 1$;
- If $x = 1$ then $y = 10$, because $10^1 = 10$;
- If $x = 2$ then $y = 100$, because $10^2 = 100$;
- If $x = 3$ then $y = 1000$, because $10^3 = 1000$;
- $\cdots$

Now we rephrase them:

- the exponent we must give to 10 in order to get 1 is 0, or, the logarithm of 1, if the basis is 10, is 0;
- the exponent we must give to 10 in order to get 10 is 1, or, the logarithm of 10, if the basis is 10, is 1;
- the exponent we must give to 10 in order to get 100 is 2, or, the logarithm of 100, if the basis is 10, is 2;
- the exponent we must give to 10 in order to get 1000 is 3, or, the logarithm of 1000, if the basis is 10, is 3;
- $\cdots$

In general, the logarithm of a number x with basis a (positive and different from 1):

$$\log_a x$$

is nothing but the exponent y we must give to a in order that:

$$a^y = x \text{ or } x = a^y$$

The number x is called *argument* of the logarithm. It is *always positive*[8].

The shape of the graph of a logarithmic function crucially depends on whether the basis a is smaller or greater than 1. The Fig. 2.7 contains the logarithmic curves with different bases:

---

[8]This depends on the fact that, as we have written above, an exponential function a^x can take only *positive* values. A common mistake consists in maintaining that logs are always positive: this is wrong. For instance, if the log basis is 10 and we consider:

$$\log_{10} 0.1$$

we see that the exponento to be given to 10, in order to get 0.1, is -1, which is negative.

Fig. 2.7 Plots of $\log_a x$

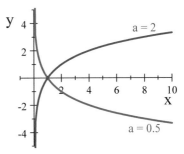

Fig. 2.8 Logs with bases e and 10

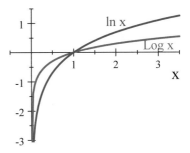

The logarithmic curve with basis $1/2$ shows the fact that if the basis is smaller than 1, when the argument increases, its logarithm decreases. Logs with basis 2 are crucial in information theory and relevant in some social contexts.

Logs with basis e are the most important ones: they are the only logs we will use systematically. They are also called *Napierian*[9] *logs* or *natural logs*. Instead of indicating them with:

$$\log_e x$$

we will systematically use the international official notation:

$$\ln x$$

Logs with basis 10, often denoted as:

$$\text{Log } x$$

have mere historical relevance in making possible complex computations before calculators and computers were available. The Fig. 2.8 illustrates the difference between them.

The main relevance of logs today depends on their inverse nature w.r.t. exponentials.

---

[9]From the name of the Scottish John Napier (1550–1617), whose scientific nickname was Neper.

It is important that the reader becomes familiar with some properties of logs. We summarize them immediately, being obviously p, q positive numbers:

- $\log_a (pq) = \log_a p + \log_a q$;
- $\log_a (p/q) = \log_a p - \log_a q$;
- $\log_a (p^q) = q \log_a p$;
- $\log_a a^x = \log_a (\exp_a x) = x$, in particular: $\ln e^x = x$ (any log kills the corresponding exponential):
- $\exp_a \log_a (x) = x$, in particular: $e^{\ln x} = x$ (any exponential kills its corresponding log).

Remark 2.1.2 An important consequence of these properties, which will turn out to be crucial in the applications, is that, given any exponential function:

$$f(x) = a^x$$

we can think of it as an exponential function, exactly, with basis e:

$$f(x) = e^{x \ln a}$$

The idea conveyed in the last formula is very simple.

Example 2.1.5 You have a population, whose size $P(t)$ grows 10% per year. The population size $P(t)$ at t is trivially:

$$P(t) = P(0) \, 1.1^t$$

Well, introducing:

$$\delta = \ln 1.1 \approx 0.095\,31$$

we can rewrite the evolution equation for P as follows:

$$P(t) = P(0) \, e^{\delta t} \approx P(0) \, e^{0.095\,31t}$$

2.1.4 Continuous Functions

Let us start with a simple economic example.

Example 2.1.6 A firm produces a good. The produced quantity is $x \geq 0$. The costs the firm incurs are:

(1) a fixed amount 5000, think of plants;
(2) a variable amount proportional to the production volume x, say: $3x$, think of raw materials, labor, energy.

In Economics, the *total cost function* is:

$$f(x) = 5000 + 3x$$

and its graph is a straight line with intercept[10] 5000 and slope[11] 3.

In real world things are not so simple as, in general, a plant has a well defined and fixed production capacity: if the production volume x does non exceed — say — 1000 units, no problem, but if you want to push the production volume over 1000 units, you have to modify the plant and fixed costs jump from 5000 to 7000. The cost function which takes into account this fact is:

$$f(x) = \begin{cases} 5000 + 3x \text{ for } x \leq 1000 \\ 7000 + 3x \text{ for } x > 1000 \end{cases}$$

Its graph is given by two half-lines (Fig. 2.9).

The graph suggests that at $x = 1000$ the cost function f "makes an obvious jump", switching from $5000 + 3x$, for $x \leq 1000$ to $7000 + 3x$ for $x > 1000$. Practically this means that if $x < 1000$ is close to 1000, well, the cost of such production volume will be close to that of producing exactly 1000 units:

$$f(1000) = 5000 + 3 \cdot 1000 = 8000$$

while, if $x > 1000$, but close to 1000, will induce costs of amount:

$$f(x) = 7000 + 3 \cdot x$$

very different from $f(1000) = 8000$. For instance $f(1001) = 7000 + 3 \cdot 1001 = 10\,003 >> 8000$. Of course this depends on the variation at $x = 1000$ in fixed costs.

Fig. 2.9 Total cost discontinuous function

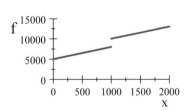

Functions that do not "jump" at some point x^*, as the f in the Example above at $x^* = 1000$, are said to be *continuous* at x^*. Otherwise they are labelled as *discontinuous* at x^*. The concrete interpretation is that if you're dealing with a continuous function and you are sufficiently close to some point x^*, the values that f takes are not far from the one f takes at x^*.

---

[10]The mathematical notion of intercept can have interesting concrete interpretations. Like in this case.

[11]See the previous note, changing intercept with slope.

Remark 2.1.3 The idea of continuity is relevant for Social Sciences. In Chap. 4 we will see striking examples when dealing with dynamic systems; a slight variation in the starting point of some process will reveal to have dramatic consequences on the process itself. Many people think that functions in the world are continuous: 'Change slightly your behavior and the effects will change only slightly'. We think to share with the readers the idea that in reality there are thresholds, characterizing opposite effects.

2.1.5 An Annoying Detail and a Tribute to L.D. Landau

The core of this chapter is Differential Calculus. It has been constructed using intuition a lot of time before its rigorous setting. The pioneers of Calculus found fantastic results and made significant errors. We will work as pioneers, basing many arguments on intuition, hopefully without paying the tax of errors.

The cost of constructing a rigorous calculus has a name: the *theory of limits*. Such a cost is rather high. As we will move only in a restricted part of calculus, it is not necessary for us a complete development of the theory of limits. We will need only a simulacrum, due to Lev Davidovič Landau[12], who proposed a series of shorthand notation symbols, which will allow us to construct Calculus incurring only in a reasonable small cost.

We will use only one of these symbols :

$$\boxed{o\,(h)}$$

to be read "small o of heich".

We will need it to handle two cases:

- C1 — The quantity h is an infinitesimal: h is a variable quantity approaching 0 (we will write $h \to 0$);
- C2 — The quantity h is an infinite: h is a variable quantity, which grows indefinitely (we will write $h \to +\infty$).

In both cases a problem arises. That of comparing quantities, depending on h, having the same qualitative behavior as: $\begin{cases} h \to 0 \\ h \to +\infty \end{cases}$.

The symbol $o\,(h)$ is nothing but a very efficient way to say that the ratio between some quantity $o\,(h)$ and h itself:

$$\left| \frac{o\,(h)}{h} \right|$$

can be made as small as we want: it's sufficient to take h sufficiently close to 0 (case C1) or sufficiently large (case C2).

---

[12] Born in Baku 1908, he passed away in Moscow in the Spring of 1968.

In general, such an occurrence is technically handled as follows. You choose a positive number $\varepsilon > 0$, which can be arbitrarily small (for instance $\varepsilon = 1/10^3$, $1/10^4$, ...): well, there is a

$$\begin{cases} h \text{ sufficiently close to } 0 \\ h \text{ sufficiently large} \end{cases}$$

such that:

$$\left| \frac{o(h)}{h} \right| < \varepsilon \tag{2.1.1}$$

What will frequently turn out to be crucial is that there could be no need to specify analytically $o(h)$. We will see later why this opportunity is practically relevant.

The concrete importance of (2.1.1) is that in several cases, when you have to handle quantities proportional to h together with quantities $o(h)$, you are allowed to ignore the last ones.

Let us consider some examples.

Example 2.1.7 Take, for instance:

$$4h^2$$

to be compared with h when $h \to 0$. The ratio:

$$\frac{4h^2}{h} = 4h$$

in absolute value, is smaller than ε if:

$$|4h| < \varepsilon \Leftrightarrow |h| < \frac{\varepsilon}{4}$$

Therefore:

$$4h^2 = o(h) \text{ if } h \to 0$$

Take, for instance:

$$4h^2 + h^3$$

For sufficiently small h we have that:

$$h^3 < h^2$$

therefore:

$$\left| 4h^2 + h^3 \right| < 5h^2$$

As, obviously, $5h^2 = o(h)$ we get:

$$4h^2 + h^3 = o(h) \text{ if } h \to 0$$

Example 2.1.8 Consider now the comparison between:

$$1000h$$

and

$$\frac{h^2}{1000}$$

when $h \to +\infty$. The absolute value of the ratio:

$$\frac{1000h}{h^2/1000} = \frac{1000000}{h}$$

can be made smaller of ε for h sufficiently large:

$$\frac{1000000}{h} < \varepsilon \Leftrightarrow h > \frac{1000000}{\varepsilon}$$

2.1.6   The Small o Algebra

Economics is a Social Science.

Its principles and its results can be disputable. Probably the most robust principle is that, in the presence of efficient markets, free lunches are not allowed. The world of Bengodi[13] does not exist.

Incredibly a Social Science tells us an interesting story about os.

We have several $o\,(h)$, and we do not want to distinguish their names[14] and we would like to handle them jointly. It could appear sensational that, without specifying their expression, we can write:

$$a \cdot o\,(h) + b \cdot o\,(h) + \cdots + z \cdot o\,(h) = o\,(h)$$

Bengodi impossibility simply tells us that the accumulation of (a finite number of) negligible quantities w.r.t. h produces a quantity with the same negligibility property.

Take two $o\,(h)$ and multiply them with each other. What you get is another (but stronger!) $o\,(h)$:

$$o\,(h) \cdot o\,(h) = o\left(h^2\right)$$

The impossibility of free lunches is waiting for us around the corner. Consider h^2 and h^3, with $h \to 0$. Both of them are $o\,(h)$, but their quotient has different natures

---

[13] Originally due to Giovanni Boccaccio, *Decameron*, 3rd count, during the 8th day. In Bengodi realm, free lunches were allowed.

[14] May be that what we know is simply that they are $o\,(h)$ and that we reasonably suspect that such partial information is sufficient, at least for some purposes.

according to what you put at the numerator and at the denominator:

$$\frac{h^3}{h^2} \to 0, \text{ while } \frac{h^2}{h^3} \to \infty$$

therefore, in general, the behavior of an expression of the type:

$$\frac{o\,(h)}{o\,(h)}$$

cannot be elementarily detected.

Now it's the time to bridge the notion of continuous function and the symbol $o\,(h)$.

Remark 2.1.4 Take a function $f : \mathbb{R} \to \mathbb{R}$ and consider a point x^* at which f is continuous: no jumps at x^*! Well, we can state this fact simply maintaining that:

$$f\left(x^* + h\right) - f\left(x^*\right) = o\,(1)$$

This simply means that if h is sufficiently small, then $f\,(x^* + h)$ is sufficiently close to $f\,(x^*)$. The expression $f\,(x^* + h) - f\,(x^*)$ will play a crucial role above. It is called *increment of f when x moves from x^* to $x^* + h$*.

Example 2.1.9 Take:

$$f\,(x) = x^2$$

Let $x^* = 1$. We have:

$$f\,(1 + h) - f\,(1) = (1 + h)^2 - 1 = 2h + h^2$$

For h sufficiently small:

$$h^2 < |h| \Longrightarrow \left|2h + h^2\right| < 3\,|h|$$

If we want that $|f\,(1 + h) - f\,(1)| < \varepsilon$ it suffices to choose:

$$|h| < \frac{\varepsilon}{3}$$

2.1.7 *Some Rankings*

We will use some comparisons. We collect here, without proof, some simple results, which later will turn out to be useful.

2.1.7.1   Powers of Quantities Going to 0

We consider (power) functions of the type:

$$f(h) = h^\alpha$$

Assume first that $\alpha > 0$. It's intuitive[15] that if h goes to 0 the same holds for f. We can translate this fact in Landau's language, writing that:

$$h^\alpha = o(1) \text{ when } h \to 0$$

Consider now two (power) functions of the type:

$$f(h) = h^\alpha; g(h) = h^\beta$$

If $\beta > \alpha$ then:
$$g(h) = h^\beta = o[f(h)] = o(h^\alpha)$$

As an example:
$$h^3 = o(h^2) \text{ when } h \to 0$$

2.1.7.2   Quantities Going to Infinity (and Soccer)

Think of soccer. There are both National Tournaments and the Champions League. In the first ones you compare homogeneous teams (same country), in the second one you compare heterogeneous teams (different countries).

- Power infinities:
$$h^\alpha \text{ with } \alpha > 0 \text{ and } h \to +\infty$$

  In these tournament what happens is that:

$$\text{if } \alpha > \beta \text{ then } h^\beta = o(h^\alpha)$$

- Exponential infinities:

$$a^h \text{ with } a > 1, \text{ and } h \to +\infty$$

---

[15]Please, give a glance to the "power onion" at p. 104.

In this tournament what happens is that:

$$\text{if } a > b \text{ then } b^h = o\left(a^h\right)$$

- Champions League.[16] If we compare power functions with exponential functions, the former ones (power) always loose:

$$h^\alpha = o\left(a^h\right)$$

for every positive α and every $a > 1$. In some cases a comparison is useful between logs and powers. Well, logs loose systematically[17]:

$$\ln^\alpha h = o\left(h^\beta\right) \text{ with } \alpha > 0 \text{ and } \beta > 0$$

The following numerical examples support our assertions.

Example 2.1.10 Let's consider the following functions:

$$f(x) = x^3 \qquad g(x) = e^{2x} \qquad h(x) = \ln^{10} x$$

When x goes to infinity, i.e.: $x \to +\infty$, we have the following statements:

$$x^3 = o(e^{2x}) \qquad \ln^{10} x = o\left(x^3\right) \qquad \ln^{10} x = o\left(e^{2x}\right)$$

Meanwhile for the function $f(x) = x^k$, with $k > 0$, we have:
(1) - $x^3 = o(x^k)$ for any $k > 3$ and $x^k = o(x^3)$ for any $0 < k < 3$; and
(2) - $x^k = o(e^{2x})$ and $\ln^{10} x = o\left(x^k\right)$, for any positive k, in fact, the following ratios:

$$\frac{x^k}{e^{2x}} \to 0 \qquad \text{and} \qquad \frac{\ln^{10} x}{x^k} \to 0$$

go to 0 as $x \to +\infty$, for any positive k.

---

[16] At last!

[17] There is a dramatic problem concerning the notation to be used for powers of logs: some people like to indicate the square of the natural logarithm of x as:

$$(\ln x)^2$$

They are right, but, when handling such quantities, a minority prefer the shortest notation:

$$\ln^2(x) \text{ or, even courageously, } \ln^2 x$$

The Authors are wrong, but, within this book they do not think there are too many misunderstanding possibilities. Therefore, the will use the lighter notation.

2.2 Local Behavior and Global Behavior

2.2.1 Local Behavior, Derivative and Differential

The key concept of infinitesimal calculus, which constitutes one of its pillars, is the *differential of a function at some point* x^*.

The main question consists in understanding that the behavior, near some point x^*, of (almost) every function $y = f(x)$, we are interested in, is substantially the same as that of some easy function:

$$y = g(x) = mx + q$$

whose behavior can be diagnosed very easily.

Consider two different functions:

$$f(x) = e^x \text{ and } g(x) = x + 1$$

and compare them on some "large interval" around $x^* = 0$, for instance in the Fig. 2.10 the interval $[-1, +1]$ is considered.

Fig. 2.10 Large interval

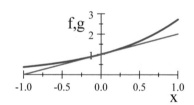

If now we restrict the observation interval to $[-0.1, +0.1]$ we get the Fig. 2.11:

Fig. 2.11 Small interval

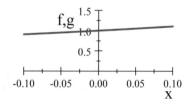

It is clear that under this "local" perspective, the two functions are practically indistinguishable.

The differential calculus exploits such an intuition and systematically draws information about the behavior near x^* of a "difficult function" like $f(x) = e^x$, through the "easy function" $g(x) = x + 1$, which locally mimics it.

In order to make comparisons between the behavior of some "difficult function" $f(x)$ and an "easy" one $g(x) = mx + q$, we must first require that at the point x^*, they take the same value. What we require is that:

$$g(x^*) = f(x^*)$$

As far as we are concerned with the two parameters characterizing g, this requirement boils down to:

$$mx^* + q = f(x^*)$$

hence:

$$y = g(x) = f(x^*) + m \cdot (x - x^*) \tag{2.2.1}$$

and we are left only to find the appropriate slope m.

Definition 2.2.1 We call *increment*[18] *of the function* f, when its argument x moves from x^* to $x^* + h$ the difference between the new value of f and the old one (we have already anticipated this notion at p. 115):

$$f(x^* + h) - f(x^*)$$

The quantity h is called *increment of the independent variable.*

Remark 2.2.1 A popular notation for the increment of f is $\Delta f(x^*)$. It is simpler, but it does not specify h. A popular substitute for h is the more cumbersome symbol Δx.

Let us try to understand differences and similarities in the behavior of difficult functions and easy functions. This will turn out to be crucial in constructing differential calculus. We do that comparing their increments. We start with some simple examples, that will suggest us general notions.
Take:

$$f(x) = x^2$$

Its increment is:

$$f(x^* + h) - f(x^*) = (x^* + h)^2 - x^{*2} = (2x^*)h + h^2 = (2x^*)h + o(h)$$

For an easy function:

$$g(x) = mx + q$$

---

[18]Normally speaking "increment" means positive variation of some quantity; e.g.; "my income has recently had an increment of $10,000$ €/year ". In the technical language we are introducing, an increment could also be zero or even negative.

we have:

$$g\left(x^* + h\right) - g\left(x^*\right) = m\left(x^* + h\right) + q - mx^* - q = mh = mh + o\left(h\right)$$

As 0 is $o\left(h\right)$ too, it should not be too disturbing that, instead of writing simply mh, we propose to write $mh + o\left(h\right)$.

Let us compare the structure of the two increments:

- they are different, at least because of the two $o\left(h\right)$s;
- both of them are the sum of a quantity proportional to h with a quantity which is negligible w.r.t. h;
- the coefficient of h for easy functions is the constant m, it does not depend on x^*, while in the case of $f\left(x\right) = x^2$, such a coefficient turns out to depend on x^*: in particular it is $m\left(x^*\right) = 2x^*$.

Take now another difficult function:

$$f\left(x\right) = x^3$$

and workout the same exercise. Its increment turns out to be:

$$\left(x^* + h\right)^3 - x^{*3} = \left(3x^{*2}\right)h + 3x^*h^2 + h^3 = \left(3x^{*2}\right)h + o\left(h\right)$$

Also in this case, the increment of f can be written as:

$$m\left(x^*\right)h + o\left(h\right)$$

Here:

$$m\left(x^*\right) = 3x^{*2}$$

Should we consider several other difficult functions, then what we would recover is the same structure of their increments:

$$f\left(x^* + h\right) - f\left(x^*\right) = m\left(x^*\right)h + o\left(h\right)$$

Functions f for which this happens could be called *smooth*, because of the smoothness of their graph near x^*.

We can think of $m\left(x^*\right)$ as the local slope of the curve representing f.

The linear function, which, at best approximates f near x^* is obtained from Eq. (2.2.1), inserting the appropriate local slope.

The graph of such a linear function is the tangent to the graph of f at x^*.

Assume that:

$$f\left(x\right) = x^2$$

that $x^* = 1$. The Fig. 2.12 collects the graphs of f and of its tangent straight line at x^*.

Example 2.2.1 Let's consider the function defined by $f(x) = x - 2x^2$ and $x^* \in \mathbb{R}$. Its increment $f\left(x^* + h\right) - f\left(x^*\right)$, with h different from 0, is

Fig. 2.12 Tangent line of x^2 at $x^* = 1$

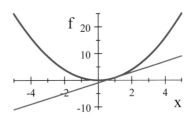

$$f\left(x^* + h\right) - f\left(x^*\right) = \left[\left(x^* + h\right) - 2\left(x^* + h\right)^2\right] - \left[x^* - 2\left(x^*\right)^2\right]$$
$$= x^* + h - 2\left(x^*\right)^2 - 4x^*h - 2h^2 - x^* + 2\left(x^*\right)^2$$
$$= -4x^*h + h - 2h^2 = \left(1 - 4x^*\right) \cdot h - 2h^2$$
$$= \left(1 - 4x^*\right) \cdot h + o(h)$$

where we have indicated by $o(h)$ the amount $-2h^2$. It's clear that $m\left(x^*\right) = 1 - 4x^*$. For the function $\phi(x) = 3x^3 - 2$ and $x^* \in \mathbb{R}$, following the same above procedure, we have

$$\phi\left(x^* + h\right) - \phi\left(x^*\right) = \left[3\left(x^* + h\right)^3 - 2\right] - \left[3\left(x^*\right)^3 - 2\right]$$

$$= 9\left(x^*\right)^2 \cdot h + o(h)$$

where $o(h) = 9x^*h^2 + 3h^3$.

It should be clear that, for a smooth function, its behavior can be detected from the one of the approximating *easy* function.

The following example illustrates why there are also non-smooth functions, even if this aspect is not so crucial for us.

Example 2.2.2 Take[19]:

$$f(x) = |x| = \begin{cases} x \text{ if } x \text{ is positive or null} \\ \\ -x \text{ if } x \text{ is negative} \end{cases}$$

and $x^* = 0$. Its graph is plotted in Fig. 2.13.

---

[19]Two perfectly equivalent definitions of this functions are at hand:

$$f(x) = |x| = \begin{cases} x \text{ if } x \text{ is positive} \\ \\ -x \text{ if } x \text{ is negative or null} \end{cases}$$

or $f(x) = |x| = \sqrt{x^2}$.

In the experience of the Authors, even if $|.|$ has (at least) three simple possible equivalent definitions, it continues to be some "mysterious object" for many undergraduate students.

Fig. 2.13 $f(x) = |x|$

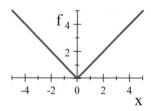

It "picks" at 0 (if touched from below). Let us look at the structure of its increment:

$$f(h) - f(0) = |h| - 0 = |h| = \begin{cases} +1 \times h = h, \text{ if } h \text{ is positive} \\ \\ -1 \times h = -h, \text{ if } h \text{ is negative} \end{cases}$$

If we try to write this increment in the form we have found for smooth functions, what we get is:

$$f(h) - f(0) = m(h)h + o(h)$$

The second addendum in the r.h.s. is artificial (in the sense it is 0), but in the first addendum, what we see is that the coefficient of h irreducibly is not independent of h, as it is ± 1, according to the sign of h.

We are ready to solemnize some notions we have met in looking at the examples we have analyzed.

The official name of smooth functions is different.

Definition 2.2.2 Consider a function f. If its increment near x^* can be represented as:

$$f(x^* + h) - f(x^*) = m(x^*)h + o(h)$$

the function f is said to be *differentiable* at x^*. The first addendum in the r.h.s. $m(x^*)h$ is called *differential of* f *at* x^*, The coefficient of h in the differential is called *derivative* of f at x^*.

It is important to understand the geometric interpretation of the notions of *differential* and of *derivative.* Here are their geometric interpretations. The derivative of a function at x^* is the slope of the tangent line to the graph of the function at x^*, meanwhile the differential at x^* is most of the increment of the function when it moves from x^* to $x^* + h$. It is the increment measured on the tangent line, instead of on the graph of the function.

Example 2.2.3 Let's consider the function $f : \mathbb{R} \to \mathbb{R}$ defined by $f(x) = 2x^2 + x$ and a point $x^* \in \mathbb{R}$. We have

$$f(x^* + h) - f(x^*) = 2(x^* + h)^2 + (x^* + h) - 2x^{*2} - x^*$$

$$= 2x^{*2} + 4x^*h + 2h^2 + x^* + h - 2x^{*2} - x^*$$
$$= 4x^*h + 2h^2 + h = (4x^* + 1) \cdot h + 2h^2$$
$$= (4x^* + 1) \cdot h + o(h)$$

therefore f is differentiable at x^*, where $m(x^*) = 4x^* + 1$ and $o(h) = 2h^2$, and the product $(4x^* + 1) \cdot h$ is the differential of f at x^*.

An important connection between differentiability and continuity of a function $f : \mathbb{R} \to \mathbb{R}$ is contained in the following:

Proposition 2.2.1 *For $f : \mathbb{R} \to \mathbb{R}$, if f is differentiable at x^*, then f is continuous at x^*.*

The formal proof is trivial and we spare the readers. The intuition is rather obvious: if some function locally behaves as a linear (affine) one, it can't jump.

Remark 2.2.2 With reference to the examples we have considered before, we suggest our readers to examine the following statements:

(1) A linear function $y = mx + q$ is differentiable everywhere, its differential is mh, its derivative is constant and equal to m;

(2) The function $y = x^2$ is differentiable everywhere, its differential at x^* is $2x^*h$, its derivative at x is $2x$;

(3) The function $y = x^3$ is differentiable everywhere, its differential at x^* is $3x^{*2}h$, its derivative at x is $3x^2$.

Example 2.2.4 Let us anticipate the use of derivatives that can be done. Take the function:

$$f(x) = x^2$$

Its graph has been represented above, at page 121. It tells us that the function f decreases when x is negative and increases if x is positive. For a linear function $y = mx + q$ these properties are revealed by the sign of the parameter m: if m is negative, the function decreases, if m is positive the function increases. We have seen above that the derivative of a function is the local slope of its graph. We have already ascertained that the local slope for f in this case is $2x$. If we want to find for which values of x the function f decreases, we could look for the values of x such that the local slope is negative. This would bring us to the inequality:

$$2x < 0$$

equivalent to:

$$x < 0$$

Analogously we could look for the points at which f increases, studying the condition:

$$2x > 0$$

which would provide us with:

$$x > 0$$

For a function like this one, the information provided by the derivative is not exciting, as we already know how f behaves, but it could become precious in more complex cases. Pay attention to the next

Example 2.2.5 Consider this function:

$$f(x) = 5x^2 - x^3$$

We have no prior precise idea about the behavior of this function, but let's see how we can extract information from the derivative of f with appropriate computations, we have substantially already made, we could reach the conclusion that f is everywhere differentiable:

$$f(x + h) - f(x) = m(x)h + o(h)$$

with derivative[20]:

$$m(x) = 10x - 3x^2 = x(10 - 3x)$$

Such derivative is the product of two factors. The first factor (x) changes its sign at 0, while the second factor $(10 - 3x)$ at $10/3$. The Table 2.1 resumes the information about signs and allows us to decide in which intervals f increases or decreases.

---

[20]In most cases the information we want to extract from a derivative is confined in its sign. If the derivative expression is of the type "sum":

$$a + b$$

even if we know the sign of a and b, we cannot say anything about the sign of the sum, because the absolute value of the addenda can be relevant. If the expression of the derivative is of the type "product":

$$ab$$

and we know the signs of the two fators, we can establish the sign of the derivative.

Therefore, **once you have at hand a derivative, the first thing to do is to try to factorize it**. This is the reason why we have replaced $10x - 3x^2$ with its factorized versione $x(10 - 3x)$.

Table 2.1 Derivative sign of $5x^2 - x^3$

Interval\Factor→	x	$(10 - 3x)$	Sign of m	Behavior of f
$(-\infty, 0)$	$-$	$+$	$-$	↘
$(0, 10/3)$	$+$	$+$	$+$	↗
$(10/3, +\infty)$	$+$	$-$	$-$	↘

Therefore the function decreases on the left of 0 and on the right of 10/3, while it is increasing between 0 and 10/3. The function attains a (local) minimum at 0 and a (local) maximum at 10/3. Rewriting f as a product:

$$f(x) = x^2(5 - x)$$

it is clear that when x goes to $-\infty$, the function f diverges to $+\infty$ and *vice-versa* at $+\infty$. The plot of f (see Fig. 2.14), you will learn to draw later in detail, is absolutely compatible with the information we have collected until now.

Fig. 2.14 $f(x) = 5x^2 - x^3$

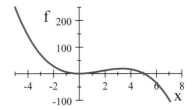

2.2.2 Notation for Derivatives

There are several ways to indicate derivatives. We mention the most frequent:

- **Our preferred**: $f'(x)$ for the derivative of f at a generic point f (it is a function of x) and $f'(x^*)$ for the derivative at the specific point x^* (it is a number, once x^* has been fixed). For instance, if $f(x) = x^2$, then $f'(x) = 2x$ and $f'(5) = 10$.
- **Newton original notation**: $\dot{f}$. A dot above instead of a prime $'$. It is used in Physics, essentially when x is time. It does not present any advantage w.r.t. the other possibilities.
- **Cauchy notation**: $D[f(x)]$, to denote the function $f'(x)$, which is the *derivative function* of $f(x)$. It works well when computing derivative functions, as we will see later, but if we want to specify that we are talking about the *number* derivative of f at x^*, the notation should become a bit cumbersome:

$$D[f(x)]_{x=x^*}$$

- **Leibnitz notation**: very interesting and practically useful, even if sometimes it can be misleading. The derivative function of $y = f(x)$ can be denoted with[21]:

$$\frac{dy}{dx} \text{ or with } \frac{d}{dx} f(x) = \frac{df(x)}{dx}$$

The first symbol is simple but a bit reticent, because it is not clear whether we are dealing with the derivative function or with the value of the derivative, as a number, at some point. The second symbol is fully equivalent to Cauchy's one:

$$D = \frac{d}{dx} \tag{2.2.2}$$

under all respects.

Remark 2.2.3 Looking at the notation display, common sense would suggest:

(1) Discard Newton notation (and the Authors agree: no sensible advantage);
(2) Discard Leibnitz notation: at best (2.2.2) it is replicated by Cauchy notation, otherwise it is unclear. Here the Authors do not agree, as we will see, that the somehow relaxed Leibnitz notation presents some practical advantages.

2.2.3 Derivative and Differential: What's the Most Important Notion

The aim of this (sub)section is to settle an irrelevant problem.

A natural question is: "Why we could not skip it?".

People, potentially interested in this book, come from different education systems, where, different traditions turn out to imply divergent perspectives.

Readers, already familiar with differential calculus, could have been surprised by the previous pages,[22] where they find no trace of the (almost religious) myth of introducing derivatives as the limit of the increment ratio (or, may be, difference quotient):

$$f'(x) = \lim_{h \to 0} \frac{f(x+h) - f(x)}{h} \tag{2.2.3}$$

In our presentation $f'(x)$ appears only as local slope of the graph of f at the point x.

We avoid the notion of limit in this textbook, simply because its time-cost is high and its benefits, in this type of treatises, are not so important.

---

[21] If you are thinking that $\frac{dy}{dx}$ is a fraction, you are wrong. You must accept that this mock-fraction is not a fraction, but sometimes it behaves like a true one.

[22] The Authors can provide empirical evidence about.

Anyway, to help some readers to link their past with their present, we state that the notion of *differentiable function* and the one of *derivable* function, i.e.; a function for which (2.2.3) holds are equivalent.

Theorem 2.2.1 *The function* f *is differentiable at* x^* *if and only if it is derivable at* x^*.

What is relevant is that the existence of the derivative (no matter how we define it) is crucial for functions of (exactly) one variable, but when we will turn our attention to functions of several variables, we will discover that the existence of derivatives is a substantially weak information, but the notion of differentiability will continue to be crucial also in this wider context.

Example 2.2.6 Let's consider the function $f(x) = x^3 - 2x$ with domain $\mathbb{R}$. We observe that its difference quotient is:

$$\frac{f(x+h) - f(x)}{h} = \frac{(x+h)^3 - 2(x+h) - (x^3 - 2x)}{h} = 3x^2 - 2 + o(h)$$

and if $h \to 0$ we get $f'(x) = 3x^2 - 2$, therefore f is both derivable and differentiable on $\mathbb{R}$.

Remark 2.2.4 If you look on any Anglo-American text about these topics, may be that you do not find the word "derivable". It simply occurs because such a notion (of derivable function) is overwhelmed by the one of "differentiable function". On the other side, if you look at the French Bible of mathematics (the unfinished monumental treatise by N. Bourbaki[23]) and you search for the "dirty" notion of differential, which would be "*différentielle*" you do not find it: simply because, for them, it was the devil. This simple mathematical fact should be useful for European Political Scientists.

Our opinion is that the crucial notion of calculus is that of *differential*, a structurally simple quantity that captures most of the increment of a function:

$$f(x+h) - f(x) \approx f'(x)h \text{ for } h \text{ sufficiently small}$$

Please, note that, using Leibnitz notation, calling dy the differential of $y = f(x)$ and dx the increment of the independent variable, we get:

$$dy = \frac{dy}{dx}dx$$

The "false" fraction $\frac{dy}{dx}$, formally behaves like a true fraction!
    To sum up:

---

[23] Bourbaki is not a person, but the label for a group. Independently of Mathematics, it is interesting to look at its history.

- Differential calculus is based on the opportunity to replace, in the small, the increment:

$$f(x+h) - f(x)$$

  with the corresponding differential:

$$f'(x)h$$

- most of the useful information, provided by the differential df, is contained in its sign (and in the behavior) of the derivative f'.

2.2.4 The Computation of Derivatives, also of Order >1

You have already learned some derivatives:

$$\begin{cases} D\left[x^2\right] = 2x, \text{ by direct experience} \\ D\left[x^3\right] = 3x^2, \text{ by direct experience} \\ D\left[5x^2 - x^3\right] = 10x - 3x^2, \text{ trusting in the Authors} \end{cases}$$

A natural question arises: apart from these special cases, is it possible to compute derivatives? The answer is YES, and such a task is not too difficult.

The computation strategy for derivatives is based on two pillars:

- a list of derivatives for elementary functions;
- a set of rules telling us how to differentiate complex functions, constructed using the elementary ones.

2.2.4.1 Table of Elementary Derivatives

The Table 2.2 contains the first pillar:

Table 2.2 Elementary derivatives

Type of function	$f(x)$	$f'(x)$
Constant	k	0
Linear	$mx + q$	m
Power	x^α, with $\alpha \in \mathbb{R}, x > 0$	$\alpha x^{\alpha-1}$
General exponential	a^x with $a > 0$	$a^x \ln a$
Special exponential	e^x	e^x
Natural log	$\ln x, x > 0$	$1/x$

Some comments are useful.

1. The fact that the derivative of a *linear function* is constant depend on the fact that the local slope does not change: it is both local and global at the same time. Note that the identity function:

$$f(x) = x$$

has unitary derivative:

$$f'(x) \equiv 1$$

therefore its differential turns out to be:

$$d[x] = 1 \cdot dx = dx$$

2. The derivative of a *constant function* can be seen as a special case of derivative of a linear function: it corresponds to the special case $m = 0$.
3. About the differentiation of a *power function* we note that we had already met the two special cases $\alpha = 2, 3$. We stress also the fact that the expression of the derivative does not hold only for integer and positive exponents (like the two cases already experienced), but also for negative exponents, for non-integer exponents. Let, for instance:

$$f(x) = \sqrt{x} = x^{\frac{1}{2}}$$

Its derivative turns out to be:

$$f'(x) = \frac{1}{2}x^{\frac{1}{2}-1} = \frac{1}{2}x^{-\frac{1}{2}} = \frac{1}{2} \cdot \frac{1}{x^{\frac{1}{2}}} = \frac{1}{2\sqrt{x}}$$

Another interesting example could be this one::

$$f(x) = \frac{1}{x^{37}} = x^{-37}$$

we have:

$$f'(x) = -37x^{-37-1} = -37x^{-38} = -37 \cdot \frac{1}{x^{38}} = -\frac{37}{x^{38}}$$

In words, the differentiation rule for a power function would be:

$$\text{derivative} = \text{old exponent} \cdot x^{\text{old exponent} - 1}$$

Remark 2.2.5 In one of his previous lives of one of the Authors,[24] a kaki-tree turned out to play a significant role: a lot of kaki fruit were regularly falling to ground in summer and only few reached their winter maturity. Thinking of a power x^{α} as a kaki tree:

---

[24]This happened in Moncucco Torinese (Italy).

$$x^\alpha \rightarrow \begin{array}{l} \alpha \leftarrow \text{ potential number of lost fruits} \\ x \leftarrow \text{ the tree} \end{array}$$

you can easily remember the differentiation rule of a power, thinking of the exponent becoming coefficient as a kaki-fruit at ground, and of the new exponent as the new number on the branches: the old number diminished of 1 . The label "*kaki-rule for powers*" can help in remembering.

4. The case of the *exponential function* is a special case of the general case a^x:

$$D\left[e^x\right] = e^x \cdot \ln e$$

By definition the second factor in the r.h.s. is unitary and therefore the derivative is $e^x \cdot 1 = e^x$. It is interesting to note that the computation of the derivative of an exponential function simply boils down to multiply the function by the constant $\ln a$. In words the differentiation rule for the exponential function with base a is:

$$\text{derivative } = \text{exponential function } \cdot \ln \text{ of the basis}$$

In general, the necessary differential implication will turn out to be useful in modeling exponential variations.

5 We do not consider the differentiation of logarithmic function with basis different from e simply because we will never need them.

2.2.4.2   Differentiation Rules

Now we switch to the second pillar. We have already met a special case:

$$D\left[5x^2 - x^3\right] = D\left[5x^2\right] - D\left[x^3\right] = 10x - 3x^3$$

We have informed you about the idea of computing derivatives of "complex" functions obtained via a linear combination of the two elementary functions x^2 and x^3. The derivative of their linear combination is the same linear combination of their derivatives.

First we will give a panoramic glance to the set of the rules (Table 2.3) we will learn, then we will examine each of them in detail.

1 Differentiating a multiple of a function f — Let:

$$\phi(x) = \alpha f(x)$$

Its increment is[25]:

---

[25] We use the fact that:

$$\alpha o(h) = o(h).$$

Table 2.3 Rules of differentiation

Name of the rule	Mathematical question	example
Multiple	$\alpha f(x)$ with $\alpha \in \mathbb{R}$	$5x^2$
Sum	$f(x) + g(x)$	$x^2 + x^3$
Difference	$f(x) - g(x)$	$x^2 - x^3$
Linear combination	$\alpha f(x) + \beta g(x)$ with $\alpha, \beta \in \mathbb{R}$	$5x^2 - x^3$
Product	$f(x) \cdot g(x), g(x) \neq 0$	$x^3 e^x$
Quotient	$f(x)/g(x)$	x^3/e^x
Chain	$f[g(x)]$	e^{x^2}

$$\phi(x+h) - \phi(x) = \alpha f(x+h) - \alpha f(x) = \alpha[f(x+h) - f(x)] =$$
$$= \alpha[f'(x)h + o(h)] = [\alpha f'(x)]h + o(h)$$

Also the function ϕ turns out to be differentiable and its derivative is:

$$\phi'(x) = \alpha f'(x)$$

Example 2.2.7 Let:
$$\phi(x) = 5x^2$$

We have:
$$\phi'(x) = 5 \cdot D[x^2] = 5 \cdot 2x = 10x$$

2 Differentiating the sum of two functions f, g — Let:

$$\phi(x) = f(x) + g(x)$$

Its increment is[26]:

$$\phi(x+h) - \phi(x) = f(x+h) + g(x+h) - [f(x) + g(x)] =$$
$$= f(x+h) - f(x) + g(x+h) - g(x) =$$
$$= f'(x)h + o(h) + g'(x)h + o(h)$$
$$= [f'(x) + g'(x)]h + o(h)$$

therefore also ϕ is differentiable and:

$$\phi'(x) = f'(x) + g'(x)$$

---

[26]We continue use the "dirty" fact that:

$$o(h) + o(h) = o(h).$$

Example 2.2.8 Let:

$$\phi(x) = x^2 + x^3$$

We have:

$$\phi'(x) = D\left[x^2 + x^3\right] = D\left[x^2\right] + D\left[x^3\right] = 2x + 3x^2 = x(2 + 3x)$$

3 Differentiating the difference of two functions f, g — Let:

$$\phi(x) = f(x) - g(x)$$

Its increment is[27]:

$$
\begin{aligned}
\phi(x+h) - \phi(x) &= f(x+h) - g(x+h) - [f(x) - g(x)] = \\
&= f(x+h) - f(x) - [g(x+h) - g(x)] = \\
&= f'(x)h + o(h) - g'(x)h - o(h) \\
&= \left[f'(x) - g'(x)\right]h + o(h)
\end{aligned}
$$

therefore also ϕ is differentiable and:

$$\phi'(x) = f'(x) - g'(x)$$

Example 2.2.9 Let:

$$\phi(x) = x^2 - x^3$$

We have:

$$\phi'(x) = D\left[x^2 - x^3\right] = D\left[x^2\right] - D\left[x^3\right] = 2x - 3x^2$$

4 Differentiating a linear combination of two functions f, g — Let:

$$\phi(x) = \alpha f(x) + \beta g(x)$$

Note that this case embraces all the three previous ones: the multiple is obtained letting $\beta = 0$, the sum appears with $\alpha = \beta = 1$, the difference with $\alpha = 1$ and $\beta = -1$.

Its increment is[28]:

---

[27] We use the "dirty" fact that:

$$o(h) - o(h) = o(h).$$

[28] We use the "dirty" facts that:

$$\alpha o(h) = o(h); \quad \beta o(h) = o(h) \text{ and}$$
$$o(h) \pm o(h) = o(h).$$

$$\phi(x+h) - \phi(x) = \alpha f(x+h) + \beta g(x+h) - [\alpha f(x) + \beta g(x)] =$$
$$= \alpha[f(x+h) - f(x)] + \beta[g(x+h) - g(x)] =$$
$$= \alpha f'(x)h + o(h) + \beta g'(x)h + o(h) =$$
$$= [\alpha f'(x) + \beta g'(x)]h + o(h)$$

therefore also ϕ is differentiable and:

$$\phi'(x) = \alpha f'(x) + \beta g'(x)$$

Example 2.2.10 Let:

$$\phi(x) = 5x^2 - x^3$$

We have:

$$\phi'(x) = D[5x^2 - x^3] = 5 \cdot D[x^2] - D[x^3] = 10x - 3x^2$$

Remark 2.2.6 The rule we have just seen tells us that the derivative of a linear combination of functions is equal to the linear combination of the derivatives. Mathematicians describe this fact saying that the operator D is *linear*. Readers have already met a linear operator when studying linear applications from — say — $\mathbb{R}^n$ to $\mathbb{R}^m$ (see the section linear application in the linear algebra part, p. 67). Later we will meet another important linear operator. The structure will be the same.

The first four differentiation rules could generate false hopes: the most common case concerns the product of two functions. Some people could expect that:

$$D[f(x)g(x)] = f'(x)g'(x) \tag{2.2.4}$$

Well, this is simply false (its true, if and only if, at least one of the two factors is constant). To convince ourselves about it is sufficient to consider a simple example and to look at what should be true in the case formula (2.2.4) would be true:

$$D[x^2] = D[x \cdot x] = D[x] \cdot D[x] = 1 \cdot 1 = 1$$

which would contradict what we already know:

$$D[x^2] = 2x$$

5 Differentiating the product of two functions f, g — Let:

$$\phi(x) = f(x)g(x)$$

Its increment is:

$$\phi(x+h) - \phi(x) = f(x+h)g(x+h) - f(x)g(x)$$

With some manipulations, we would like to be reticent about, it is possible to see that also ϕ is differentiable and:

$$\phi'(x) = f'(x)g(x) + f(x)g'(x)$$

Example 2.2.11 Let:

$$\phi(x) = x^3 e^x$$

We have:

$$\phi'(x) = D\left[x^3 e^x\right] = D\left[x^3\right]e^x + x^3 D\left[e^x\right] = 3x^2 e^x + x^3 e^x = x^2(3+x)e^x$$

Remark 2.2.7 There is a simple trick to remember this rule. It is not politically correct, but somehow efficient. Two old ladies f, g, gossip eager,[29] usually visit their haidresser for their perms. The (somehow country) hairdresser has only one workstation. Therefore while Ms. f is under the hair dryer (labelled f'), Ms. g sits nearby, in order to start the gossip. Once Ms. g is under the hairdryer (labelled g'), Ms. f sits nearby, in order to usefully continue their gossip. The differentiation rule;

$$(fg)' = f'g + fg'$$

deserves the title of "*hairdresser rule* ", and we will occasionally refer to.

6 Differentiating the quotient of two functions f, g — Let:

$$\phi(x) = \frac{f(x)}{g(x)}, \text{ with } g(x) \neq 0$$

Its increment is:

$$\phi(x+h) - \phi(x) = \frac{f(x+h)}{g(x+h)} - \frac{f(x)}{g(x)}$$

With some manipulations, we omit, it is possible to see that also ϕ is differentiable and:

$$\phi'(x) = \frac{g(x)f'(x) - f(x)g'(x)}{g^2(x)}$$

There is a very simple way to remember the preceding formula. Two steps are crucial:

(1) The new denominator is nothing but the square of the original one: $g(x)$ becomes $g^2(x)$.

---

[29]Thought of not so common among males..., at least by males.

(2) The construction of the numerator starts from the denominator $g(x)$ and runs like in the hairdresser case, simply replacing the sign "+" with "− " .

Example 2.2.12 Example 24 Let:

$$\phi(x) = \frac{x^3}{e^x}$$

We have:

$$\phi'(x) = \frac{e^x D\left[x^3\right] - x^3 D\left[e^x\right]}{e^{2x}} = \frac{3x^2 e^x - x^3 e^x}{e^{2x}} = \frac{x^2(3-x)}{e^x} = x^2(3-x)e^{-x}$$

7 Differentiation of a composite functions, the chain rule. This is the spice of life. Here is the problem. Think of the income x of the citizen Mr. X. A part of it — say — $z = g(x)$ is taxable. The amount of taxes y, due by Mr. X, is a function f of the taxable income z:

$$y = f(z) = f[g(x)]$$

If the income of Mr. X varies, the tax return will vary. This is a standard fiscal policy scheme. The variation of the income return for Government depends on both two functions:

- the tax rules of some income x, which are specified by g;
- the taxation rules, which associate to any taxable income an amount due to the Government, specified by f.

In very simple (and unrealistic) cases the taxation rules could be the following: the taxable income is some fraction τ of the income itself:

$$z = \tau x$$

the amount of taxes is some fraction α of the taxable amount:

$$y = \alpha z$$

therefore:
$$y = \alpha(\tau x) = (\alpha \tau) x$$

The simple case rule is: "the amount of taxes can be computed from the income x, simply multiplying the two percentages":

- τ the percentage of taxable income;
- α, the tax rate."

We will discover that such a simple rule works also in the general case, when f, g are non-linear but only differentiable, i.e; locally approximable with linear functions.

Let:
$$\phi(x) = f[g(x)]$$

its increment is:

$$\phi(x+h) - \phi(x) = f[g(x+h)] - f[g(x)] =$$
$$= f'[g(x)] \cdot [g(x+h) - g(x)] + o[g(x+h) - g(x)] =$$
$$= f'[g(x)]g'(x)h + o(h)$$

therefore:

$$\phi'(x) = f'[g(x)]g'(x)$$

The computation rule for derivatives of composite functions is often called *chain rule*.

Example 2.2.25 Let:

$$\phi(x) = e^{x^2}$$

This function stems from the composition of: $f(y) = e^y$ and $y = g(x) = x^2$:

$$\phi(x) = \exp\left(x^2\right)$$

We a think of the exponential function as the "external" function and of the quadratic function as the internal one. The chain rule tells us that we must first differentiate the external function and then multiply the derivative we have found by the derivative of the internal function:

$$D\left[\exp\left(x^2\right)\right] = \exp\left(x^2\right) \cdot 2x = 2xe^{x^2}$$

Remark 2.2.8 **Onion rule** — A simple rule, stems from the idea that, if you want to peel an onion, you must start from its external leaves, and proceed with the internal ones. If you want to peel (pardon, to differentiate $\phi(x) = e^{x^2}$), you first state that its external leave is $\exp(.)$ and that the internal one is $(.)^2$. The derivative of the external is a simple photocopy: e^{x^2} and in must be multiplied with the derivative of the internal leave (pardon, of the internal function) $D(x^2) = 2x$. What the onion rule produces is: $e^{x^2} \cdot (2x) = 2xe^{x^2}$.

Sometimes we will need to differentiate the derivative of some function f. What we will obtain is the *second derivative* of f.

Definition 2.2.3 Given some function $f : \mathbb{R} \to \mathbb{R}$, differentiable at any relevant point x, with derivative $f'(x)$, we call *second derivative* of the function:

$$D\left[f'(x)\right]$$

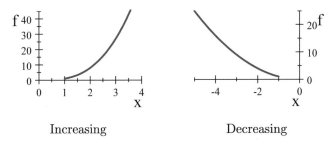

Increasing   Decreasing

Fig. 2.15 Increasing and decreasing functions

Such a derivative can be indicated with symbols like:

$$f''(x); \quad D^2[f(x)]; \quad \frac{d^2}{dx^2}f(x)$$

Example 2.2.26 The function we consider is the same of the previous example:

$$\phi(x) = e^{x^2}$$

We have already computed its (first) derivative:

$$\phi'(x) = 2xe^{x^2}$$

If we want its second derivative, we must simply differentiate the first one:

$$\phi''(x) = D\left[2xe^{x^2}\right] = 2 \cdot e^{x^2} + 2x \cdot 2xe^{x^2} = 2e^{x^2}\left(1 + 2x^2\right)$$

2.2.4.3 Common Uses of Derivatives

Increasing and decreasing functions — Take an interval and assume that some function f is differentiable at each point of the interval and that its derivative f' is positive. This implies that the value of f is necessarily *increasing* over that interval. This precisely means that if we take two points in the interval: $x_1 < x_2$, then $f(x_1) < f(x_2)$.

A similar conclusion will hold in the case f' is negative: the value of the function will decrease. If $x_1 < x_2$, then $f(x_1) > f(x_2)$. In this case, we will say that f is *decreasing*. The Fig. 2.15 illustrates the two cases.

2.2.4.4 Extrema

Think now of a derivative which, close to some point x^*, is positive on the left of x^* and negative on the right. This implies that the function is increasing on the left of

Fig. 2.16 $f(x) = x^2 e^{-x}$

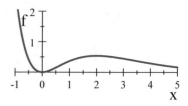

x^* and decreasing on the right. Look for instance at Fig. 2.16 containing the graph of the function $f(x) = x^2 e^{-x}$, close to 2:

Its derivative is:

$$f'(x) = x(2 - x)e^{-x}$$

The third factor has no influence on the sign of f', which close to 2 on the left is positive, but its sign switches to $-$ on the right of 2.

We say that the point 2 is a *local maximum point*. ("local max point" for short is authorized).

If we look at the point 0, the situation is somehow symmetric, on the left f' is negative and on the right it is positive; the function f decreases on the left of 0 and increases on the right. We say that 0 is a *local minimum point*. ("local min point", in a hurry).

Definition 2.2.4 Solemnly, if $f(x^*) \geq f(x^* + h)$ for sufficiently small $|h|$ we say that x^* is a *local maximum point*. If we reverse "$\geq$" with "$\leq$" we get the notion of *local minimum point*.

We will use the word *local extremum* to refer to a point which is a (local) max or min point.

Remark 2.2.9 At extrema points functions switch from increasing to decreasing or vice versa. If functions are differentiable at that points, then there the derivative f' must be 0.

Definition 2.2.5 If at a point x^* the derivative of f is 0, we say that x^* is a *stationary point* for f. In other words stationary points are nothing but the solutions of the equation:

$$f'(x) = 0$$

If at an internal point x^* of some interval over which f is defined and differentiable there is an extremum, then necessarily $f'(x^*) = 0$. This fact is intuitively trivial: if $f'(x^*) > 0$, in the small, f looks like an increasing linear affine function; if $f'(x^*) < 0$, in the small, f looks like a decreasing linear affine function. These possibilities are canceled by the assumption that x^* is an extremum.

Necessarily $f'(x^*) = 0$, or it does not exist.[30]

---

[30]We have assumed that f is differentiable everywhere, but this framework is politically interesting: $f'(x^*) \neq 0$ could imply some engagement, the inexistence of $f'(x^*)$ looks like abstention.

Fig. 2.17 $f(x) = x^5 - 3x^4 + 2x^3$

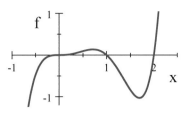

The fact that some function is stationary at some point does not guarantee that such a point is an extremum.

Consider the function:

$$f(x) = x^5 - 3x^4 + 2x^3$$

The Fig. 2.17 shows its graph[31]:

Its derivative is:

$$f'(x) = 5x^4 - 12x^3 + 6x^2$$

The stationary points for f solve:

$$5x^4 - 12x^3 + 6x^2 = 0$$

There are three stationary points:

$$0; \quad \frac{6 \mp \sqrt{6}}{5}$$

or, less cryptically:

$$x_1 = 0;$$

$$x_2 = \frac{6 - \sqrt{6}}{5} \approx 0.7101$$

$$x_3 = \frac{6 + \sqrt{6}}{5} \approx 1.6899$$

The point $x_1 = 0$ is not an extremum: f is increasing both on the left and on the right of 0.

The point x_2 is a max point: f increases on the left and decreases on the right. Finally, the point x_3 is a min point: f decreases on the left and increases on the right.

The Table 2.4 summarizes what we have seen until now about the information contents of first derivatives.

---

[31]For the moment, believe in the Authors. Later you will be able to construct such graphs autonomously.

Table 2.4 Information contents of 1st order derivatives

Behavior of f'	Then for f
$f'(x^*) = 0$	x^* is a stationary point for f
$f' > 0$ on an interval	f increases on that interval
$f' < 0$ on an interval	f decreases on that interval
f' switches from $+$ to $-$ at x^*	x^* is a max point
f' switches from $-$ to $+$ at x^*	x^* is a min point

Fig. 2.18 $f(x) = -x^2$

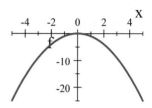

2.2.4.5 2nd Order Derivatives, Convexity and Concavity

Take the function $f(x) = -x^2$ whose graph is in the Fig. 2.18. Its (first) derivative is $f'(x) = -2x$ and it is positive if x is negative and *vice versa*. What we have just seen tells us that the function attains a maximum at 0.

Its second derivative is -2 and it tells us that the *local slope* (f') of the graph has negative derivative, which implies that the *local slope is decreasing*. In fact, roughly speaking, we can say that:

- if x is "very negative" then f' is "very positive": f very rapidly increases;
- if x is "slightly negative" then f' is "slightly positive": f slowly increases;
- if x is "slightly positive" then f' is "slightly negative": f slowly decreases;
- if x is "very positive" then f' is "very negative": f very rapidly decreases.

Functions of this type are characterized by the fact that any portion of their graph, on an interval, stays above the straight line connecting the endpoints of the interval. These functions are called *concave functions*. The Fig. 2.19 illustrates two examples.

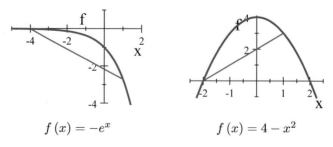

$$f(x) = -e^x \qquad\qquad f(x) = 4 - x^2$$

Fig. 2.19 Concave functions

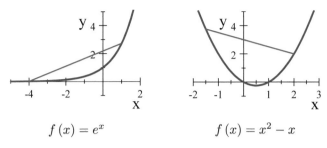

$$f(x) = e^x \qquad\qquad f(x) = x^2 - x$$

Fig. 2.20 Convex functions

For functions like $f(x) = x^2$ the situation is symmetrically reversed: for them the graph stays everywhere under the straight line. These functions are called *convex functions*. The Figs. 2.20 provides us with a pair of convex functions.

These properties have relevant implication for Social Sciences (not only in Economics, in which they play a central role).

Second derivatives are extremely useful in detecting the presence/absence of these properties.

If on some interval f'' is negative, than on that interval f is concave, while if f'' is positive, then f turns out to be convex.

Definition 2.2.6 The points at which the second derivative changes its sign are called *inflexion points*.

For instance, if you go back to p. 139 and you look at the graph of the function, you can see that the stationary point 0, which was not an extremum is an inflection point, because the function is concave on the left of 0 and convex on the right.

For twice differentiable functions, values x at which there could be an inflexion point can be detected via the condition:

$$f''(x) = 0$$

Example 2.2.27 Let's consider the function defined by $f(x) = xe^x$. Its first and second derivatives are:

$$f'(x) = e^x(x+1) \qquad \text{and} \qquad f''(x) = e^x(x+2)$$

therefore $f''(x) = 0$ where $e^x(x+2) = 0$ and we obtain $x = -2$ (since e^x is always positive). At $x = -2$ the function shows a change of concavity (in this case form concave to convex).

Example 2.2.28 Let's take the function $f(x) = \ln(x+1)$, which is defined for every $x > -1$. Its first and second derivatives are:

$$f'(x) = \frac{1}{x+1} \qquad \text{and} \qquad f''(x) = -\frac{1}{(x+1)^2}$$

Table 2.5 Information contents of 1st and 2nd order derivatives

Behavior of f', f''	Then for f
$f'(x^*) = 0$ and $f''(x^*) < 0$	x^* is a max point for f
$f'(x^*) = 0$ and $f''(x^*) > 0$	x^* is a min point for f
$f'' > 0$ on an interval	f is convex on that interval
$f'' < 0$ on an interval	f is concave on that interval
f'' switches from $+$ to $-$ at x^* or vice versa	x^* is an inflection point

Its first derivative is positive and it's clear that its second derivative keeps the same negative sign for every $x > -1$, therefore f is increasing over $(-1, +\infty)$, has no inflexion points and is concave everywhere.

The Table 2.5 summarizes the information content[32] of 1st and 2nd derivatives of a function.

2.2.4.6 Drawing Graphs of Functions $f : \mathbb{R} \to \mathbb{R}$

How can we get the graph of a function, knowing its analytic definition? In principle the task could not be not easy, if we are confined to use the few tools presented above. For some relevant applications in Social Sciences, the functions encountered are rather simple and therefore the task can be fulfilled rather easily.

The strategy we suggest for drawing the graph of some function $f : \mathbb{R} \to \mathbb{R}$ is articulated in the following steps:

1. Determine the domain E of the function f: it is the subset of $\mathbb{R}$ in which the function is defined. The points of $\mathbb{R}$ that must be **excluded** are the ones at which:

    (a) **denominators do vanish**;
    (b) **the argument of even radicals are negative**;
    (c) **the arguments of a logarithm is non-positive**. Frequently, for our purposes, it will be sufficient to work only on a subset[33] of the "mathematical" domain E.

2. Try to identify the **behavior** of f **at the boundary** of E. If you know the theory of limits,[34] use it. If not, try numerically.

---

[32]The first two rows under the description are important, because they provide us with cheap criteria for finding extrema. We saw above that looking at the sign of f' close to some point, we can detect the presence of an extremum. But they are expensive because they require us to find the sign of a function (f'), while what we are learning is that, using second derivatives, it is sufficient to find their value only at a point.

[33]Think that $f(x)$ connects the electoral consensus x of a party to, say, the number $f(x)$ of members of a Parliament, chosen by that party. It is clear that the variation interval, which is relevant for politics is the interval $[0\%, 100\%] = [0, 1]$ or: $0 \leq x \leq 1$. A negative consensus or a consensus above 100% would be void of sense.

[34]Which is omitted in this text.

3. Find the points at which **the graph of f crosses or touches the Cartesian axes**:

   (a) the intersection with the y axis exists iff $0 \in E$ is at the point whose abscissa is 0 and whose ordinate is $f(0)$;
   (b) the points of the graph belonging to the abscissas axis solve the equation $f(x) = 0$. They are called "zeroes of f".

4. Find the points of the **graph above the x axis**: they satisfy the inequality $f(x) > 0$ and the points of the **graph under the x axis**: they satisfy the inequality $f(x) < 0$.
5. Compute the **first derivative** of f and

   (a) find where $f' > 0$, there f will be **increasing**;
   (b) find where $f' < 0$, there f will be **decreasing**;
   (c) find the **stationary points** for f and decide whether they are **extrema** or not, maybe using later the second derivative.

6. Compute the **second derivative** of f and

   (a) find where $f'' > 0$, there f will be **convex**;
   (b) find where $f'' < 0$, there f will be **concave**;
   (c) find the 0 **points** for f'' and decide whether they are **inflexion points**, where concavity changes.

   Here is an example.

Example 2.2.29 The function is the one we have already met above at p. 139:

$$f(x) = x^5 - 3x^4 + 2x^3$$

We will show how the strategy we have designed allows us to detect the behavior of f.

1. The domain E of f is $\mathbb{R}$: for every $x \in \mathbb{R}$, we can successfully compute $f(x)$.
2. The addenda are functions. We know that, among power functions, there is a strict hierarchy determined by the exponents. The first addendum x^5 dominates the others, so, when $x \to -\infty$, the same is made by f and when $x \to +\infty$ also f does. Therefore $f(x) \to +\infty$ if $x \to +\infty$ and $f(x) \to -\infty$ if $x \to -\infty$. The Fig. 2.21 illustrates our first conclusions.
3. (a) The intersection with the y axis is at the origin itself is

$$f(0) = 0$$

   while (b) the intersections with the x axis are at the solutions of the equation

$$x^5 - 3x^4 + 2x^3 = 0$$

   The left hand side can be factored:

$$x^3 \left(x^2 - 3x + 2 \right)$$

Table 2.6 Function sign

Interval\Factor	x^3	$x^2 - 3x + 2$	f
$(-\infty, 0)$	$-$	$+$	$-$
$\left(0, \left(6 - \sqrt{6}\right)/5\right)$	$+$	$+$	$+$
$\left(\left(6 - \sqrt{6}\right)/5, \left(6 + \sqrt{6}\right)/5\right)$	$+$	$-$	$-$
$\left(\left(6 + \sqrt{6}\right)/5, +\infty\right)$	$+$	$+$	$+$

so that we detect three intersection points: $x_1^* = 0$, $x_2^* = \left(6 - \sqrt{6}\right)/5 \approx$ 0.710 1 and $x_3^* = \left(6 + \sqrt{6}\right)/5 \approx 1.689\,9$.

4. The points of the graph above the x satisfy the inequality $f(x) > 0$ and the points of the graph under the x axis: they satisfy the inequality $f(x) < 0$. In order to single out the various intervals, it is useful to construct a table, providing us with the sign of the three factors in the various intervals.

   From the Table 2.6 we argue that the graph of f stays under the x axis if $x < 0$ or $x_2^* < x < x_3^*$, while it stays above if $0 < x < x_2^*$ or $x > x_3^*$.

5. The first derivative of f is:

$$f'(x) = \frac{\mathrm{d}}{\mathrm{d}x}\left(x^5 - 3x^4 + 2x^3\right) = 5x^4 - 12x^3 + 6x^2 = x^2\left(5x^2 - 12x + 6\right)$$

It does not change its sign at 0, where the first factor x^2 (so to say) "passes" from a positive sign on the left, to a positive sign on the right. The origin is not an extremum for f. The second factor $5x^2 - 12x + 6$ has a more exciting behavior. It is null at the solutions of the equation $5x^2 - 12x + 6 = 0$. Such solutions are $x_4^* = \left(6 - \sqrt{6}\right)/5 \approx 0.710\,1$ and $x_5^* = \left(6 + \sqrt{6}\right)/5 \approx 1.689\,9$. It is negative for $x_4^* < x < x_5^*$ and positive outside such interval. The Table 2.7 summarises the results.

Table 2.7 First derivative sign

Where $\pm$ for f'	1st fact. x^2	2nd fact. $5x^2 - 12x + 6$	f'	f
$\left(-\infty, \dfrac{6 - \sqrt{6}}{5}\right)$	$+$ or 0	$+$	$+$ or 0	↗
$\left(\dfrac{6 - \sqrt{6}}{5}, \dfrac{6 + \sqrt{6}}{5}\right)$	$+$	$-$	$-$	↘
$\left(\dfrac{6 + \sqrt{6}}{5}, +\infty\right)$	$+$	$+$	$+$	↗

Table 2.8 Second derivative sign

Where $\pm$ for f''	1st fact. $4x$	2nd fact. $5x^2 - 9x + 3$	f''	f
$(-\infty, 0)$	$-$	$+$	$-$	$\cap$
$\left(0, \dfrac{9 - \sqrt{21}}{10}\right)$	$+$	$+$	$+$	$\cup$
$\left(\dfrac{9 - \sqrt{21}}{10}, \dfrac{9 + \sqrt{21}}{10}\right)$	$+$	$-$	$+$	$\cap$
$\left(\dfrac{9 + \sqrt{21}}{10}, +\infty\right)$	$+$	$+$	$+$	$\cup$

Fig. 2.21 Limit behavior

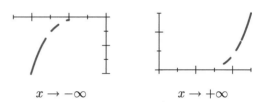

$$x \to -\infty \qquad\qquad x \to +\infty$$

6. The second derivative of f is:

$$f''(x) = \frac{d}{dx}\left(5x^4 - 12x^3 + 6x^2\right) = 20x^3 - 36x^2 + 12x = 4x\left(5x^2 - 9x + 3\right)$$

It vanishes at $x_6^* = 0$ and at the roots of:

$$5x^2 - 9x + 3 = 0$$

which are $x_7^* = \left(9 - \sqrt{21}\right)/10 \approx 0.44174$ and $x_8^* = \left(9 + \sqrt{21}\right)/10 \approx 1.$
3583. At each of such points the concavity of the graph changes: all of them are inflexion points. The Table 2.8 summarizes our conclusions.
At this point the graph Fig. 2.17 at p. 139 should not be a mystery.

Example 2.2.30 (Demand Elasticity) — Consider the case of a monopolistic producer. Call x the quantity of a good to be produced and sold at the unitary price p. It is obvious that the choice of p will determine the amount $x(p)$ that will be sold. At a cheap price p the demand x will be high, while for large values of p (the good is expensive), the demand will be low. The function:

$$x = f(p)$$

is usually called *demand function*. We will assume that it is differentiable and that $f'(p) < 0$. Call $I(p)$ the amount of money collected by the producer as a function of the sale price p. We have:

$$I(p) = px(p)$$

we want to understand how the total sale volume $I(p)$ depends on price. The question is not trivial because $I(p)$ is the product of two factors, which do move in opposite directions: if p increases then $x(p)$ decreases and vice versa. We differentiate I:

$$I'(p) = x(p) + px'(p)$$

The function I will be increasing if:

$$x(p) + px'(p) > 0$$

or, equivalently:

$$p\frac{|x'(p)|}{x(p)} < 1$$

The function:

$$\varepsilon(p) = p\frac{|x'(p)|}{x(p)}$$

is usually called *elasticity* of the demand function. Note that it could also be written as:

$$\varepsilon(p) = p\,|D[\ln x(p)]|\tag{2.2.5}$$

and that it can be thought of as an approximation of:

$$|(\Delta x/x)/(\Delta p/p)|$$

If it is smaller than 1 the demand is said to be *rigid* (or *inelastic*), because if the price increases, the demand contraction is so small that the total amount of money collected by the producer increases. If it is greater than 1, it is said to be *elastic* as the demand contraction determined by an increase in the price prevails. If ε is 1 the demand is said to be *anelastic*: the two effects compensate each other. Obviously the notion of elasticity is referred to some point p. The same demand function can have different elasticities, even $\gtrless 1$, at different points. An interesting case is the following one:

$$x(p) = \frac{A}{x^\alpha}$$

where $A, \alpha > 0$. Using Eq. (2.2.5), it can be checked very simply that $\varepsilon(p) \equiv \alpha$.

Using tools we will see later it can be shown that these demand functions are the only ones with constant elasticity.

Fig. 2.22 $f(p) = \frac{1000}{p^{0.8}}$

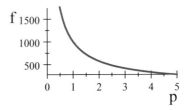

In fact they are frequently called *CE functions*, where CE stays for "Constant Elasticity".

Let us look at a numerical example.

Example 2.2.31 We are given a demand function for some public service

$$f(p) = \frac{A}{p^\alpha} \tag{2.2.6}$$

where A, α are positive parameters. The positive variable p is the unitary price of the service. Formula (2.2.6) provides us with the demand volume as a function of unitary tariff p for the service (f decreases as p increases). The Fig. 2.22 shows such a function in the case $A = 1000$ and $\alpha = 0.8$.

The elasticity of the demand function $\varepsilon\,(p)$, is:

$$\varepsilon\,(p) = -p \cdot \frac{d}{dp} \ln f(p)$$

Considering (2.2.5) we have:

$$\varepsilon\,(p) = -p \cdot \frac{d}{dp} \ln \left(\frac{A}{p^\alpha} \right) = p \cdot \frac{d}{dp} \ln p^\alpha = \alpha \,(\text{constant})$$

We want to explore the effect of a change in the tariff on the total expenditure for that public service. The total expenditure E, is:

$$E(p) = p \cdot f\,(p)$$

It is the product of unitary tariff with the required volume of public service. Considering the demand function (2.2.6), the total expenditure is

$$E(p) = p \cdot \frac{A}{p^\alpha} = A \cdot p^{1-\alpha}$$

The Fig. 2.23 illustrates the behavior of the total expenditure E in three particular cases: (i) *strong sensitivity* to price variations ($\alpha = 1.2$); (ii) *unitary sensitivity* to price variations ($\alpha = 1$); (iii) *moderate sensitivity* to price variations ($\alpha = 0.8$).

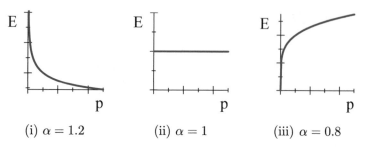

(i) $\alpha = 1.2$ (ii) $\alpha = 1$ (iii) $\alpha = 0.8$

Fig. 2.23 Demand sensitivity to price and expenditure

Exercise 2.2.1 We leave to the reader the case of a linear affine demand function, for instance:

$$f(p) = 1000 - 10p$$

We suggest to compute the demand elasticity $\varepsilon(p)$ and to explore the behavior of the expenditure $E(p)$ as a function of price.

*Example 2.2.32 **Elasticity again** —* We complete the model we have met talking about elasticity, taking into account costs incurred by the producer. This is the well-known optimum problem for a monopolist. A monopolist produces the quantity x a good and must sell it at the unitary price $p(x)$ in order to place all the production in the market. The function $p(x)$ is usually called *inverse demand function*. Of course the collected amount is:

$$I(x) = xp(x)$$

In order to produce the quantity x, the monopolist incurs in costs, whose amount is $C(x)$. We assume that both p and C are differentiable. The profit for the monopolist is:

$$P(x) = I(x) - C(x)$$

In order to maximize the profit, special attention should be given to stationary points x^* for P, i.e., points at which P' is zero. As:

$$P'(x) = I'(x) - C'(x)$$

at such points the following equality must hold:

$$I'(x^*) - C'(x^*)$$

It is nothing but the famous classic economic principle that at an optimum marginal returns (I') and marginal costs (C') must coincide.

Look now at this:

Example 2.2.33 The inverse demand function is:

$$p(x) = \frac{B}{x^\beta}; \qquad x > 0$$

with $B, \beta > 0$. The cost function is linear affine (fixed costs + variable costs), with v positive:

$$C(x) = F + vx$$

The profit function is:

$$P(x) = xp(x) - F - vx = Bx^{1-\beta} - F - vx$$

The marginal profit is:

$$P'(x) = (1 - \beta)Bx^{-\beta} - v$$

If $\beta \geq 1$, $P'(x)$ is negative for every value of x and the best decision for the monopolist would be to stop the production. If $\beta < 1$, the marginal profit is (very) positive for x sufficiently close to 0 and it becomes negative for values of x sufficiently large. Consequently, there is an $x^* > 0$, which turns out to be optimal. We get it solving the equation:

$$(1 - \beta)Bx^{-\beta} - v = 0$$

hence:

$$x^* = \left[\frac{(1 - \beta)B}{v}\right]^{1/\beta}$$

The expression for x^* we have obtained is interesting also from a political point of view. Assume that a further tax could be introduced that increases the unitary variable cost v. The expression of x^* tells us that if v increases then x^* decreases and may be that such a contraction is able to cancel the expected increment in tax return. A reduction in x^* would bring to a higher sale price p of the good and, consequently, the tax formally paid by the monopolist would be at least partially transferred to other people. Last detail: if we compute the second derivative of P we get:

$$P''(x) = -\beta(1 - \beta)Bx^{-\beta-1}$$

which is negative if $\beta < 1$. In this case P is concave and therefore the local maximum x^* we have detected is a global one.

Example 2.2.34 An NGO offers a service. The size of the service offered[35] is $x \geq 0$. The total cost $C(x)$ the NGO incurs has the rather standard structure:

---

[35]For instance, the number of people assisted by the NGO.

$$C(x) = F + xv(x)$$

where F is the fixed cost of the service and $v(x) \geq 0$ the unitary variable cost, assumed to be differentiable. The efficiency of the service can be obviously measured through the average service cost:

$$A(x) = \frac{C(x)}{x} = \frac{F}{x} + v(x)$$

We are interested in understanding how A varies with x. We get:

$$A'(x) = -\frac{F}{x^2} + v'(x)$$

The first addendum is negative. If $v' \leq 0$ there is no stationary point for A and, consequently, no efficient service size. If $v' > 0$, there can be stationary points. We introduce the further assumption, that v' does not decrease if x increases. It's a natural assumption in the case there are scale diseconomies. Not only the unitary variable cost increases with the service volume, but it does that with increasing speed. It is easy to understand that in this case there is a unique $x^* > 0$ at which A' is null, and that this is a minimum point. It solves the equation:

$$v'(x) = \frac{F}{x^2}$$

We are interested in understanding the relationship between the average cost A and the marginal cost $M(x) = C'(x)$ at the maximum efficiency point. A classical result in Microeconomics is that, at x^*, $A = M$. In formulas:

$$A(x^*) = M(x^*)$$

We check this general result in the more familiar terms of fixed and variable costs, which, likely, are more understandable from an applied viewpoint. The marginal cost is:

$$M(x) = C'(x) = xv'(x) + v(x)$$

Let's look at this interesting example.

Example 2.2.35 Let:
$$v(x) = \alpha + \beta x, \text{ with } \alpha, \beta > 0$$

The simple idea, behind this example, is that the unitary variable cost increases linearly with the volume of the service the NGO provides. The total cost specification turns out to be:
$$C(x) = F + (\alpha + \beta x)x = F + \alpha x + \beta x^2$$

The average cost results:

$$A(x) = \frac{C(x)}{x} = \frac{F}{x} + \alpha + \beta x$$

If we look for the maximum efficiency size of the service, we must look for stationary points for A. We obtain easily:

$$A'(x) = -\frac{F}{x^2} + \beta$$

The only concretely relevant solution solves positively:

$$-\frac{F}{x^2} + \beta = 0$$

hence[36]:

$$x^* = \sqrt{\frac{F}{\beta}}$$

The minimized average cost turns out to be:

$$\frac{F}{\sqrt{\frac{F}{\beta}}} + \alpha + \beta\sqrt{\frac{F}{\beta}} = \alpha + 2\sqrt{F\beta}$$

Now, we move to the marginal cost:

$$M(x) = \beta x + \alpha + \beta x = \alpha + 2\beta x$$

At the efficient point:

$$M\left(\sqrt{\frac{F}{\beta}}\right) = \alpha + 2\beta\sqrt{\frac{F}{\beta}} = \alpha + 2\sqrt{F\beta}$$

there is coincidence between average and marginal cost. The dependence on F of the optimal solutions is absolutely relevant. If an NGO wants efficiency, the scale of basic investments (F) must have an adequate scale. A natural political question is "What follows?". The answer is not difficult: if an NGO offers a service to too few people the average cost of the intervention is too high. If it offers a service to too many people, again the average cost of the intervention is too high. Therefore

---

[36]We ignore the algebraic negative solution, which is irrelevant in this framework.

there is an optimal way for sizing the service. It turns out to depend (non-linearly) on the amount F. this has a influence on the effects of an optimal political use of resources F.

2.3   What's a Function of a Vector

In hard sciences the idea that you have one cause which determines one effect turns out to be common. All of us, at the high school, have met the gravity model;

$$y = \frac{1}{2}gx^2$$

informing us that if a body falls because of gravity, in the absence of frictions, the space it covers (y) depends on time (x), according to the formula above, where g is the "gravity acceleration": interesting, simple, illuminating.

The scheme is clean and appealing:

$$y = f(x)$$

We can think of f as a black box, which transforms the input x into the output y:

$$x \longrightarrow \boxed{f} \longrightarrow y$$

As x is a real number (an element of $\mathbb{R}$) and the same is true for y as well, we have arguments to maintain the simplicity of the scheme.

But this is nothing new. Please go back to p. 99.

We have also considered a few lines above a problem of interest Economic Policy.

In Social Sciences, such a simple scheme is often too simplistic. For instance, think you want to study some election process. In order to forecast the number of the votes some candidate will collect you must take into account several variables: profile, reputation, electoral strategy,...

To study these problems, we must necessarily move to more complex schemes, in which the input is non-necessarily a single variable, but a vector $\mathbf{x}$ of n variables:

$$\mathbf{x} = \begin{bmatrix} x_1 \\ x_2 \\ \cdots \\ x_n \end{bmatrix}$$

We want to extend the basic scheme as follows :

$$\mathbf{x} \longrightarrow \boxed{f} \longrightarrow y$$

or:

$$y = f(\mathbf{x}) = f(x_1, x_2, \ldots, x_n)$$

Also the output could be a vector

$$\mathbf{y} = \begin{bmatrix} y_1 \\ y_2 \\ \ldots \\ y_m \end{bmatrix}$$

This means simply that instead of having only one "cause" and one "effect" y we will have several "causes" and several "effects":

$$\mathbf{y} = \mathbf{f}(\mathbf{x}) \text{ or, more explicitly: } \begin{cases} y_1 = f_1(\mathbf{x}) \\ y_2 = f_2(\mathbf{x}) \\ \ldots \\ y_m = f_m(\mathbf{x}) \end{cases}$$

or, also, with maximum detail:

$$\begin{cases} y_1 = f_1(x_1, x_2, \ldots, x_n) \\ y_2 = f_2(x_1, x_2, \ldots, x_n) \\ \ldots \\ y_m = f_m(x_1, x_2, \ldots, x_n) \end{cases}$$

Functions of this type will be called *vector functions*, because of $m > 1$, *of a vector*, because of $n > 1$.

A natural question concerns whether these new functions, whose structure can be symbolically described as: $\mathbf{f} : \mathbb{R}^n \to \mathbb{R}^m$ are "more difficult" because the scalar x has become the n-vector $\mathbf{x}$ or because the scalar y has become the m-vector $\mathbf{y}$. The answer is that the most expensive extension concerns the number of independent variables. The fact that you have to handle simultaneously the m components of $\mathbf{y} \in \mathbb{R}^m$ does not create serious problems: we have simply to deal with m functions of the type:

$$y_r = f_r(\mathbf{x}) = f_r(x_1, x_2, \ldots, x_n), \text{ with } r = 1, 2, \ldots, m$$

What will turn out to be new is the fact that, instead of a single independent variable x, we must handle $n > 1$ independent variables:

$$\mathbf{x} = \begin{bmatrix} x_1 \\ x_2 \\ \ldots \\ x_n \end{bmatrix}$$

This fact will bring us to face new problems. We compare two simple cases:

$$f : \mathbb{R} \to \mathbb{R} \text{ or } y = f(x)$$

and:

$$f : \mathbb{R}^2 \to \mathbb{R} \text{ or } y = f(x_1, x_2)$$

In particular, if $n = 1$, and we explore what happens if we move from some point $x_1^* = x^*$ to a new point $x^* + h$, we move along the x-axis and it is natural that what f makes is adequately featured by $f'(x^*)$.

If $n > 1$, for instance, $n = 2$, and we move from $\mathbf{x}^* = \begin{bmatrix} x_1^* \\ x_2^* \end{bmatrix}$ to

$$\mathbf{x}^* + \mathbf{h} = \begin{bmatrix} x_1^* + h_1 \\ x_2^* + h_2 \end{bmatrix}$$

we have to handle the fact that both the two arguments x_1, x_2 do vary. We will learn quickly how to differentiate f w.r.t. each arguments. When we will introduce the derivatives of f using, we will have to specify with respect to which variable (x_1 or x_2) we are working. When differentiating w.r.t. x_1, we will explore explore the variation direction if x_1 varies alone. Analogously, differentiating w.r.t. x_2, we will explore the variation direction of x_2 alone. But what happens if both x_1 and x_2 simultaneously will vary? For a social scientist it should turn out to be clear that if, jointly, two variables do move, there is not only a direct effect of each of them, but also their interaction can be relevant. Simple examples can be exhibited, in which the behavior of f when one of the arguments does vary, is not informative about the overall behavior of the function. The next example provides us with an extreme case with political relevance.

Example 2.3.1 Candidates are evaluated with respect to their attitude in two directions: (1) democratic attitude, measured by x_1 and (2) national identity defense, measured by x_2. The simplified world we consider is modeled as follows. There is a benchmark attitude:

$$\mathbf{x}^* = \begin{bmatrix} x_1^* \\ x_2^* \end{bmatrix}$$

and a voter provides her/his consensus in the presence of positive increments in *both* the values of $\mathbf{x}$ w.r.t. the benchmark position. We can think of consensus as a Boolean variable:

$$\begin{cases} 1 \text{ stays for " yes, you have my consensus"} \\ \\ 0 \text{ stays for " no, sorry, you haven't it"} \end{cases}$$

At $\mathbf{x}^*$ the consensus function $c(x_1, x_2)$ is null:

$$c\left(x_1^*, x_2^*\right) = 0$$

The behavior of such a function is:

$$c\,(x_1, x_2) = \begin{cases} 1 \text{ if } x_1 > x_1^* \text{ and } x_2 > x_2^* \\ \\ 0 \text{ if the variation of some component is non-positive} \end{cases}$$

Starting from $\mathbf{x}^*$, if we increase the value of the first component, or the one of the second, the function c is still null. If increase both, its value jumps to 1. The information conveyed by the variation of a single component is not sufficient.

2.3.1 Graphic Representation of a Function of $n \geq 2$ Variables

We hope the reader has appreciated the role of the graph of $f : \mathbb{R} \to \mathbb{R}$ in understanding the behavior of f. When $f : \mathbb{R}^n \to \mathbb{R}$, things become more difficult for a simple reason: human beings live in a 3D world and cannot easily conceive a world with dimension $n > 3$. To represent a function of n variables, we would need an $n + 1$ space. If $n = 3$, we're already done.

In the case $f : \mathbb{R}^2 \to \mathbb{R}$, we can always use the ground plane to represent the couple (x_1, x_2) and a vertical axis to represent f as a *surface*.

Look, for instance, at this example:

$$f\,(x_1, x_2) = 10 - x_1^2 - x_2^2 \tag{2.3.1}$$

The graph of the function, i.e., the set of the points in $\mathbb{R}^3$, with coordinates:

$$\begin{bmatrix} x_1 \\ x_2 \\ 10 - x_1^2 - x_2^2 \end{bmatrix}$$

is called "paraboloid". The Fig. 2.24 illustrates such type of surface.

Even if such a representation is clear and easy (if using appropriately a computer and some software), it is not very manageable because it is a 3D object. In this case $f : \mathbb{R}^2 \to \mathbb{R}$, there is an escape opportunity. If we accept to loose some information, we can get a 2D geometric representation of f with some dignity.

Fig. 2.24 Paraboloid surface

The idea is far from being new. All of us have been trained to use geographic maps. The 3D behavior of the earth surface (planes, mountains, hills, valleys,…) is frequently described using *contour lines*, which are constructed on the (x_1, x_2) plane fixing some value k for f and connecting in the line all the points (x_1, x_2) such that:

$$f(x_1, x_2) = k$$

We will call such a line also *level curve* . In case of need we will add the label "with quote k".

Let us look at a simple example:

$$f(x_1, x_2) = x_1 x_2$$

From the equation of the curve of level k:

$$x_1 x_2 = k$$

we get:

$$x_2 = \frac{k}{x_1}$$

We can choose some values for k and plot the corresponding contour lines. The Fig. 2.25 collects some contour lines of f:

The hyperbolas branches living in the first and in the third quadrant constitute the level curves of quotas 1, 2, 3, while the ones in the second and fourth quadrant constitute the contour lines for the quotas $-1, -2, -3$. The set of a number of contour lines is called *contour map*.

The Fig. 2.26 shows the countour map of the function considered in the example (2.3.1).

Fig. 2.25 Contour lines

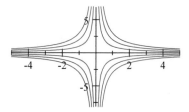

Fig. 2.26 Contour map

2.3.2 How Big is a Vector?

When studying Linear Algebra we have cautiously avoided to answer this question. For numbers the "magic" notion of absolute value helps:

$$|x| = \begin{cases} x \text{ if } x \geq 0 \\ -x \text{ if } x < 0 \end{cases} \quad \text{for instance } |-5| = |+5| = (+)\,5$$

has already proved to be interesting, at least providing us with an example of a non-differentiable function (at 0) (see p. 122).

If $\mathbf{x} \in \mathbb{R}^2$, our old friend Pythagoras suggests us to use the length of the segment connecting $\mathbf{o}$ with $\mathbf{x}$:

$$\|\mathbf{x}\| = \sqrt{x_1^2 + x_2^2}$$

as measure.[37]

Example 2.3.2 Let's consider point $\mathbf{x} = (3, 6)$ on the plane. The norm of $\mathbf{x}$ is equal to

$$\|\mathbf{x}\| = \sqrt{3^2 + 6^2} = \sqrt{45} = 3\sqrt{5}$$

that is the distance of $\mathbf{x}$ from the origin of the plane is $3\sqrt{5}$.

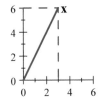

The same could be made in the case of a 3-vector, where the norm turns out to be the length of the diagonal of a parallelepiped.

Example 2.3.3 Let's consider point $2 = (4, 4, 2)$ on $\mathbb{R}^3$. The norm of $\mathbf{x}$ is equal to

$$\|\mathbf{x}\| = \sqrt{4^2 + 4^2 + 2^2} = \sqrt{36} = 6$$

that is the distance of $\mathbf{x}$ from the origin of the space $\mathbb{R}^3$ is 6.

---

[37] According to widespread uses, the length of a 1-vector x is denoted with $|x|$, while, for $n > 1$ the double bars are repeated and we will change our notation, replacing the "absolute value" of $x \in \mathbb{R}$, with its *norm* of $\mathbf{x}$ in $\mathbb{R}^n$.

Definition 2.3.1 Given a vector $\mathbf{x} \in \mathbb{R}^n$, we call *(Euclidean) norm* of $\mathbf{x}$ the non-negative number:

$$\|\mathbf{x}\| = \left\| \begin{bmatrix} x_1 \\ x_2 \\ \cdots \\ x_n \end{bmatrix} \right\| = \sqrt{x_1^2 + x_2^2 + \cdots + x_n^2}$$

The Euclidean norm is not the only one we could imagine. But, for our purposes, it works exactly as the other possible ones.

Note that $\|\mathbf{x}\| = 0$ iff $\mathbf{x} = \mathbf{o}$.

Remark 2.3.1 If $n = 1$ the notion of norm boils down to coincide with the notion of absolute value:

$$\|[x]\| = \sqrt{x^2} = |x|$$

Example 2.3.4 Let $\mathbf{x} \in \mathbb{R}^4$:

$$\mathbf{x} = \begin{bmatrix} 1 \\ -2 \\ 3 \\ -4 \end{bmatrix}$$

We have:

$$\|\mathbf{x}\| = \sqrt{1^2 + (-2)^2 + 3^2 + (-4)^2} = \sqrt{30} \approx 5.4772$$

2.3.3 Derivatives of Functions of a Vector

Let $f : \mathbb{R}^n \to \mathbb{R}$:

$$y = f(x_1, x_2, \ldots, x_n) = f(\mathbf{x}) \text{ being } \mathbf{x} \in \mathbb{R}^n$$

for instance, in the case $f : \mathbb{R}^3 \to \mathbb{R}$, we could meet this function:

$$y = x_1^2 + x_1 x_2 - x_2 x_3^2 \tag{2.3.2}$$

The number y turns out to depend on the 3-vector:

$$\begin{bmatrix} x_1 \\ x_2 \\ x_3 \end{bmatrix}$$

If we want to differentiate a function of a vector, i.e., of several variables, we must choose the variable w.r.t. which we proceed. For functions of scalars: $f : \mathbb{R} \to \mathbb{R}$ there was no choice, because we had only one independent variable. Here, for functions $f : \mathbb{R}^n \to \mathbb{R}$, we have n possibilities.

In principle, a function of n variables could have until n derivatives of the first order. The adverb "until" is appropriate because differentiation means "looking for a linear approximation" and we know there are functions that cannot be linearly approximated.

This is not the case for (2.3.2).

For such functions the differentiation process is simple: if you differentiate w.r.t. some variable x_s, the other variables must be thought of as constant.

In practice, if we differentiate y w.r.t. x_1, we get:

$$2x_1 + x_2$$

thinking of x_2, x_3 as constant quantities.

If we differentiate y w.r.t. x_2, we get:

$$x_1 - x_3^2$$

while, if we choose x_3, we find:

$$-2x_2x_3$$

Even starting from this simple example, it is clear that we need to introduce an appropriate notation and some framework.

Given some function $f : \mathbb{R}^n \to \mathbb{R}$, we indicate its *(partial) derivatives* w.r.t. some independent variable x_s with one of the following symbols[38]:

$$f'_{x_s}(\mathbf{x}) \text{ or } f_{x_s}(\mathbf{x}) \text{ or } \frac{\partial}{\partial x_s} f(\mathbf{x}) \text{ or } \frac{\partial f(\mathbf{x})}{\partial x_s} \text{ or } D_{x_s} f(\mathbf{x})$$

In principle a function $f : \mathbb{R}^n \to \mathbb{R}$ has n first partial derivatives w.r.t. each of the n independent variables. We can host them in a (row) n-vector:

$$f'(\mathbf{x}) = \left[f'_{x_1}(\mathbf{x}) \quad f'_{x_2}(\mathbf{x}) \quad \cdots \quad f'_{x_n}(\mathbf{x}) \right]$$

called *gradient* of f and commonly denoted also as:

$$\nabla f(\mathbf{x}) = \left[f'_{x_1}(\mathbf{x}) \quad f'_{x_2}(\mathbf{x}) \quad \cdots \quad f'_{x_n}(\mathbf{x}) \right]$$

where the symbol "∇" is usually read "nabla".[39]

For instance, in the example we started from:

$$f(\mathbf{x}) = f(x_1, x_2, x_3) = x_1^2 + x_1x_2 - x_2x_3^2 \qquad (2.3.3)$$

---

[38] The multiplicity of symbols is not beloved by the Authors, but it is a matter of fact in practice.

[39] No Author is responsible for this name, which comes from the Phoenician name "nabla" for "harp". All of us can admit that ∇ recalls the shape of a harp.

we get:

$$\nabla f(\mathbf{x}) = \begin{bmatrix} 2x_1 + x_2 & x_1 - x_3^2 & -2x_2x_3 \end{bmatrix} \qquad (2.3.4)$$

We already know that for functions $f : \mathbb{R} \to \mathbb{R}$ the points at which the derivative is null, called stationary points, are interesting (for instance for optimization problems). The situation does not change for functions $f : \mathbb{R}^n \to \mathbb{R}$. In this more general case, *stationary points* are those at which the gradient of f is null:

$$\nabla f(\mathbf{x}) = \mathbf{0}$$

ore, more explicitly:

$$\begin{cases} f'_{x_1}(\mathbf{x}) = 0 \\ f'_{x_2}(\mathbf{x}) = 0 \\ \quad \dots \\ f'_{x_n}(\mathbf{x}) = 0 \end{cases}$$

Example 2.3.5 Take the function defined above at (2.3.3). From its gradient (2.3.4) we get the non-linear system

$$\begin{cases} 2x_1 + x_2 = 0 \\ x_1 - x_3^2 = 0 \\ -2x_2x_3 = 0 \end{cases}$$

The third equation provides us with the alternative $x_2 = 0$ or $x_3 = 0$ or both. Replacing with 0 or x_2 or x_3 in the other two equation we find always the same stationary point: $\mathbf{0}$, or $x_1 = x_2 = x_3 = 0$, the origin in $\mathbb{R}^3$. We can conclude that, for this function there is only one stationary point, which is[40]:

$$\mathbf{0} = \begin{bmatrix} 0 \\ 0 \\ 0 \end{bmatrix}$$

We know that for functions of a scalar $f : \mathbb{R} \to \mathbb{R}$ a crucial role is played by the differential:

$$df(x) = f'(x) h$$

being h the increment of the independent variable, passing from x to $x + h$.

In the current more general context $f : \mathbb{R}^n \to \mathbb{R}$, we can fully recover the notion of *differential*:

---

[40]We encourage the reader to meditate that this point is not stationary because all of its components are 0, but because, if we give to x_1, x_2, x_3 value 0, then all of the three derivatives of f, or f'_{x_1}, f'_{x_2}, f'_{x_3} take value 0.

$$df(\mathbf{x}) = \begin{bmatrix} f'_{x_1}(\mathbf{x}) & f'_{x_2}(\mathbf{x}) & \cdots & f'_{x_n}(\mathbf{x}) \end{bmatrix} \begin{bmatrix} h_1 \\ h_2 \\ \cdots \\ h_n \end{bmatrix} =$$

$$= \nabla f(\mathbf{x}) \cdot \mathbf{h} = \sum_{s=1}^{n} f'_{x_s}(\mathbf{x}) h_s =$$

$$= f'_{x_1}(\mathbf{x}) h_1 + f'_{x_2}(\mathbf{x}) h_2 + \cdots + f'_{x_n}(\mathbf{x}) h_n$$

Remark 2.3.2 Note that if $\mathbf{x}^*$ is a stationary point for f, then, no matter which is $\mathbf{h}$ the differential is null: $df(\mathbf{x}^*) = 0$.

Example 2.3.6 Let's take the function $f : \mathbb{R}^2 \to \mathbb{R}$ defined by:

$$f(x_1, x_2) = 2x_1^2 + 3x_2^2 + 10$$

The variation of f when we move from the starting point $\mathbf{x}^* = (1, 1)$ to $\mathbf{x} = (1.05, 0.95)$ is:

$$f(x_1, x_2) - f\left(x_1^*, x_2^*\right) = 2 \cdot 1.05^2 + 3 \cdot 0.95^2 + 10 - 15 \approx -0.087\,5$$

The gradient of f at $\mathbf{x}$ is the row vector

$$\nabla f(\mathbf{x}) = \begin{bmatrix} 4x_1 & 6x_2 \end{bmatrix}$$

Introducing the vector of the increments of the independent variables

$$\mathbf{h} = \begin{bmatrix} 0.05 \\ -0.05 \end{bmatrix}$$

we can calculate the differential of f at $\mathbf{x}^*$:

$$df(\mathbf{x}) = \nabla f(1, 1) \cdot \begin{bmatrix} 0.05 \\ -0.05 \end{bmatrix} = \begin{bmatrix} 4 & 6 \end{bmatrix} \cdot \begin{bmatrix} 0.05 \\ -0.05 \end{bmatrix} = -0.1$$

We note that $df(\mathbf{x}) \approx \Delta f(\mathbf{x})$.

The notion of differential turns out to be crucial in defining differentiability for functions of several variables.

Now we are ready to do that.

Definition 2.3.2 We say that a function $f : \mathbb{R}^n \to \mathbb{R}$ is *differentiable* at $\mathbf{x}^* \in \mathbb{R}^n$ if its increment:

$$f(\mathbf{x}^* + \mathbf{h}) - f(\mathbf{x}^*) = f\left(x_1^* + h_1, x_2^* + h_2, \ldots, x_n^* + h_n\right) - f\left(x_1^*, x_2^*, \ldots, x_n^*\right)$$

can be represented as follows:

$$f\left(\mathbf{x}^* + \mathbf{h}\right) - f\left(\mathbf{x}^*\right) = \mathbf{m}\left(\mathbf{x}^*\right)\mathbf{h} + o\left(\|\mathbf{h}\|\right)$$

being $\mathbf{m}\left(\mathbf{x}^*\right)$ an n-(row) vector.

For functions of a scalar: $f : \mathbb{R} \to \mathbb{R}$ we have already met the notion of second derivative:

$$f''(x) = D\left[f'(x)\right]$$

what happens in the case $f : \mathbb{R}^n \to \mathbb{R}$, with $n > 1$?

We have seen that the (first) derivative of a function $f : \mathbb{R} \to \mathbb{R}$ at some point is a number, while, when we move to functions $f : \mathbb{R}^n \to \mathbb{R}$, with $n > 1$, the analogous of the first derivative "explodes" and becomes a vector: the gradient of f, at that point. For analogous reasons, what we can expect when looking for second partial derivatives is an explosion even larger.

It's crucial to understand that if we differentiate twice a function, non-necessarily the variables w.r.t. which we differentiate in the two steps are the same. The standard second derivative of a function $f : \mathbb{R}^n \to \mathbb{R}$ is of the type:

$$D_{x_s}\left[f'_{x_r}(\mathbf{x})\right]$$

Practically, starting from the first derivative w.r.t. to the independent variable x_r, we can generate n second partial derivatives, in which the second differentiation is conducted w.r.t. any variable x_s, $(s = 1, 2, \ldots, n)$ with $s = r$ or not.

We have learned above that a good place to store the first derivatives of some $f : \mathbb{R}^n \to \mathbb{R}$ is the n - (row) vector ∇f. It should seem natural for the reader to think of the possibility of storing the second partial derivatives in a square matrix, usually called *Hessian*[41] matrix of f, where each row is nothing but the gradient of a partial first derivative. The first row is generated by the first partial derivative of f w.r.t. x_1, and differentiating it w.r.t. x_1 again or, $x_2,\ldots, x_n$.

first row: $\nabla f'_{x_1}(\mathbf{x}) = \left[D_{x_1} f'_{x_1}(\mathbf{x}) \quad D_{x_2} f'_{x_1}(\mathbf{x}) \quad \cdots \quad D_{x_n} f'_{x_1}(\mathbf{x}) \right]$

for the second row we would have the gradient of $f'_{x_2}(\mathbf{x})$:

second row: $\nabla f'_{x_2}(\mathbf{x}) = \left[D_{x_1} f'_{x_2}(\mathbf{x}) \quad D_{x_2} f'_{x_2}(\mathbf{x}) \quad \cdots \quad D_{x_n} f'_{x_2}(\mathbf{x}) \right]$

At this point, the matrix structure should appear rather natural. The matrix, which collects all of the (possible) second derivatives of f, is:

---

[41] As it has been introduced in the XIX century by the German mathematician Ludwig Otto Hesse (1811–1874).

$$f''(\mathbf{x}) = \nabla^2 f(\mathbf{x}) = \begin{bmatrix} f''_{x_1,x_1}(\mathbf{x}) & f''_{x_1,x_2}(\mathbf{x}) & \cdots & f''_{x_1,x_n}(\mathbf{x}) \\ f''_{x_2,x_1}(\mathbf{x}) & f''_{x_2,x_2}(\mathbf{x}) & \cdots & f''_{x_2,x_n}(\mathbf{x}) \\ \cdots & \cdots & \cdots & \cdots \\ f''_{x_n,x_1}(\mathbf{x}) & f''_{x_n,x_2}(\mathbf{x}) & \cdots & f''_{x_n,x_n}(\mathbf{x}) \end{bmatrix}$$

Example 2.3.7 We have seen above that the gradient of:

$$f(\mathbf{x}) = f(x_1, x_2, x_3) = x_1^2 + x_1 x_2 - x_2 x_3^2$$

is:

$$\nabla f(\mathbf{x}) = \begin{bmatrix} 2x_1 + x_2 & x_1 - x_3^2 & -2x_2 x_3 \end{bmatrix}$$

We can now compute the Hessian matrix of f:

$$f''(\mathbf{x}) = \nabla^2 f(\mathbf{x}) = \begin{bmatrix} 2 & 1 & 0 \\ 1 & 0 & -2x_3 \\ 0 & -2x_3 & -2x_2 \end{bmatrix}$$

Remark 2.3.3 Please, note that the Hessian matrix is symmetric, in the sense that entries in 'symmetric' position, w.r.t. the principal diagonal are equal. It's not a case, because, under general conditions, the order with which you compute derivatives does not affect the result:

$$f''_{x_r,x_s}(\mathbf{x}) = f''_{x_s,x_r}(\mathbf{x})$$

This proposition is known as *Schwarz Theorem*.[42]

The principal diagonal of the matrix collects second derivatives, for which the differentiation variable is the same in the two differentiation steps. Such derivatives are said to be *pure* second derivatives. Off the principal diagonal, the differentiation variable changes across the two steps. They are usually called *mixed* second derivatives.

2.3.4 *Unconstrained Extrema for Functions $f : \mathbb{R}^n \to \mathbb{R}$*

A common problem for functions $f : \mathbb{R}^n \to \mathbb{R}$ consists in finding their extrema (maximum or minimum points).

We already met it for functions $f : \mathbb{R} \to \mathbb{R}$, but the case of several independent variables is largely more exciting (and useful in Social Sciences).

What Mathematics can offer is a scissor (and we often experiment how a scissor is good). The strength of a scissor is that it has two blades. The strength of Mathematics

---

[42] Schwarz is not a colour, but the name of the German mathematician Karl Hermann Amandus Schwarz (1843–1921).

in these problems is that it has two blades. A good scissor cuts paper and that's all. The blades, Mathematics provides, are not the ones of a standard scissor we would buy, but that's life;

- sometimes, using $\nabla f(\mathbf{x}^*), \nabla^2 f(\mathbf{x}^*)$, we can decide that at some point $\mathbf{x}^* \in \mathbb{R}^n$ the function f attains a (local) maximum or a (local) minimum;
- sometimes, using $\nabla f(\mathbf{x}^*), \nabla^2 f(\mathbf{x}^*)$, we can decide that at some point $\mathbf{x}^* \in \mathbb{R}^n$ the function f attains no (local) maximum or no (local) minimum;
- in some very special case[43], the information provided by gradient and Hessian matrix are not sufficient. For scissors it would embarrassing, but not for Maths as the non-decidible cases turn out to be somehow exceptional.

First we have to look for stationary points. In fact if $\mathbf{x}^* \in \mathbb{R}^n$ is an extremum point for f and f admits the gradient at $\mathbf{x}^*$, then surely it has to be null:

$$\nabla f(\mathbf{x}^*) = \mathbf{0}$$

In other words all maximum and minimum points are stationary points of $f : \mathbb{R}^n \to \mathbb{R}$. The stationarity condition is only *necessary* and it is called *first order condition* because it involves only first order partial derivatives .

To decide whether a stationary point actually is an extremum point, one needs more information about the (local) behavior of f.

The hessian matrix $\nabla^2 f(\mathbf{x}^*)$, computed at any stationary point $\mathbf{x}^*$, may provide us with all information we need. In fact, some conditions about the hessian matrix, called *second order conditions*, are sufficient to assess the nature of a stationary point.

However it is better to recall first some ancillary information about matrices, we already met before.

Consider the (n, n)-matrix:

$$f''(\mathbf{x}^*) = \nabla^2 f(\mathbf{x}^*) = \begin{bmatrix} f''_{x_1,x_1}(\mathbf{x}^*) & f''_{x_1,x_2}(\mathbf{x}^*) & \cdots & f''_{x_1,x_n}(\mathbf{x}^*) \\ f''_{x_2,x_1}(\mathbf{x}^*) & f''_{x_2,x_2}(\mathbf{x}^*) & \cdots & f''_{x_2,x_n}(\mathbf{x}^*) \\ \cdots & \cdots & \cdots & \cdots \\ f''_{x_n,x_1}(\mathbf{x}^*) & f''_{x_n,x_2}(\mathbf{x}^*) & \cdots & f''_{x_n,x_n}(\mathbf{x}^*) \end{bmatrix}$$

It's the Hessian matrix of f, computed at the point $\mathbf{x}^*$. It's a (square & symmetric) matrix of numbers: $h_{r,s} = f''_{x_r,x_s}(\mathbf{x}^*)$

---

[43]Typically, if you are working on empirical data this ambiguous situation does not appear. We find it in "artificial cases", that can be of interest for some mathematician, but absolutely irrelevant in most of the applications.

$$\nabla^2 f \left(\mathbf{x}^* \right) = H = \begin{bmatrix} h_{1,1} & h_{1,2} & \cdots & h_{1,n} \\ h_{2,1} & h_{2,2} & \cdots & h_{2,n} \\ \cdots & \cdots & \cdots & \cdots \\ h_{n,1} & h_{n,2} & \cdots & h_{n,n} \end{bmatrix}$$

We already know the notion and the relevance of principal NW minors for a square matrix H (see the first chapter).

Let $H_1, H_2, H_3, \ldots, H_n$ be the sequence of the NW principal minors of H, the Hessian matrix computed at $\mathbf{x}^*$. Construct the corresponding sign sequence.

Two sequences are *sufficient* to state the extremum nature of a stationary point $\mathbf{x}^*$:

(i) at $\mathbf{x}^*$ there is a local maximum[44] if the signs do alternate starting from $-$:

$$-, +-, +, \ldots, (-)^n \tag{2.3.5}$$

(ii) at $\mathbf{x}^*$ there is a local minimum if the signs are all positive:

$$+, +, +, +, \ldots, + \tag{2.3.6}$$

A further immediate diagnosis is possible:

- if the sequence of signs is different from any of the two preceding ones, no local maximum or minimum point stays there.
- If there is some zero instead of $+, -$ then only two cases may occur:

1. a suitable replacement of 0's with $+$ or $-$ signs can bring to one of the two sequences (2.3.5) or (2.3.6). Nothing can be said without a further and deeper analysis: if the sequence (2.3.5) can be reached, there could be a local maximum; if the sequence (2.3.6) can be reached, there could be a local minimum, and both these possibilities are compatible with the possibility that $\mathbf{x}^*$ be not an extremum point.
2. no possible replacement of 0's with $+$ or $-$ signs can bring to one of the two sequences (2.3.5) or (2.3.6). In such a case the existence of any extremum can be excluded. For instance:

$$0, -, 0 \tag{2.3.7}$$

is incompatible with both (2.3.5) and (2.3.6).

The Table 2.9 collects some examples of sign sequences and should help the reader in understanding

---

[44]The symbol $(-)^n$ means $+$ if n is even or $-$ if n is odd.

Table 2.9 Sign examples

Sequence	Diagnosis
$-, +, -, +$	max
$+, +, +, +$	min
$\pm, -, \pm, +$	No extremum: the 2nd sign is incompatible both with (2.3.5) and (2.3.6)
$0, +, \pm, -$	No extremum: the 4th sign is incompatible both with (2.3.5) and (2.3.6)
$0, +, 0, +$	There could be a max or a min or no extremum
$+, +, 0, +$	There cannot be a max, there could be a min or no extremum

Let us look at the following:

Example 2.3.8 Consider the function :

$$f(x_1, x_2) = ax_1^2 + bx_2^2 \text{ with } a, b \in \mathbb{R}$$

Its gradient is:

$$\nabla f(x_1, x_2) = \begin{bmatrix} 2ax_1 & 2bx_2 \end{bmatrix}$$

for every value of $a, b \in \mathbb{R}$, the origin:

$$\mathbf{x}^* = \mathbf{0} = \begin{bmatrix} 0 \\ 0 \end{bmatrix}$$

is a stationary point. The Hessian matrix is:

$$\begin{bmatrix} 2a & 0 \\ 0 & 2b \end{bmatrix}$$

The Table 2.10 summarizes the possible conclusions.

The diagnoses between parentheses are obtained in general with a deeper analysis. Only in this very special case they are possible at sight. We add that, in the case "no

Table 2.10 Possible sequences and conclusions

a, b	Sign sequence	Diagnosis
$a, b > 0$	$+, +$	min
$a, b < 0$	$-, +$	max
$a \cdot b < 0$	$+, -$ or $-, -$	no extremum
$a = 0, b > 0$	$0, 0$	??? (actually it's a weak min)
$a > 0, b = 0$	$+, 0$	??? not a max (actually it's a weak min)
$a = 0, b < 0$	$0, 0$	??? (actually it's a weak max)
$a < 0, b = 0$	$-, 0$	??? not a min (actually it's a weak max)

extremum", the point we have met is called *saddle point*: it is an important notion we will encounter later at p. 174.

Example 2.3.9 We have seen above that the gradient of:

$$f(\mathbf{x}) = f(x_1, x_2, x_3) = x_1^2 + x_1 x_2 - x_2 x_3^2$$

is:

$$\nabla f(\mathbf{x}) = \begin{bmatrix} 2x_1 + x_2 & x_1 - x_3^2 & -2x_2 x_3 \end{bmatrix}$$

It is null at the triplets $\begin{bmatrix} x_1 \\ x_2 \\ x_3 \end{bmatrix}$ such that:

$$\begin{cases} 2x_1 + x_2 = 0 \\ x_1 - x_3^2 = 0 \\ -2x_2 x_3 = 0 \end{cases}$$

There is only one stationary point:

$$x_1^* = x_2^* = x_3^* = 0$$

the origin. The Hessian matrix of f at the origin is:

$$f''(\mathbf{o}) = \nabla^2 f(\mathbf{o}) = \begin{bmatrix} 2 & 1 & 0 \\ 1 & 0 & 0 \\ 0 & 0 & 0 \end{bmatrix}$$

The sequence of its NW principal minors is:

$$H_1 = 2, \; H_2 = -1, \; H_3 = 0$$

The negativity of H_2 excludes that the origin is an extremum point. The principle is the same illustrated above by (2.3.7).

Example 2.3.10 An important example of unconstrained minimum is highly relevant for all sciences, and, in particular, for the Social ones. We use an economic scheme, but there are others in social sciences. Think of some country and call Y_t its GNP[45] obtained during the year $t = 1, 2, \ldots, T$. We have data about the GNP for the relevant period and we know also the consumption level C_t, that has been observed (Table 2.11). A classic macroeconomic problem consists in looking for some relationship between GNP and consumption. The simplest model one could imagine (the so-called "consumption function theory") boils down to assume that,

---

[45] *Gross National Product* = value of goods and services produced in that system in a certain time interval. Typically an annual period.

Year	Consumption	GNP
1	C_1	Y_1
2	C_2	Y_2
3	C_3	Y_3
...	...	...
T	C_T	Y_T

Table 2.11 Consumption and GNP data

apart from disturbs (ε_t):

$$C_t = a + bY_t + \varepsilon_t \tag{2.3.8}$$

in other words: the total consumption is (dirty) linear affine function of the GNP. the dirty nature of the relationship is determined by the disturbing terms (ε_t). An important problem consists in extracting reliable estimates of the parameters a, b from the two time series of consumptions and GNPs we are given.

In particular b is known as *marginal propensity to consumption* and is absolutely relevant in many economic policy decisions. A naïve idea could be that of looking for the values of a, b which minimize the *total error* $\sum_{t=1}^{T} \varepsilon_t$. We do not share this approach as it is compatible with compensations of large errors of opposite signs. The best approach consists in considering the *total quadratic error* $\sum_{t=1}^{T} \varepsilon_t^2 = f(a, b)$ and looking for the values of a, b that bring it to a minimum. This idea is known as minimum least squares and is labelled as OLS principle (=Ordinary Least Squares principle). From (2.3.8) we get $\varepsilon_t = C_t - a - bY_t$ and hence:

$$f(a, b) = \sum_{t=1}^{T} (C_t - a - bY_t)^2$$

First, let us compute the gradient of f:

$$\nabla f(a, b) = \left[f'_a(a, b) \quad f'_b(a, b) \right] =$$
$$= \left[-2 \sum_{t=1}^{T} (C_t - a - bY_t) \quad -2 \sum_{t=1}^{T} Y_t (C_t - a - bY_t) \right]$$

The stationarity conditions boil down to:

$$\begin{cases} -2 \sum_{t=1}^{T} (C_t - a - bY_t) = 0 \\ -2 \sum_{t=1}^{T} Y_t (C_t - a - bY_t) = 0 \end{cases}$$

They can be rewritten as:

$$\begin{cases} \sum_{t=1}^{T} C_t - aT - b\sum_{t=1}^{T} Y_t = 0 \\ \sum_{t=1}^{T} Y_t C_t - a\sum_{t=1}^{T} Y_t - b\sum_{t=1}^{T} Y_t^2 = 0 \end{cases}$$

Dividing by T both sides of the equations, and introducing the symbols $\overline{C} = \dfrac{\sum_{t=1}^{T} C_t}{T}$ for the average consumption, $\overline{Y} = \dfrac{\sum_{t=1}^{T} Y_t}{T}$ for the average GNP, $\overline{YC}$ for the average product (GNP consumption), and $\overline{Y^2}$ for the average of the squared income, we get the system:

$$\begin{cases} \overline{C} - a - b\overline{Y} = 0 \\ \overline{YC} - a\overline{Y} - b\overline{Y^2} = 0 \end{cases}$$

Multiplying both sides by $\overline{Y}$ and subtracting, side by side, we get:

$$b\left(\overline{Y^2} - \overline{Y}^2\right) = \overline{YC} - \overline{Y} \cdot \overline{C}$$

hence:

$$b = \frac{\overline{YC} - \overline{Y} \cdot \overline{C}}{\overline{Y^2} - \overline{Y}^2}$$

The denominator is known as the *variance* of the GNP, while the numerator is known as the *covariance between* C_t *and* Y_t. This is nothing but an anticipation of what the readers are expected to learn in Statistics courses. The economic interpretation of a is an economic puzzle: "the consumption level you would have with a null GNP". We confine it to a simple formula:

$$a = \overline{C} - \overline{Y} \cdot \frac{\overline{YC} - \overline{Y} \cdot \overline{C}}{\overline{Y^2} - \overline{Y}^2}$$

Are we sure, we have found a minimum? We can hope in an answer coming from the Hessian matrix of f:

$$\nabla^2 f(a, b) = \begin{bmatrix} 2T & 2T\overline{Y} \\ 2T\overline{Y} & 2T\overline{Y^2} \end{bmatrix}$$

Its NW principal minors are $2T > 0$ and $4T^2\overline{Y^2} - 4T^2\overline{Y}^2 = 4T^2\left(\overline{Y^2} - \overline{Y}^2\right)$. The last factor is nothing but the s.c. *variance* of the GNP. It is sufficient that in a pair of years, at least, the GNPs have taken different levels, in order that $\overline{Y^2} - \overline{Y}^2 > 0$. We're certainly in the presence of a minimum. It's local, but also global.

2.3.5 Constrained Extrema

What we have seen above is that if you catch the net:

$$\nabla f(\mathbf{x}) = \mathbf{0}$$

you have the possibility to find points at which some function $f : \mathbb{R}^n \to \mathbb{R}$ attains a maximum or a minimum. we have also seen criteria allowing us to say whether at some stationary point $\mathbf{x}^*$ there is an extremum.

Frequently the optimization problems encountered in social sciences are slightly more difficult.

Take the following:

Example 2.3.11 A politician has to design an electoral campaign. The politician can invest in two different media (1 and 2). The invested amounts are, respectively, x_1, x_2. The politician knows as his/her investment will provide him/her with votes. This means we are given a function:

$$V(x_1, x_2)$$

which tells us how many votes the candidate will collect for any pair of investments $\begin{bmatrix} x_1 \\ x_2 \end{bmatrix}$ in the two media. Unfortunately our candidate is not the Donald, who manages an unbounded budget, but our candidate must respect the obvious budget constraint:

$$x_1 + x_2 = b \tag{2.3.9}$$

asking to spend exactly the assign budget b. It is reasonable to think that V increases, both with x_1 and x_2. The constraint (2.3.9) does not allow for unlimited expenses: here's the core of the current topic: the statement of the problem $\mathcal{P}$:

$$\mathcal{P} \begin{cases} \max_{x_1 x_2} y = V(x_1, x_2) \\ \text{sub} \\ x_1 + x_2 = b \end{cases}$$

In principle, this problem is non-trivial and a mathematically sound proof of the correctness of the solution procedure is not completely easy at least at the technical level of this textbook.

What we know is that there is the possibility to provide an economic interpretation of the procedure, which is more intuitive than the purely mathematical one.

We will use this strategy, but, before, we sketch the path we will follow.

Our exposition strategy will be organized in some steps:

1. we will state the constrained optimization $\mathcal{P}$ in its generality;

2. we will provide an immediate economic interpretation[46] for $\mathcal{P}$;
3. we will analyze a very special case of $\mathcal{P}$, labelled as $\mathcal{P}'$, and we will obtain, through its economic interpretation, a necessary condition for a solution, which is widely used in practice;
4. we will intuitively extend such necessary condition to the general case labelled as $\mathcal{P}''$, which turns out to be equivalent to the original $\mathcal{P}$ and we will offer also useful sufficient conditions for an extremum.

The problem of constrained optimization was successfully solved by the Italian mathematician J.L. (*comte*) de Lagrange[47] and his name will recur frequently in the vocabulary of such problems.

2.3.5.1   The Problem

We are given a function $f : \mathbb{R}^n \to \mathbb{R}$,

$$y = f(\mathbf{x}) = f(x_1, x_2, \dots, x_n)$$

which associates to vectors $\mathbf{x}$ in $\mathbb{R}^n$ a real number y. We would like to maximize y, through the appropriate choice $\mathbf{x}^*$ of $\mathbf{x}$:

$$\mathbf{x}^* = \begin{bmatrix} x_1^* \\ x_2^* \\ \dots \\ x_n^* \end{bmatrix}$$

or:

$$f(\mathbf{x}^*) \geq f(\mathbf{x})$$

for $\mathbf{x}$ not too far from $\mathbf{x}^*$. However, the choice of $\mathbf{x}$ is not free. We are also given $m(< n)$ constraints, that must be respected:

$$\begin{cases} g_1(x_1, x_2, \dots, x_n) = b_1 \\ g_2(x_1, x_2, \dots, x_n) = b_2 \\ \qquad \dots \\ g_m(x_1, x_2, \dots, x_n) = b_m \end{cases}$$

---

[46]The reason is that using the economic intuition it is possible to find heuristically the solution for $\mathcal{P}$.

[47]Born in Turin 1736, passed away in Paris 1813.

We can summarize this *set of constraints* as:

$$\mathbf{g}\,(\mathbf{x}) = \mathbf{b}$$

being $\mathbf{g} : \mathbb{R}^n \to \mathbb{R}^m$ and $\mathbf{b} \in \mathbb{R}^m$.

The preceding example falls in this general scheme with $n = 2$, number of media and $b =$ budget for the electoral campaign.

2.3.5.2   Economic Interpretation (the Optimal Production Mix)

A firm produces n goods. They are numbered with $r = 1, 2, \ldots, n$. The quantities, the firm produces, of these goods are:

$$\mathbf{x} = \begin{bmatrix} x_1 \\ x_2 \\ \ldots \\ x_n \end{bmatrix}$$

The firm makes a margin $y = f\,(\mathbf{x})$ to be maximized. The objective function f is of the type $f : \mathbb{R}^n \to \mathbb{R}$. The production process requires the use of m resources, for instance, m machines (with $m < n$). In order to produce the n-vector $\mathbf{x}$ of quantities the use of the m resources is quantified by m *absorption functions*:

$$\begin{bmatrix} g_1\,(\mathbf{x}) \\ g_2\,(\mathbf{x}) \\ \ldots \\ g_m\,(\mathbf{x}) \end{bmatrix} = \mathbf{g}\,(\mathbf{x})$$

We specify that $\mathbf{g}\,(\mathbf{x})$ is an m-vector collecting the amount of time of each of the m machines absorbed by the production plan described by the n-vector $\mathbf{x}$.

Therefore, the (vector) absorption function $\mathbf{g}$ is of the type: $\mathbf{g} : \mathbb{R}^n \to \mathbb{R}^m$. For each of the m machines we are assigned a time budget b_s for the machine # s. The vector $\mathbf{b} \in \mathbb{R}^m$ collects all these time budgets and the choice of $\mathbf{x}$ is required to respect the full use of each of the m machines:

$$\mathbf{g}\,(\mathbf{x}) = \mathbf{b} \tag{2.3.10}$$

or, more explicitly, the constraints to be respected are:

$$\begin{cases} g_1\,(x_1, x_2, \ldots, x_n) = b_1 \\ g_2\,(x_1, x_2, \ldots, x_n) = b_2 \\ \qquad \ldots \\ g_m\,(x_1, x_2, \ldots, x_n) = b_m \end{cases}$$

2.3.5.3 A Very Special Case ($n = 2$ and $m = 1$)

We call $\mathcal{P}_1$ the following problem, where only two goods are produced and only 1 machine is needed:

$$\mathcal{P}_1 = \begin{cases} \max_{x_1, x_2} y = f(x_1, x_2) \\ \text{sub} \\ g(x_1, x_2) = b \end{cases}$$

Our strategy will consist in:

- Constructing the variant $\mathcal{P}_2$ of the problem $\mathcal{P}_1$ where we will allow for violations of the equality constraint $g(x_1, x_2) = b$, taking into account their economic consequences;
- Constructing the variant $\mathcal{P}_3$ of the variant $\mathcal{P}_2$, which will turn out to be equivalent to $\mathcal{P}_1$; as we will be able to provide a first order condition (FOC) for $\mathcal{P}_3$, consequently inherited by $\mathcal{P}_1$.

Our itinerary could appear a bit tortuous, but we think it is the best to reach almost costlessly Lagrange's central result.

Let's now construct $\mathcal{P}_2$. Now, we allow for the possibility to violate the constraint: if the firm wants to use more machine-time than the amount b it is endowed, it can purchase the extra time needed: $g(x_1, x_2) - b$ at its market price λ (the letter is a "Greek" homage to J.L. Lagrange), but, if the firm want to use less time than b, the left extra time $b - g(x_1, x_2)$ can be sold at the same price. Of course these violations have an incidence on the margin. The corrected margin $L(\lambda; x_1 x_2)$ ("L" as Lagrange) turns out to depend both on the quantities x_1, x_2 and on the exogenous market price of the resource λ. In $\mathcal{P}_2$ the price λ is given.

FOC for an optimal decision can be easily constructed as $\mathcal{P}_2$ is an unconstrained max problem:

$$\mathcal{P}_2 = \begin{cases} \max_{x_1, x_2} L(\lambda; x_1 x_2) = f(x_1, x_2) + \lambda[b - g(x_1, x_2)] \\ \lambda \text{ given} \end{cases}$$

They boil down to:

$$\nabla_{x_1, x_2} L = \begin{bmatrix} f'_{x_1}(x_1, x_2) - \lambda g'_{x_1}(x_1, x_2) & f'_{x_2}(x_1, x_2) - \lambda g'_{x_2}(x_1, x_2) \end{bmatrix} =$$
$$= \begin{bmatrix} 0 & 0 \end{bmatrix}$$

or:

$$\begin{cases} f'_{x_1}(x_1, x_2) - \lambda g'_{x_1}(x_1, x_2) = 0 \\ f'_{x_2}(x_1, x_2) - \lambda g'_{x_2}(x_1, x_2) = 0 \end{cases}$$

Two equations, two unknowns. Of course, the optimal production policy:

$$\mathbf{x}^* (\lambda) = \begin{bmatrix} x_1^* (\lambda) \\ x_2^* (\lambda) \end{bmatrix}$$

will turn out to depend[48] on the exogenous price λ.

In this framework, $\mathcal{P}_2$, the market price is neutrally fixed and the firm adjusts its decision to λ.

To construct $\mathcal{P}_3$, we remove the neutrality assumption about λ, and we assume that λ is chosen by some "evil subject" in order to minimize the margin for the firm: if the total resource amount b is not fully used, the firm will be heavily fined for that, in the case an extra-time beyond b is needed, its price will be unaffordable. This situation can be formalized as follows:

$$\min_{\lambda} \max_{x_1, x_2} L (\lambda; x_1 x_2) = f (x_1, x_2) + \lambda [b - g (x_1, x_2)] \qquad (2.3.11)$$

For the problem $\mathcal{P}_3$, the function L can be called *auxiliary function* or *Lagrangean function* or, more simply, *Lagrangean*. The point $(\lambda^*; x_1^*, x_2^*) \in \mathbb{R}^3$, solving the system (2.3.11) must be a *saddle point*[49], in the sense that at it L attains a maximum w.r.t. $(x_1, x_2) \in \mathbb{R}^2$ and a minimum w.r.t. λ. As the FOC for maxima and minima are the same, they simply suggest us to look for stationary points for L:

$$\nabla L (\lambda; x_1, x_2) = \mathbf{0}$$

or, more explicitly:

$$\begin{cases} L'_{\lambda} (\lambda; x_1, x_2) = 0 \\ L'_{x_1} (\lambda; x_1, x_2) = 0 \\ L'_{x_2} (\lambda; x_1, x_2) = 0 \end{cases}$$

Let's look at the following:

Example 2.3.12 The function to be maximized is:

$$f (x_1, x_2) = 10 - x_1^2 - x_2^2$$

under the constraint:

$$x_1 + x_2 = 10$$

---

[48]To help intuition: if λ is very high an intelligent policy turns out to consist in producing directly small quantities of the two goods, in order to be able to sell (at a very high price) the machine time directly unused, while, if λ is small, it should be convenient to produce directly a lot, using completely the available amount of resource and potentially purchase at cheap conditions machine time from others.

[49]Think to sit on a horse saddle: in the standard direction of the horse motion you are at a minimum, while in the orthogonal direction you are at a maximum (and you have to pay attention...).

The Lagrangean for the problem is:

$$L(\lambda; x_1, x_2) = 10 - x_1^2 - x_2^2 + \lambda(10 - x_1 - x_2)$$

Its gradient turns out to be:

$$\nabla L(\lambda; x_1, x_2) = \begin{bmatrix} 10 - x_1 - x_2 & -2x_1 - \lambda & -2x_2 - \lambda \end{bmatrix}$$

The FOCs produce the linear system:

$$\begin{cases} 10 - x_1 - x_2 = 0 \\ -2x_1 - \lambda = 0 \\ -2x_2 - \lambda = 0 \end{cases}$$

Its solution is:

$$\begin{cases} \lambda^* = -10 \\ x_1^* = 5 \\ x_2^* = 5 \end{cases}$$

Remark 2.3.4 The first equation (stationarity w.r.t. λ) inserts in the FOCs the respect of the constraint, a pillar of P_1, temporarily turned off in P_2 and finally re-born in P_3.

A concrete interpretation of λ^* is natural in Economics and generally useful in Social Sciences. A widespread perception of the auxiliary variable λ as a simple technical tool to solve an optimization problem and of its optimal value λ^* as something not useful are well consolidated. Our opinion is absolutely orthogonal. The value of λ^* hides important information, that could be relevant also for Social Sciences different from Economics. Assume you are an entrepreneur[50] and that you want to maximize your margin function $y = f(x_1, x_2)$, being x_1, x_2 the quantities of two goods you are producing. The equation:

$$g(x_1, x_2) = b$$

describe very well a, so called, bottleneck in your production process: maybe the amount b of the resource at your disposal confines (maybe dramatically) the set of the admissible production policies x_1, x_2. Assume that, using this method, you find the best way to exploit the b units of the resource, i.e., the optimal production mix:

$$\mathbf{x}^*(b) = \begin{bmatrix} x_1^*(b) \\ x_2^*(b) \end{bmatrix}$$

which, obviously turns out to depend on b. The same holds for the optimized margin

---

[50]This assumption about the reader will be removed soon.

$$y^* (b) = f \left[x_1^* (b), x_2^* (b) \right] = L \left[\lambda^* (b), x_1^* (b), x_2^* (b) \right] =$$
$$= f \left[x_1^* (b), x_2^* (b) \right] + \lambda^* (b) \left[b - g \left[x_1^* (b), x_2^* (b) \right] \right]$$

In the case you have the opportunity to purchase some extra quantity h of the resource, you could re-optimize your strategy, allocating the new resource amount $b + h$ for producing the two goods. Under reasonable regularity assumptions, $y^* (b)$ turns out to be differentiable: the optimized margin variation $y^* (b + h) - y^* (b)$ would be:

$$y^* (b + h) - y^* (b) = y^{*\prime} (b) h + o (h)$$

and it would be of some interest to know $y^{*\prime} (b)$. Indeed, assume that you could purchase an amount h of the resource, paying for it ph, so that you can read p as the unitary market price of the resource. The comparison between p and $y^{*\prime} (b)$ becomes crucial: if $p > y^{*\prime} (b)$ it is not convenient to purchase additional doses of the resource, but it would be better to sell (at least a part of) the resource on the market. If $p < y^{*\prime} (b)$, the acquisition of the resource is convenient, because the differential margin (thinking of $h > 0$): $y^{*\prime} (b) h$ is greater that the differential cost ph. At this point the relevance of $y^{*\prime} (b)$ should be clear. How could we compute it? It is astonishing, but true: if we have solved the FOCs system, no computation is needed, because[51]:

$$y^{*\prime} (b) = \lambda^* (b)$$

The Lagrange multiplier at the optimum tells us how h, a small variation of b, would have an impact on the optimized objective function. Such a multiplier is usually called *shadow price* of the resource.

We can recycle such interpretation in Politics.

Example 2.3.13 Let's retake under consideration the problem introduced in the Example 2.3.11:

$$\mathcal{P} \begin{cases} \max_{x_1 x_2} y = V (x_1, x_2) \\ \text{sub} \\ x_1 + x_2 = b \end{cases}$$

The Lagrangean for the problem is:

$$L (\lambda; x_1, x_2) = V (x_1, x_2) + \lambda (b - x_1 - x_2)$$

The FOCs are:

$$\begin{cases} b - x_1 - x_2 = 0 \\ V'_{x_1} (x_1, x_2) - \lambda = 0 \\ V'_{x_2} (x_1, x_2) - \lambda = 0 \end{cases}$$

---

[51] Analytically wrong, but useful for remembering:

$$L^* = f^* + \lambda^* (b - g^*)$$

The derivative of L^* w.r.t. b is... λ^*.

At the optimum point we must have:

$$V'_{x_1}\left(x_1^*, x_2^*\right) = V'_{x_1}\left(x_1^*, x_2^*\right) = \lambda^* \text{ and } x_1^* + x_2^* = b$$

The common value of $V'_{x_1}\left(x_1^*, x_2^*\right)$, $V'_{x_1}\left(x_1^*, x_2^*\right)$ is roughly telling us the number of extra votes produced by 1€ more in the budget. Such value coincides with λ^* and it is the shadow price of the resource (budget) in terms of votes .

Example 2.3.14 **The optimal consumption** — Let's consider a basket of two goods, A and B. The utility function $u : \mathbb{R}^2 \to \mathbb{R}$ defined by:

$$u(a, b) = a^\alpha b^{1-\alpha} \qquad \alpha \in (0, 1)$$

measures the benefit for a consumer of a basket containing a units of good A and b units of good B. The Fig. 2.27 shows the plot of u in the case $\alpha = 1/3$.

Fig. 2.27 $u(a, b) = a^{1/3}b^{2/3}$

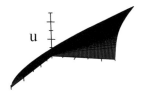

Let's indicate with $c \geq 0$ a given satisfaction[52] level for a consumer. It is interesting to investigate how a and b must vary in order that the level c remains constant. Starting from:

$$a^\alpha b^{1-\alpha} = c$$

we get the family of functions depending on the parameter $c \geq 0$

$$b(c, a) = \left(\frac{c}{a^\alpha}\right)^{1/(1-\alpha)} = c^{1/(1-\alpha)} \cdot a^{\alpha/(\alpha-1)}$$

The family of curves in the plane, corresponding to different levels for c, is the contour map for u (see at p.156).

We observe that

$$b'_a(c, a) = c^{1/(1-\alpha)} \frac{\alpha}{\alpha - 1} a^{1/(\alpha-1)} < 0$$

therefore b decreases as a increases in order to keep the same level of satisfaction c. Let's indicate with p_A the unitary price of good A, p_B the unitary price of good B and β the budget available. We can consider the following constrained optimization problem

---

[52]Frequently called in Economics "utility" or, more recently, "benefit".

$$\begin{cases} \max_{a,b} u(a, b) = a^{\alpha} b^{1-\alpha} \\ \text{sub} \\ ap_A + bp_B = \beta \end{cases}$$

The Lagrangean $L : \mathbb{R}^3 \to \mathbb{R}$, associated to the problem, is:

$$L (\lambda; a, b) = u(a, b) + \lambda (\beta - ap_A - bp_B)$$

where λ is the Lagrange multiplier. The gradient of L is the row vector

$$\nabla L (\lambda; a, b) = \begin{bmatrix} \beta - ap_A - bp_B & \alpha a^{\alpha-1} - \lambda p_A & (1 - \alpha) b^{-\alpha} - \lambda p_B \end{bmatrix}$$

The FOC is $\nabla L (\lambda, a, b) = \mathbf{0}$, or:

$$\begin{cases} \beta - ap_A - bp_B = 0 \\ \alpha a^{\alpha-1} - \lambda p_A = 0 \\ (1 - \alpha) b^{-\alpha} - \lambda p_B = 0 \end{cases}$$

and we get the optimal basket components:

$$a^* = \beta \frac{\alpha}{p_A} \quad \text{and} \quad b^* = \beta \frac{1 - \alpha}{p_B}$$

which obviously satisfy the budget constraint:

$$\left(\beta \frac{\alpha}{p_A} \right) \cdot p_A + \left(\beta \frac{1 - \alpha}{p_B} \right) \cdot p_B = \beta$$

The Lagrange's multiplier, generated, for instance, from the second equation, is

$$\lambda^* = \alpha^{\alpha} \cdot (1 - \alpha)^{1-\alpha} \left(\frac{p_B}{p_A} \right)^{\alpha}$$

Its economic interpretation is straightforward. Should the budget β be increased of a (small) amount h, the optimized utility level would have the increment:

$$\lambda^* h + o (h)$$

Exactly for this reason, λ^* is labelled "marginal utility of money".

2.3.6 The General Case

In the general case the Lagrangean becomes a function of $m + n$ variables:

$$L : \mathbb{R}^m \times \mathbb{R}^n \to \mathbb{R}$$

or:

$$L (\lambda_1, \lambda_2, \ldots, \lambda_m; x_1, x_2, \ldots, x_n)$$

defined as:

$$f (x_1, x_2, \ldots, x_n) + \lambda_1 [b_1 - g_1 (x_1, x_2, \ldots, x_n)] +$$
$$+ \lambda_2 [b_2 - g_2 (x_1, x_2, \ldots, x_n)] + \ldots + \lambda_m [b_m - g_m (x_1, x_2, \ldots, x_n)]$$
$$= f (\mathbf{x}) + \lambda_1 [b_1 - g_1 (\mathbf{x})] + \lambda_2 [b_2 - g_2 (\mathbf{x})] + \ldots + \lambda_m [b_m - g_m (\mathbf{x})]$$

Introducing a new row m-vector:

$$\boldsymbol{\lambda} = \begin{bmatrix} \lambda_1 & \lambda_2 & \cdots & \lambda_m \end{bmatrix}$$

and coupling it with the constraints system (2.3.10) it is truly nice to be able to rewrite the Lagrangean in the very compact form:

$$L (\boldsymbol{\lambda}, \mathbf{x}) = f (\mathbf{x}) + \boldsymbol{\lambda} \big[\mathbf{b} - \mathbf{g} (\mathbf{x})\big]$$

which perfectly mimics the simpler case we examined above. The economic interpretation of L does not change.

The FOCs:

$$\nabla L (\boldsymbol{\lambda}, \mathbf{x}) = \mathbf{0} \in \mathbb{R}^{m+n}$$

consist of a system of $m + n$ equations:

$$\begin{cases} \nabla_{\boldsymbol{\lambda}} L (\boldsymbol{\lambda}, \mathbf{x}) = \mathbf{0} \in \mathbb{R}^m \\ \nabla_{\mathbf{x}} L (\boldsymbol{\lambda}, \mathbf{x}) = \mathbf{0} \in \mathbb{R}^n \end{cases}$$

or:

$$\begin{cases} \left. \begin{array}{l} b_1 - g_1 (x_1, x_2, \ldots, x_n) = 0 \\ b_2 - g_2 (x_1, x_2, \ldots, x_n) = 0 \\ \quad \cdots \\ b_m - g_m (x_1, x_2, \ldots, x_n) = 0 \end{array} \right\} \leftarrow m \text{ equations} \\[2em] \left. \begin{array}{l} f'_{x_1} (x_1, x_2, \ldots, x_n) - \sum_{s=1}^{m} D_{x_1} g_s (x_1, x_2, \ldots, x_n) = 0 \\ f'_{x_2} (x_1, x_2, \ldots, x_n) - \sum_{s=1}^{m} D_{x_2} g_s (x_1, x_2, \ldots, x_n) = 0 \\ \quad \cdots \\ f'_{x_n} (x_1, x_2, \ldots, x_n) - \sum_{s=1}^{m} D_{x_n} g_s (x_1, x_2, \ldots, x_n) = 0 \end{array} \right\} \leftarrow n \text{ equations} \end{cases}$$

$$(2.3.12)$$

Note that the first m equations require the respect of the constraints, so that it is true that $\mathcal{P}_3$ brings us back to $\mathcal{P}_1$.

Any solution of the system (2.3.12) is an $(m + n)$-vector:

$$\begin{bmatrix} \boldsymbol{\lambda}^* \\ \mathbf{x}^* \end{bmatrix}$$

at which the gradient of L is the null vector:

$$\nabla L\left(\boldsymbol{\lambda}^*, \mathbf{x}^*\right) = \mathbf{o} \in \mathbb{R}^{m+n}$$

could identify an extremum.

Let's look at a numerical example:

Example 2.3.15 The function to be maximized is:

$$f\left(x_1, x_2, x_3\right) = 10 - x_1^2 - x_2^2 - x_3^2$$

under the constraints:

$$\begin{cases} x_1 + x_2 + x_3 = 10 \\ x_1 + 2x_2 + 3x_3 = 20 \end{cases}$$

The Lagrangean for the problem is:

$$L\left(\lambda_1, \lambda_2; x_1, x_2, x_3\right) = 10 - x_1^2 - x_2^2 - x_3^2 + \lambda_1\left(10 - x_1 - x_2 - x_3\right) + \\ + \lambda_2\left(20 - x_1 - 2x_2 - 3x_3\right)$$

Its gradient turns out to be:

$$\nabla L\left(\lambda_1, \lambda_2; x_1, x_2, x_3\right) = \begin{bmatrix} 10 - x_1 - x_2 - x_3 \\ 20 - x_1 - 2x_2 - 3x_3 \\ -2x_1 - \lambda_1 - \lambda_2 \\ -2x_2 - \lambda_1 - 2\lambda_2 \\ -2x_3 - \lambda_1 - 3\lambda_2 \end{bmatrix}^T$$

The FOCs boil down to the following linear system:

$$\begin{cases} x_1 + x_2 + x_3 = 10 \\ x_1 + 2x_2 + 3x_3 = 20 \\ -2x_1 - \lambda_1 - \lambda_2 = 0 \\ -2x_2 - \lambda_1 - 2\lambda_2 = 0 \\ -2x_3 - \lambda_1 - 3\lambda_2 = 0 \end{cases}$$

in matrix form:

$$\begin{bmatrix} 0 & 0 & 1 & 1 & 1 \\ 0 & 0 & 1 & 2 & 3 \\ -1 & -1 & -2 & 0 & 0 \\ -1 & -2 & 0 & -2 & 0 \\ -1 & -3 & 0 & 0 & -2 \end{bmatrix} \begin{bmatrix} \lambda_1 \\ \lambda_2 \\ x_1 \\ x_2 \\ x_3 \end{bmatrix} = \begin{bmatrix} 10 \\ 20 \\ 0 \\ 0 \\ 0 \end{bmatrix}$$

hence:

$$
\begin{bmatrix} \lambda_1^* \\ \lambda_2^* \\ x_1^* \\ x_2^* \\ x_3^* \end{bmatrix} = \begin{bmatrix} 0 & 0 & 1 & 1 & 1 \\ 0 & 0 & 1 & 2 & 3 \\ -1 & -1 & -2 & 0 & 0 \\ -1 & -2 & 0 & -2 & 0 \\ -1 & -3 & 0 & 0 & -2 \end{bmatrix}^{-1} \begin{bmatrix} 10 \\ 20 \\ 0 \\ 0 \\ 0 \end{bmatrix} = \begin{bmatrix} -20/3 \\ 0 \\ 10/3 \\ 10/3 \\ 10/3 \end{bmatrix} \qquad (2.3.13)
$$

Remark 2.3.5 In the case of several constraints, we have several resources and several Lagrange multipliers. Their concrete interpretation as *shadow prices* of the resources is straightforward: the value of — say — λ_s^* approximates the variation of the optimized objective function $y^* = y^*$ (**b**) of 1 extra-unit of the sth resource[53]: $b_s \rightarrow b_s + 1$.

FOCs provide us with necessary optimality conditions: they simply reduce to the stationarity conditions of L.

How can we get *second order conditions* (SOCs), that combined with the FOCs provide us with a (local) maximum?

First of all we introduce the notion of *degrees of freedom*. The notion is pretty intuitive. If we have a function of 2 variables and 1 constraint, the "true" number of independent variables is 1, because once we have chosen an independent variable, the other will be fixed by the constraint. If we have a function of 4 variables and 1 constraint, the "true" number of independent variables is 3, because once we have chosen 3 independent variable, the fourth one will be fixed by the constraint. We call *number of degrees of freedom* for a problem of optimization of a function of n variables, under $m \, (< n)$ constraints, the difference δ between the number of variables and the number of constraint equations:

$$
\delta = n - m
$$

For instance, in the case of the Example 2.3.15 we have:

$$
\delta = 3 - 2 = 1
$$

In the case of unconstrained optima we have enjoyed the useful role played by all of the principal NW minors of $\nabla^2 f$, the Hessian matrix of the function to be optimized.

In the constrained maxima setting, we will work again successfully with NW principal minors. The matrix involved will be once again a Hessian matrix, the one of L, at the stationary point. Here is its structure:

---

[53] To be more precise: an increment h_s in b_s determines a variation in y^* which is of the type:

$$
\lambda_s^* h_s + o(h_s).
$$

$$\nabla^2 L\left(\boldsymbol{\lambda}^*, \mathbf{x}^*\right) = \begin{bmatrix} \mathbf{O} & -\nabla \mathbf{g}\left(\mathbf{x}^*\right) \\ \left[-\nabla \mathbf{g}\left(\mathbf{x}^*\right)\right]^{\mathrm{T}} & \nabla^2_{\mathbf{x},\mathbf{x}} L \end{bmatrix}$$

In this matrix $\mathbf{O}$ is a null matrix[54] of order m, $\nabla \mathbf{g}\left(\mathbf{x}^*\right)$ is the Jacobian (m, n)-matrix of $\mathbf{g}$, and $\nabla^2_{\mathbf{x},\mathbf{x}} L$ is the partial Hessian (n, n)-matrix of L w.r.t. the decision variables $\mathbf{x}$.

A less synthetic description of the square matrix of order $m + n$ follows, for each block:

$$\mathbf{O}_{(m,m)} = \begin{bmatrix} 0 & 0 & \cdots & 0 \\ 0 & 0 & \cdots & 0 \\ \cdots & \cdots & \cdots & \cdots \\ 0 & 0 & \cdots & 0 \end{bmatrix}$$

$$-\nabla \mathbf{g}\left(\mathbf{x}^*\right)_{(m,n)} = \begin{bmatrix} -D_{x_1} g_1\left(\mathbf{x}^*\right) & -D_{x_2} g_1\left(\mathbf{x}^*\right) & \cdots & -D_{x_n} g_1\left(\mathbf{x}^*\right) \\ -D_{x_1} g_2\left(\mathbf{x}^*\right) & -D_{x_2} g_2\left(\mathbf{x}^*\right) & \cdots & -D_{x_n} g_2\left(\mathbf{x}^*\right) \\ \cdots & \cdots & \cdots & \cdots \\ -D_{x_1} g_m\left(\mathbf{x}^*\right) & -D_{x_2} g_m\left(\mathbf{x}^*\right) & \cdots & -D_{x_n} g_m\left(\mathbf{x}^*\right) \end{bmatrix}$$

$$\left[-\nabla \mathbf{g}\left(\mathbf{x}^*\right)\right]^{T}_{(n,m)} = \begin{bmatrix} -D_{x_1} g_1\left(\mathbf{x}^*\right) & -D_{x_2} g_1\left(\mathbf{x}^*\right) & \cdots & -D_{x_n} g_1\left(\mathbf{x}^*\right) \\ -D_{x_1} g_2\left(\mathbf{x}^*\right) & -D_{x_2} g_2\left(\mathbf{x}^*\right) & \cdots & -D_{x_n} g_2\left(\mathbf{x}^*\right) \\ \cdots & \cdots & \cdots & \cdots \\ -D_{x_1} g_m\left(\mathbf{x}^*\right) & -D_{x_2} g_m\left(\mathbf{x}^*\right) & \cdots & -D_{x_n} g_m\left(\mathbf{x}^*\right) \end{bmatrix}^{T} =$$

$$= \begin{bmatrix} -D_{x_1} g_1\left(\mathbf{x}^*\right) & -D_{x_1} g_2\left(\mathbf{x}^*\right) & \cdots & -D_{x_1} g_m\left(\mathbf{x}^*\right) \\ -D_{x_2} g_1\left(\mathbf{x}^*\right) & -D_{x_2} g_2\left(\mathbf{x}^*\right) & \cdots & -D_{x_2} g_m\left(\mathbf{x}^*\right) \\ \cdots & \cdots & \cdots & \cdots \\ -D_{x_n} g_1\left(\mathbf{x}^*\mathbf{x}\right) & -D_{x_n} g_2\left(\mathbf{x}^*\right) & \cdots & -D_{x_n} g_m\left(\mathbf{x}^*\mathbf{x}\right) \end{bmatrix}$$

and, finally:

$$\nabla^2_{\mathbf{x},\mathbf{x}} L_{(n,n)} = \begin{bmatrix} D^2_{x_1,x_1} L\left(\boldsymbol{\lambda}^*, \mathbf{x}^*\right) & D^2_{x_1,x_2} L\left(\boldsymbol{\lambda}^*, \mathbf{x}^*\right) & \cdots & D^2_{x_1,x_n} L\left(\boldsymbol{\lambda}^*, \mathbf{x}^*\right) \\ D^2_{x_2,x_1} L\left(\boldsymbol{\lambda}^*, \mathbf{x}^*\right) & D^2_{x_2,x_2} L\left(\boldsymbol{\lambda}^*, \mathbf{x}^*\right) & \cdots & D^2_{x_2,x_n} L\left(\boldsymbol{\lambda}^*, \mathbf{x}^*\right) \\ \cdots & \cdots & \cdots & \cdots \\ D^2_{x_n,x_1} L\left(\boldsymbol{\lambda}^*, \mathbf{x}^*\right) & D^2_{x_n,x_2} L\left(\boldsymbol{\lambda}^*, \mathbf{x}^*\right) & \cdots & D^2_{x_n,x_n} L\left(\boldsymbol{\lambda}^*, \mathbf{x}^*\right) \end{bmatrix}$$

It is pretty clear that the first m NW minors are 0. The overwhelming presence of 0's in the first m lines will likely continue to produce at the beginning null NW principal minors.

---

[54]It collects the second partial derivatives of L w.r.t. the Lagrange multiplier. As L is linear in the multipliers, this submatrix of the Hessian matrix is a null matrix.

What we are interested in is not the whole sequence of the NW minors (like in the unconstrained case), which would provide us with a sequence of non-informative 0s. We could prove that only in the *last* $\delta = n - m$ minors.

An alternative way to precise the list of relevant NW minors, simply, indicates the order of the first one we take into consideration. Its order is $2m + 1$. The next one will have order $2m + 2$, etc.

The *condition that guarantees a local maximum* is that these minors, starting from the one of order $2m + 1$ and arriving to the largest one (of order $m + n$) have *alternate signs*, starting from $(-)^{m+1}$.

Let us look at a pair of examples.

Example 2.3.16 Reconsider the Example 2.3.11. Here is the Hessian matrix for the Lagrangean $L(\lambda; x_1, x_2) = V(x_1, x_2) + \lambda(b - x_1 - x_2)$:

$$\nabla^2 L = \begin{bmatrix} 0 & -1 & -1 \\ -1 & V''_{x_1,x_1}(x_1, x_2) & V''_{x_1,x_2}(x_1, x_2) \\ -1 & V''_{x_2,x_1}(x_1, x_2) & V''_{x_2,x_2}(x_1, x_2) \end{bmatrix}$$

The first principal NW minor is the one of order $2m + 1 = 2 + 1 = 3$, or the determinant of the Hessian matrix:

$$\det \begin{bmatrix} 0 & -1 & -1 \\ -1 & V''_{x_1,x_1}(x_1^*, x_2^*) & V''_{x_1,x_2}(x_1^*, x_2^*) \\ -1 & V''_{x_2,x_1}(x_1^*, x_2^*) & V''_{x_2,x_2}(x_1^*, x_2^*) \end{bmatrix} =$$

$$= 2V''_{x_1,x_2}(x_1^*, x_2^*) - V''_{x_1,x_1}(x_1^*, x_2^*) - V''_{x_2,x_2}(x_1^*, x_2^*)$$

In the case its sign is $(-)^{m+1} = (-)^{1+1} = +$ we would be sure that the stationary point of L is a maximum point.

In order to deepen our understanding, we can borrow from scientific marketing a widely accepted specification for V:

$$V(x_1, x_2) = ax_1^{\alpha_1} x_2^{\alpha_2} \text{ with } a, \alpha_1, \alpha_2 > 0 \text{ known parameters} \qquad (2.3.14)$$

We stress that maximizing V is the same as maximizing:

$$W(x_1, x_2) = \ln V(x_1, x_2) = \ln a + \alpha_1 \ln x_1 + \alpha_2 \ln x_2$$

We can rewrite the problem as:

$$\mathcal{P} \begin{cases} \max_{x_1,x_2} y = W(x_1, x_2) \\ \text{sub} \\ x_1 + x_2 = b \end{cases}$$

or, more explicitly:

$$\mathcal{P} \begin{cases} \max_{x_1, x_2} y = \ln a + \alpha_1 \ln x_1 + \alpha_2 \ln x_2 \\ \text{sub} \\ x_1 + x_2 = b \end{cases}$$

The Lagrangean is:

$$L(\lambda; x_1, x_2) = \ln a + \alpha_1 \ln x_1 + \alpha_2 \ln x_2 + \lambda (b - x_1 - x_2)$$

Its gradient turns out to be:

$$\nabla L = \left[b - x_1 - x_2 \quad \frac{\alpha_1}{x_1} - \lambda \quad \frac{\alpha_2}{x_2} - \lambda \right]$$

It is null at:

$$\lambda^* = \frac{\alpha_1 + \alpha_2}{b}; \quad x_1^* = \frac{\alpha_1 b}{\alpha_1 + \alpha_2} \quad x_2^* = \frac{\alpha_2 b}{\alpha_1 + \alpha_2} \tag{2.3.15}$$

If we construct the Hessian matrix for the Lagrangean, we get:

$$\nabla^2 L = \begin{bmatrix} 0 & -1 & -1 \\ -1 & -\alpha_1/x_1^2 & 0 \\ -1 & 0 & -\alpha_2/x_2^2 \end{bmatrix}$$

As the number of degrees of freedom is:

$$\delta = 2 - 1 = 1$$

we need to consider only the last of $\nabla^2 L$, or its determinant:

$$\det \nabla^2 L = \frac{\alpha_1}{x_1^2} + \frac{\alpha_2}{x_2^2} > 0$$

therefore we have found a maximum.

Remark 2.3.6 If we continue to think of the example above, we can state some non-obvious considerations, may be, of some interest for political science. The specification (2.3.14) for the function V is known in the literature as the "constant elasticity function", as, if we compute the elasticity of the number of votes w.r.t. the investment in some medium, we find that such an elasticity:

$$\varepsilon_1(x_1, x_2) = x_1 \frac{D_{x_1}[V(x_1, x_2)]}{V(x_1, x_2)} = \alpha_1$$

turns out to be constant and equal to the exponent of x_1 in (2.3.14).

If we look at the Eq. (2.3.15), we learn that the optimal allocation of the budget between the two media is nothing but a proportional rule, but the weights used are non-trivial: they turn out to be nothing but the elasticities of votes w.r.t. the investment in the two media. Honestly, we are far from being sure that when politicians choose their campaign strategies are so conscious of these optimality rules.

Example 2.3.17 Now back to the previous Example 2.3.15. For that problem, with $n = 3$ and $m = 1$, we have found FOCs, see: (2.3.13):

$$\begin{bmatrix} \lambda_1^* \\ \lambda_2^* \\ x_1^* \\ x_2^* \\ x_3^* \end{bmatrix} = \begin{bmatrix} -20/3 \\ 0 \\ 10/3 \\ 10/3 \\ 10/3 \end{bmatrix}$$

Let us now look at sufficient conditions. The Hessian matrix of the Lagrangean turns out to be:

$$\begin{bmatrix} 0 & 0 & -1 & -1 & -1 \\ 0 & 0 & -1 & -2 & -3 \\ -1 & -1 & -2 & 0 & 0 \\ -1 & -2 & 0 & -2 & 0 \\ -1 & -3 & 0 & 0 & -2 \end{bmatrix}$$

We are interested in its principal minors of order $\geq 2m + 1 = 2 \cdot 2 + 1 = 5$. The only relevant minor turns out to be the largest one:

$$H_5 = \det \begin{bmatrix} 0 & 0 & -1 & -1 & -1 \\ 0 & 0 & -1 & -2 & -3 \\ -1 & -1 & -2 & 0 & 0 \\ -1 & -2 & 0 & -2 & 0 \\ -1 & -3 & 0 & 0 & -2 \end{bmatrix} = -12$$

In order to be sure that we are in the presence of a maximum point, the starting sign must be $(-)^{m+1} = (-)^{2+1} = (-)^3 = -$. Therefore the point we have found is a maximum point (at least local, but also global, after a further analysis).

The following examples should provide the reader with interesting political application of what we have seen until now.

Example 2.3.18 — ***Votes and campaign duration*** — This case deserves some attention because it mixes several topics met above. We are considering a candidate for some election. How long an electoral campaign should last? We indicate with x the campaign duration. Let $V(x)$ be the the numbers of votes the candidate can collect, as a function of x, the campaign duration. For simplicity, we assume that V is differentiable. We can realistically assume that each vote has an economic value[55],

---

[55]Think of a candidate that has used some other tool in order to gather consensus. According to such experience, every vote has had an average cost that can be assessed. The campaign to be evaluated uses new tools and new communication channels. The "value" of a vote could be thought of as its old cost. Lagrange's approach allows for an automatic comparison between the two strategies.

which can be compared with the campaign costs. We assume here that the value of a vote is independent of the number of votes that are gathered: more realistic assumptions would bring to problems, where the philosophy is the same, but some formal complications do appear. Our aim is to examine the problem in its simplest framework. Our assumption is that $W(x) = \ln V(x)$ is proportional[56] to $V(x)$ itself, through some proportionality constant $\alpha > 0$:

$$W(x) = \alpha V(x)$$

We assume that $V'(x) > 0$, that $V''(x) < 0$ and, absolutely reasonably, that V'' $(+\infty) = 0$: a long campaign is more effective than a short campaign, but every additional campaign date has a value, that decreases when the length is extended. The benefits of a very long campaign are negligible. The campaign cost $C(x)$ is absolutely standard:

$$C(x) = F + vx$$

where F is the fixed cost of the campaign (say, what the candidate pays to some consulting agency for designing the campaign) and v is the daily cost of proposing such campaign through a certain basket of media. The candidate considers his/her "political profit" as:

$$P(x) = \alpha V(x) - F - vx$$

The derivative of this *political profit* turns out to be:

$$\alpha V'(x) - v$$

We must distinguish between two possibilities:

$$\begin{cases} (1)\, \alpha V'(0) - v \leq 0 \\ (2)\, \alpha V'(0) - v > 0 \end{cases}$$

In the first case, the best policy for our candidate is… to devote his/her enthusiasm to — say — gardening, because his/her political profit decreases if the duration of the campaign increases. This simply means that the marginal profit $\alpha V'(x)$ is smaller than its marginal cost v. In the case (2) there exists a unique optimal value x^* for the campaign duration, that turns out to be optimal. Such an optimal value solves the equation:

$$\alpha V'(x) - v$$

Example 2.3.19 We continue with the preceding example. We introduce a standard specification for V. It is the already known Constant Elasticity assumption:

$$V(x) = ax^k, \text{ with } a, k > 0$$

---

[56] Such hypothesis simply boils down to asking that the growth rate of V is proportional to V itself.

In this context, we should think $k < 1$. The FOC become:

$$\alpha a k x^{k-1} = v$$

hence:

$$x^* = \left(\frac{a\alpha k}{v}\right)^{-1/(1-k)}$$

In the preceding model ($k \geq 1$), the optimal campaign duration x^* would have been x^*, terrible for us and our kids.

Example 2.3.20 — Grexit[57] **(A)** — The problem we present is inspired by the 2015 problems in the Euro area concerning Grexit. Our aim is to show that the mathematics we have considered could help to analyze some political problems. Two Countries are involved: L and B ("L" stays for lender and "B" stays for borrower). B has an enormous debt D w.r.t. L. The country B could receive liquidity from the ECB: its amount is W_0. L can claim that, at least a part of such amount is used to repay the debt D. Let x be the the amount L requires immediately to B, so that what remains to B in order to foster economic development is $u = W_0 - x$. The agreements between L and B are very simple: B will pay immediately $x = W_0 - u$, the remaining part of this liquidity will be invested over 1 year, producing new "wealth" W_1 and B will pay to L some fraction α of W_1. At this point the residual debt will be cancelled.

We examine this situation from the viewpoint of L, which is interested in extracting the best from this situation.

The economic side of the problem for L is that, asking for money today, reduces the opportunities to get money later, whose amount could be interesting because of the capabilities of B to make a good use of resources. If, in particular, we introduce the standard economic assumption of decreasing returns to scale, for L the possibility to exploit high marginal returns of B for small available liquidity amounts could be interesting for L itself.

We need to introduce a smooth function f, that, in B, transforms liquidity today u into new wealth W_1:

$$W_1 = f(u) = f(W_0 - x)$$

The assumptions we make on f are rather standard:

$$f' > 0; \qquad f'' < 0$$

This means that the final wealth increases with the initial wealth and that the scale-returns are decreasing.

Therefore, the Country L will collect these amounts:

---

[57]This example illustrates some stylized facts concerning the Grexit question of Summer 2015.

$$\begin{cases} x = W \text{ now} \\ \alpha f(u) = \alpha f(W_0 - x) \text{ at the date 1} \end{cases}$$

L wants to choose x optimally. We can describe the problem using the standard scheme of constrained maximization. We need to introduce a discount factor v, which transforms € at 1 into € now. Such a discount factor can be expressed in terms of the interest rate i, prevailing between 0 and 1:

$$v = \frac{1}{1 + i}$$

The interest rate i is nothing but the cost of the public debt of the country L.
    The (selfish) problem for L is:

$$\begin{cases} \max_{x,u} x + \alpha v f(u) \\ \text{sub} \\ x + u = W_0 \end{cases}$$

The Lagrangean is:

$$L(\lambda; x, u) = x + \alpha v f(u) + \lambda (W_0 - x - u)$$

The FOCs are:

$$\begin{cases} W_0 - x - u = 0 \\ 1 - \lambda = 0 \\ \alpha v f'(u) - \lambda = 0 \end{cases}$$

and they tell us a politically interesting story:

$$\begin{cases} \lambda^* = 1 \\ f'(u^*) = \dfrac{1 + i}{\alpha} \\ x^* = W_0 - u^* \end{cases}$$

The first equation tells us that for the country L every extra € that B will obtain has value exactly 1 € because it will be collected:

$$\begin{cases} \text{immediately in } x \\ \text{or} \\ 1 \text{ period later, with interests, as } \alpha v f(u^*) \end{cases}$$

The second equation tells us that, at the optimal point for L, there must be equality between the marginal productivity of money left to B, f' at u^*, and the ratio $(1 + i)/\alpha$: a threshold which takes into account the (opportunity) cost for L, to leave money to B and of the fraction α of the final wealth of B, that L will be allowed to extract.

The third equation is a mere budget equation.

It is important to establish whether the equation:

$$f'(u^*) = \frac{1+i}{\alpha} \tag{2.3.16}$$

has some economically significant solution.

The assumptions we have made about f are useful. The fact that f' is decreasing (implied by $f'' < 0$), tells us that, on the left of u^*, L'_u is positive and that on the right it is negative. This "$\nearrow\searrow$" pattern guarantees a max point has been found.

In order that it exists, it is necessary that $f'(0) > (1+i)/\alpha$. If we assume the so-called Inada conditions[58] about f:

$$\begin{cases} f'(0) = +\infty \\ \text{and} \\ f'(+\infty) = 0 \end{cases}$$

exactly one and only one u^* satisfying (2.3.16) does exist.

In the case:

$$u^* > W_0$$

the optimal policy for L would be to finance B, giving it more money. In the case $u^* \leq W_0$, the best policy consists in asking for the immediate payment of $x^* = W_0 - x^*$ and in collecting $\alpha f(u^*)$ one period later. This is the only way in which the ECB decision (i.e.; the choice of W_0 can affect the decisions of L: if $u^* < W_0$, the specific level of W_0 is irrelevant for L).

Example 2.3.21 — Grexit (B) — We have studied the model framing it into the constrained optimization scheme. The triviality of the constraint:

$$x + u = W_0$$

suggests to solve for x and to propose a new frame for the problem as an unconstrained one:

$$\max_u W_0 - u + \alpha v f(u)$$

this experiment allows us to evaluate pros and cons of this simplification.

The pros are represented by the fact that the FOC provides us immediately with the condition: (2.3.16).

---

[58]Inada conditions: $f'(0) = +\infty$, $f'(+\infty) = 0$, simply require that f' covers all the possible values between $+\infty$ (at 0) and 0 at $+\infty$. They are due to the Japanese mathematical economist Ken-Ichi Inada (1925-2002).

The cons depend on the fact that, avoiding the Lagrange approach, we have no access to the important economic info:

$$\lambda^* = 1$$

which illustrates the perfectly rational attitude of L.

Example 2.3.22 — Grexit (C) — We will revisit the preceding examples in order to make the model more realistic and to inform the readers about the possibility to study political problems of this kind using an important tool: *game theory*: probably the first piece of brand new mathematics created for the study of Economics, where it turned out to become pervasive, but also in other Social Sciences (Political Science included). The key point of game theory is the study of the effects of the interaction between different decision makers. A situation which is absolutely customary in Politics.

In the previous examples, the role of B was substantially passive: L decides u^*, B uses u^* fruitfully (according to the function f) and what remains in the hands of B is nothing but: $(1 - \alpha) f(u^*)$.

Now we extend the model, in order to include a decision opportunity for B. The idea is very simple: the function f, which transforms u into the wealth of B, 1 period later, before L takes its allowances, is not only a function of u, but also of the effort e, B puts on the table[59]:

$$W_1 = f(u, e)$$

We can assume that:

$$f'_e(u, e) > 0 \text{ higher effort implies higher final wealth}$$

We are proud to introduce our readers to a new problem, where the same function of two variables is involved in two interacting problems. For the country L the problem does not change: its objective is to maximize what it can extract from its decision u. For the country B the decision problem is crucially related to the cost of the effort it will engage and the return left to it. The formalization of the (double) problem is not too difficult:

$$\begin{cases} \max_u W_0 - u + \alpha v f(u, e) & \text{L decision} \\ \max_e (1 - \alpha) f(u, e) - c(e) & \text{B decision} \end{cases}$$

where $c(e)$ is the cost of the effort e. The function c is assumed to be differentiable.

We are looking for a pair (u^*, e^*), which turns out to be optimal both for L and B. Such a pair would be called *solution* of the game.

---

[59]The letter "e" is used differently in these pages. When we mean "the constant number e $\approx$ 2.718", we write "e". When we indicate with this letter the effort (a notation common in the economic literature, we write e). In the non-economic literature it is common to see e for the constant e.

The FOC for both the countries are:

$$\begin{cases} -1 + \alpha v f'_u\,(u, e) = 0 & \text{for L} \\[2mm] (1 - \alpha)\, f'_e\,(u, e) - c'\,(e) & \text{for B} \end{cases} \qquad (2.3.17)$$

Without specifying f, it is difficult to draw conclusions. Therefore, we introduce the following specification:

$$f\,(u, e) = h\,(e)\,u^k, \text{ with } 0 < k < 1$$

The quantity h increases with the effort[60] and is differentiable.
From the FOCs (2.3.17) we easily get the system:

$$\begin{cases} \alpha v k h\,(e)\,u^k = u & \text{for L} \\[2mm] (1 - \alpha)\, h'\,(e)\,u^k = c'\,(e) & \text{for B} \end{cases}$$

hence:

$$u = c'\,(e)\, \frac{\alpha k}{(1 + i)\, D\,[\ln h\,(e)]} \qquad (2.3.18)$$

which relates the potentially optimal pairs (u, e).
We can further specify our model as follows[61]:

$$\begin{cases} h\,(e) = a + be, \text{ with } a, b > 0 \\[2mm] c\,(e) = ce \end{cases} \qquad (2.3.19)$$

The FOCs become:

$$\begin{cases} -1 + \alpha v k\,(a + be)\,u^{k-1} = 0 \\[2mm] (1 - \alpha)\, b u^k = c \end{cases} \qquad (2.3.20)$$

From the second equation we get the optimal value of u for B:

$$u^* = \left[\frac{c}{(1 - \alpha)\, b}\right]^{1/k} \qquad (2.3.21)$$

Manipulating the two equations (2.3.20), we get the analogous of (2.3.18):

---

[60] A similar assumption is usual in Economic Theory when dealing with the technology innovation effects in the neoclassical framework.

[61] We admit some notation abuse. In (2.3.19) c switches from being a function name to a parameter in a linear function.

$$u = c\frac{\alpha vk\,(a + be)}{(1 - \alpha)\,b}$$

which is a linear (affine) function of e.

Fixing u at some level u^*, offered by (2.3.21), we obtain e^*:

$$e^* = a\frac{1 - \alpha\,(1 + vk)}{\alpha vk},$$

which turns out to be independent of b. At this point B can negotiate with L, both proposing u^* and proposing e^*, knowing that the pair (u^*, e^*) is the only jointly optimal for the two countries.

2.4 Exercises

Exercise 2.4.1 Find the points of maximum and/or minimum (local and global) for the function $f : \mathbb{R} \to \mathbb{R}$ defined by:

(a) $f(x) = x - \ln x$ (b) $f(x) = 4x + \dfrac{1}{x}$ (c) $f(x) = \ln x - \sqrt{x}$

Exercise 2.4.2 Consider the function $f : (0, +\infty) \to \mathbb{R}$ defined by

$$f(x) = \ln x - (\ln x)^2$$

(a) Calculate its first derivative $f'(x)$.
(b) Find the unique stationary point x^* of f.
(c) Calculate its second derivative of and establish the nature of x^*.

Exercise 2.4.3 Let α be a positive real parameter. Consider the function $f : [0, +\infty) \to \mathbb{R}$ defined by

$$f(x) = x^\alpha e^{-x}$$

(a) Calculate the first derivative of f.
(b) Find the unique stationary point $x^* > 0$.
(c) Calculate the second derivative of f and establish the nature of x^*.

Exercise 2.4.4 A company produces the physical quantity $x \geq 0$ of some good, claiming total production costs:

$$C(x) = a + bx^3$$

with a, b positive parameters.

(a) Find a, b knowing that fixed costs $C(0)$ are € 500 000 and the total production cost of 10 units, that is $C(10)$, is € 600 000.
(b) Calculate the average production cost $m(x) = C(x)/x$ and its first derivative $m'(x)$.
(c) If it exists, calculate the quantity $x^* > 0$ such that $m'(x^*)$ is null. Using $m''(x)$ say whether x^* is a maximum or a minimum point.

Exercise 2.4.5 A monopolist produces and sells a good. Call q the quantity of the good which is produced and sold. The unitary price p the monopolist must choose in order to sell quantity q is

$$p = \frac{140}{q^{0.7}}$$

(a) Construct the function $G(q)$, which describes the profit for the monopolist as a function of the production volume q, taking into account that the monopolist incurs a fixed cost € 70 and variable costs, with unitary variable cost b.
(b) Compute $G'(q)$ and find the unique stationary point q^* for G. Compute $G''(q^*)$ and explain why q^* is a maximum profit point.
(c) If the Government introduces an excise tax on the production of the good, which would be the effect on the optimal production volume?

Exercise 2.4.6 In some market the demand function for a good is

$$q = q\,(p) = 1000 - 0.1p$$

with p the unitary price and q the demanded quantity.

(a) Compute $\varepsilon\,(p)$, the elasticity of the demand function.
(b) Compute $E\,(p)$, the total expense amount in that market as a function of p. Determine also the price level p which maximizes the expense.
(c) Assuming that the supply function is

$$s = s\,(p) = 0.2p$$

compute the equilibrium price p^*.

Exercise 2.4.7 An NGO organizes a support plan for a population. Call x the number of people recruited for the support plan and call $f(x)$ the gross value of the support. Such value is modeled by the function

$$f(x) = 40x^{0.3}$$

The cost of the plan $C(x)$ is made by a fixed component $k = 27$ and a variable one, with unitary variable cost per person $v = 0.2$. The net value of the support is $N(x)$, the difference between the gross value and its cost.

(a) Construct the analytic expression of $N(x)$.

(b) Compute $N'(x)$.

(c) Show that there exists a unique stationary point x^* for N, compute x^*.

(d) Explain what you can say about the nature of x^* (global/local extremum for N) on the basis of $N''(x)$.

Exercise 2.4.8 An NGO can provide help to people in an underdeveloped country. The variable $x \geq 0$ quantifies the volume such help (for instance the number of people enjoying of fixed amount of some goods). The unitary value of the help for people is € 100. The cost $C(x)$ for the NGO of helping x people increases with the volume of the activity. It is made up of a fixed component of € 500 and of a variable unitary cost $v(x) = 40 + x$. The NGO controls the *humanitarian margin* $h(x)$, or the difference between the humanitarian result $r(x)$, determined by the value of help, and the help cost $C(x)$.

(a) Construct the humanitarian result $r(x)$ and the humanitarian margin $h(x)$.

(b) Find the values of the size of help x such that the humanitarian margin is non-negative.

(c) Find the values of the size of help such that the humanitarian margin is maximized.

Exercise 2.4.9 Let's consider the function $f : \mathbb{R}^2 \to \mathbb{R}$ defined by

$$f(x_1, x_2) = x_1^2 \ln x_2 - x_2$$

(a) Calculate the gradient of f.

(b) Find, if they exist, the stationary point of f.

(c) What is the value of f at $(2,e)$?

Exercise 2.4.10 Find, if they exist, points of maximum and minimum for $f : \mathbb{R}^2 \to \mathbb{R}$, defined by:

(a) $f(x_1, x_2) = \ln\left[1 - (x_1 - 2)^2\right] - x_2^2$

(b) $f(x_1, x_2) = 2x_1^2 + x_2^2 + 2x_1 x_2 - 2\ln x_1$

(c) $f(x_1, x_2) = 2x_1 + x_2 - \ln(x_1 x_2)$

Exercise 2.4.11 Consider the function $f : \mathbb{R}^2 \to \mathbb{R}$ defined by

$$f(\mathbf{x}) = 4x_1 x_2 - 2x_1^2 + x_2^2$$

and the constraint $3x_1 + x_2 = 5$.

(a) Construct the Lagrangean function $L(\lambda, x_1, x_2)$.

(b) Compute the gradient $\nabla L(\lambda, x_1, x_2)$ of the Lagrangean and find its stationary points.

Exercise 2.4.12 Consider the following optimization problem

$$\max_{\mathbf{x} \in \mathbb{R}^3} f(x_1, x_2, x_3) = x_2 x_3 + x_1 x_3$$

subject to the following constraints $x_2 + 2x_3 = 1$ and $x_1 + x_3 = 3$.

(a) Build the Lagrangian function L associated with the optimization problem.
(b) Calculate the gradient of L and find, if they exist, the stationary point of L.
(c) Find the Hessian matrix for L and compute the value of its (relevant) last NW principal minor, discuss the nature of the stationary point found in (b).

Chapter 3
Integral Calculus

3.1 Integrals and Areas

The following problem is frequently met in Probability and Statistics. Let's consider a function $f : \mathbb{R} \to \mathbb{R}$ and a bounded interval $[a, b]$, that is the set of real numbers x such that $a \leq x \leq b$ on which f is defined. We would like to find the area of the part of the plane bounded by $[a, b]$ on x-axis, vertical lines with equations $x = a$ and $x = b$, as sketched in Fig. 3.1a, and the graph of f.

Fig. 3.1 Area below a function

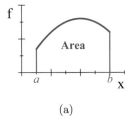
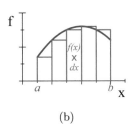

(a)   (b)

Imagine that such an area can be obtained as a sum of the areas of infinitely many thin mini-rectangles, with base dx and height $f(x)$, so the area of generic mini-rectangle is $f(x)dx$ (Fig. 3.1b). The standard symbol to indicate such an area is:

$$\int_a^b f(x)\, dx$$

that recalls precisely this way of seeing the number that represents the area. The area is obtained by adding the areas of the rectangles above. The symbol $\int$ derives from stretching the letter "s" as sum. We observe immediately that if you change the name to the dependent variable and write, in place of the above, for example:

© Springer Nature Switzerland AG 2018
L. Peccati et al., *Maths for Social Sciences*, UNITEXT - La Matematica
per il 3 + 2 113, https://doi.org/10.1007/978-3-030-02336-2_3

$$\int_a^b f(t)\,dt \tag{3.1.1}$$

the value of the area (= the integral) does not change.

Definition 3.1.1 This number is called the *definite integral of f over* $[a, b]$ and is read as follows

"definite integral of $f(t)$ in $d\,t$ over the interval $[a, b]$"

It is interesting to know how to evaluate this integral, more simply indicated by $I(a, b)$. The following section provides important information on how it can be evaluated.

3.2   Fundamental Theorem of Integral Calculus

The problem can be advantageously solved in many cases by a simple observation. Consider a function f, defined on $[a, b]$. Consider a point x of this interval.

Definition 3.2.1 We call *integral function* of f the integral:

$$F(x) = \int_a^x f(t)\,dt \tag{3.2.1}$$

Conversely, the function f is called the *integrand function*[1].

If the function f is not too much irregular, for instance, if it is continuous or if it has only a finite number of discontinuity points[2] then F does exists.

If F is known then the integral $I(a, b)$ can be evaluated[3], in fact, obviously

$$I(a, b) = F(b)$$

Finding the integral function is not always hopeless, for many cases that turn out to cover most of our interests it is even immediate. This happens by virtue of a fundamental result that we will try to guess below.

---

[1] It's a bit like mother (F) and daughter (f).

[2] Perhaps 10 trillion jumps in 1 mm!

[3] Refrain from talking about the solution of an integral. Integrals are evaluated, calculated, but not solved!

Fig. 3.2 $\int_{x_0}^{x_0+h} f(t)dt$

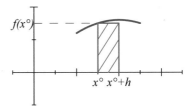

As indicated in the Fig. 3.2, let's consider the graph of a function f and suppose that it is continuous near a given point x_0. Consider the increment of F if x moves from x_0 to $x_0 + h$, that is:

$$F(x_0 + h) - F(x_0) = \int_a^{x_0+h} f(t)dt - \int_a^{x_0} f(t)dt = \int_{x_0}^{x_0+h} f(t)dt$$

If h is small enough, the area of the integral above can be approximated by the area of the rectangle with base h and height $f(x_0)$. Then we have

$$F(x_0 + h) - F(x_0) = \int_{x_0}^{x_0+h} f(t)dt = f(x_0)h + o(h)$$

The last equality authorizes us to state the *Fundamental Theorem of Integral Calculus*, usually attributed to Torricelli[4] and Barrow.[5]

Theorem 3.2.2 *Let* $f : \mathbb{R} \to \mathbb{R}$ *defined on* $[a, b]$. *Assume that* $F(x) = \int_a^x f(t)dt$ *exists for any* $x \in [a, b]$. *If at some* x_0 *the function* f *is continuous, then: (1) F is differentiable at* x_0; *(2) its derivative* F' *at* x_0 *is:*

$$F'(x_0) = f(x_0)$$

Then hopefully you can identify F, by looking at f and asking: "What function F can have f as its derivative?". For example, we want to evaluate

$$\int_0^{10} 2t\,dt$$

We could observe that this is equivalent to calculating the area of a triangle, with vertex points $(0, 0)$, $(10, 0)$, $(10, 20)$ as Fig. 3.3 illustrates. This area is $\frac{10 \cdot 20}{2} = 100$.

---

[4]The Italian guy of atmospheric pression (1608–1647).
[5](1630–1677): known for having had as a student none other than Isaac Newton, who, as a student, tried to eat apples.

Fig. 3.3 $\int_0^{10} 2t\,dt$

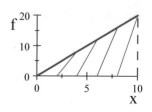

Let's see how you can achieve the same result via the Theorem of Torricelli-Barrow. We begin by noting that the integral function:

$$F(x) = \int_a^x 2t\,dt$$

has derivative $f(x) = 2x$:

$$F'(x) = 2x$$

Then the integral function will look like $F(x) = x^2 + c$, where c is any constant. By choosing $c = 0$ — in fact — we do not determine the integral function that gives the area under the graph of the function relative to the interval $[0, x]$, but the area relating to an interval $[\alpha, x]$, where α is unknown. An interesting observation, however, can be made: from the integral function for the interval $[\alpha, x]$ the integral can be deduced on $[0, 10]$, which is our target. It is sufficient to calculate the increment of x^2 over the interval $[0, 10]$, it coincides with that of $x^2 + c$

$$\left[x^2\right]_0^{10} = \left[10^2 - 0^2\right] = 100$$

This is the same result we found above.

Basically, to evaluate the definite integral:

$$\int_a^b f(x)\,dx$$

the following two steps are appropriate:

• find an *antiderivative* F of f, that is a function F, that admits f as derivative, in symbols:

$$F'(x) = f(x)$$

• and then calculate the increment of F over the integration interval $[a, b]$.

This is summarized in *Barrow's* famous *formula*:

$$\int_a^b f(x)\,dx = F(b) - F(a) \qquad (3.2.2)$$

In light of this, it becomes interesting to be able to use a methodology which finds the antiderivatives F for interesting integrands f. This is the content of the following section.

The search process of an antiderivative of f is called *indefinite integration* of f. We write, in general, in place of $F'(x) = f(x)$, the relation:

$$\int f(x)\,dx = F(x) + c \qquad c \in \mathbb{R}$$

where the integral symbol has no meaning in itself, but merely recalls the use that can be made of F to evaluate definite integrals.

3.3 Antiderivative Calculus

It is not intuitive that the goal can be reached, by reversing — in a certain sense — as seen above in respect of the calculation of derivatives. Having, for example, to integrate a linear combination of functions

$$f(x) = \alpha_1 \cdot f_1(x) + \alpha_2 \cdot f_2(x) + \cdots + \alpha_n \cdot f_n(x)$$

with $\alpha_1, \alpha_2, \ldots, \alpha_n$ real numbers, we would get:

$$\int f(x)\,dx = \alpha_1 \int f_1(x)\,dx + \alpha_2 \int f_2(x)\,dx + \cdots + \alpha_n \int f_n(x)\,dx$$

The table of basic derivatives, previously obtained, can be "inverted" to get the Table 3.1 of the basic antiderivatives. Antiderivatives:

Table 3.1 Basic antiderivatives

$f(x)$	$F(x)$
$x^\alpha,\, \alpha \neq -1,\, x > 0$	$\dfrac{x^{\alpha+1}}{\alpha+1}$
$\dfrac{1}{x}$	$\ln|x|$
$a^x,\, a > 0,\, a \neq 1$	$\dfrac{a^x}{\ln a}$

Example 3.3.1 Determine the set of the antiderivatives of $f(x) = 4x - 4^x + 3/x$ (with $x > 0$). Using the above properties we can write

$$\int \left(4x - 4^x + \frac{3}{x}\right) dx = 4\int x\,dx - \int 4^x\,dx + 3\int x^{-1}\,dx$$

from which, by calculating the integrals of elementary functions, we get:

$$\int \left(4x - 4^x + \frac{3}{x}\right) dx = 2x^2 - 4^x / \ln 4 + 3 \ln x + c \quad c \in \mathbb{R}$$

Two additional formulas, that can be adopted in practice, are known as *integration by parts* and *integration by substitution*. Let's see what these are.

3.3.1   Integration by Parts

Let's consider two differentiable functions f_1, f_2. From the differentiation rule of the product $f_1 f_2$ we have

$$D[f_1(x)f_2(x)] = f_1'(x) \cdot f_2(x) + f_1(x) \cdot f_2'(x)$$

that is

$$f_1(x) \cdot f_2'(x) = D[f_1(x)f_2(x)] - f_1'(x) \cdot f_2(x)$$

and considering the antiderivatives of both members we obtain

$$\int f_1(x)f_2'(x)\, dx = f_1(x)f_2(x) - \int f_1'(x)\, f_2(x) dx$$

known as the *integration by parts* formula.

We call $f_1(x)$ the *finite factor* and $f_2'(x)$ the *differential* one (as $f_2(x)$ is an antiderivative of $f_2'(x)$). This new formula lets us to replace a given integral with another, that could be easier one (making a good choice of a differential factor). The method of integration by parts can also be used more than once.

We suggest what we think the best way to remember this formula:

$$\int f_1(x) d f_2(x) = f_1(x)f_2(x) - \int f_2(x) d f_1(x)$$

or, eliminating unnecessary specifications:

$$\int f_1 d f_2 = f_1 f_2 - \int f_2 d f_1$$

Example 3.3.2 We want to calculate the antiderivatives of $f(x) = xe^x$. Using the method of integration by parts we get:

$$\int x e^x \, dx = \int x \, d\left[e^x\right] = x e^x - \int e^x \, d\left[x\right] = \qquad (3.3.1)$$

$$= x e^x - \int e^x \, dx = x e^x - e^x + c \quad c \in \mathbb{R}$$

In the previous example:

$$f_1(x) = x \text{ and } f_2(x) = e^x$$

A few lines above we have asserted that the formula of integration by parts could reduce the computation of some integral to that of another (hopefully easier) one. We would like to use the preceding example to show that a non-appropriate choice would bring to a more difficult task. Retake (3.3.1) and reverse the choice of the finite and of the differential factors[6]:

$$\int x e^x \, dx = \int e^x \, d\left[\frac{x^2}{2}\right] = \frac{x^2}{2} e^x - \int \frac{x^2}{2} d\left[e^x\right] =$$

$$= \frac{x^2}{2} e^x - \int \frac{x^2}{2} e^x \, dx \quad \longleftarrow \quad \text{more difficult!}$$

3.3.2 Integration by Substitution

Here we consider a special rule, already in a suitable form, for calculating definite integrals. Let's consider

$$\int_a^b f(x) \, dx$$

and suppose that the integration variable x is linked to another one t by the relationship $x = \phi(t)$, invertible with $t = \psi(x)$. Substituting in the definite integral

- $x = \phi(t)$
- $dx = \phi'(t) dt$
- the limits of integration a, b, respectively, with $\psi(a)$, $\psi(b)$,

we get

$$\int_a^b f(x) \, dx = \int_{\psi(a)}^{\psi(b)} f\left[\phi(t)\right] \phi'(t) \, dt$$

Once again we have traced the calculation of an integral with the calculation of another one. If the second is easier than the first one then we have moved forward. We illustrate this with a simple example.

---

[6]In this attempt x is the differential factor, while e^x is chosen as the finite one.

Example 3.3.3 Using the substitution method, let's calculate the following definite integral:

$$\int_0^5 e^{3x}\,dx$$

Setting $3x = t$, we have $x = t/3$ and differentiating both members we obtain $dx = dt/3$. The next step is to change the limits of integration as follows

- if $x = 0$ then $t = 3 \cdot 0 = 0$,
- if $x = 5$ then $t = 3 \cdot 5 = 15$.

Therefore, we can calculate the definite integral as follows

$$\int_0^5 e^{3x}\,dx = \frac{1}{3}\int_0^{15} e^t\,dt = \frac{1}{3}\left[e^t\right]_0^{15} = \frac{e^{15} - 1}{3}.$$

3.4 An Immediate Application: Mean and Expected Values

We have already encountered *statistic statistic variables* in the first chapter. There we have introduced the notions of both *mean value* and of *expected value*.

Concretely speaking statistic/probabilistic notions constitute an interesting application of integrals. We sketch them here, in order to see... integrals at work.

A statistical/random variable is a number which can take different values according to the outcome of some experiment.

Such a notion has been already encountered above, as an application of linear algebra, see p. 58. There statistical/random variables had the possibility to take a value in a discrete (and finite) set $\{x_1, x_2, \ldots, x_n\}$ Here we allow them to take any value in some interval (a, b), with, possibly $a = -\infty$ and $b = +\infty$.

We will work on *continuous statistical/random variables*, whose value can be any of the numbers in an interval (a, b), where a could also be $-\infty$ and/or b could be $+\infty$. In this case the frequency/probability distribution is usually described as follows:

$$\mathbf{X} \sim \begin{cases} a < x < b \leftarrow \text{ possible values} \\[2mm] f(x) \quad \leftarrow \text{ frequency/probability density function} \end{cases} \tag{3.4.1}$$

In the first line of the scheme we identify the set of possible values $\mathbf{X}$ can take, while in the second line we describe how likely $\mathbf{X}$ can take any possible value x. This task is done using $f(x)$, called *frequency/probability density function*, to be interpreted appropriately. First of all, the frequency/probability associated to some fixed possible value is 0. The appropriate perspective is to look at frequency/probability associated to some interval. The frequency/probability that $\mathbf{X}$ will fall in the interval $(x, x + h)$

is $f(x)h + o(h)$. If h is small then $f(x)h$ will be a good approximation of such a frequency/probability.

Let us look at the:

Example 3.4.1 Take this frequency/probability distribution[7]:

$$\mathbf{X} \sim \begin{cases} 0 < x < 1 \leftarrow \text{ possible values} \\ \\ 6(x - x^2) \leftarrow \text{ probability density function} \end{cases} \tag{3.4.2}$$

The frequency/probability that $\mathbf{X}$ takes a value between — say — 0.4 and 0.41, or (the same), in the interval $(0.4, 0.41)$, is approximately:

$$f(0.4) \cdot (0.41 - 0.4) = f(0.4) \cdot 0.01 = 6(0.4 - 0.4^2) \cdot 0.01 = 0.0144 = 1.44\%$$

Now, back to the general case (3.4.1). If we think to divide the interval (a, b) into many small intervals $(x, x + dx)$, we can think of $f(x)dx$ as the frequency/probability that $\mathbf{X}$ will fall in the interval $(x, x + dx)$. If we sum all these small frequencies/probabilities:

$$\int_a^b f(x)\,dx$$

we must get $1 = 100\%$. The condition:

$$\int_a^b f(x)\,dx = 1$$

is usually called *normalization condition*.

We can check whether it is satisfied in the Example (1.8.7). It is sufficient to integrate $f(x)$ over the interval $(0, 1)$ to see that the normalization condition is satisfied:

$$\int_0^1 6(x - x^2)\,dx = \left[6\left(\frac{x^2}{2} - \frac{x^3}{3}\right)\right]_0^1 = 6\left(\frac{1}{2} - \frac{1}{3}\right) = 1$$

For a continuous statistical/random variable $\mathbf{X}$, with frequency/probability distribution (3.4.1) the *expectation* $E[\mathbf{X}]$ is defined as:

$$E[\mathbf{X}] = \int_a^b xf(x)\,dx$$

In the Example (1.8.7) we would have:

---

[7]Well known as Beta distribution. We will meet it later.

$$E[X] = \int_0^1 x \cdot 6\left(x - x^2\right) dx = 6 \int_0^1 \left(x^2 - x^3\right) dx =$$

$$= 6 \left[\frac{x^3}{3} - \frac{x^4}{4}\right]_0^1 = 6 \left(\frac{1}{3} - \frac{1}{4}\right) = 0.5$$

3.4.1 Expectation and the Law of Large Numbers

Likely some readers could ask "Why such an insistence about the notion of expected value? Both in the discrete and in the continuous case?".

The question is absolutely legitimate and deserves an answer (at least a partial one).

Likely many readers will attend courses of Statistics, where they will find the completion of the answer.

Some mathematical results are very useful in Social Sciences and do generate statistical methods widely used.

Think you can repeat some random experiment in conditions you think are always the same. The random experiment generates an observation (a number) X_n for the nth experiment. For instance, the experiment could consist in picking a person in some population and asking the person, whether she/he is is in favor of a certain candidate (nothing but an opinion test). The numerical outcome could be codified as:

$$\begin{cases} 0 \text{ if not in favor} \\ \\ 1 \text{ if in favor} \end{cases}$$

For any X_s the expected value is the same:

$$E[X_s] = m$$

Compute now the (arithmetic) average of the observations:

$$Y_n = \frac{X_1 + X_2 + \cdots + X_n}{n}$$

At first sight this quotient could be appear as very wild: the numerator is the sum of n random addenda.

Well, probability theory tells us good news. If n is sufficiently large, then:

$$Y_n \approx m$$

Socially speaking this means that a sufficiently large sample will tell you "almost precisely" the truth.

This fact passes in the literature as the simplest example of the Law of Large Numbers.

Look at the following:

Example 3.4.2 This example will show concretely how the Law of Large Numbers provides us with the concrete interpretation of expected value. The AA. have run three experiments, with distribution the one of Eq. 3.4.2, with number of observations: $n = 10, 100, 1000, 10000, 100000$.

n	Y_n
10	0.603
100	0.463
1000	0.504
10000	0.499
100000	0.499

In the case of $n = 1000$ we observe a good approximation around the expectation $E(X) = 0.5$.

3.4.2 Density Function and Distribution Function

A crucial notion in Statistics and Probability is the one of *distribution function*. Its importance has essentially theoretical value as it allows for a unified treatment of discrete and continuous statistical/random variables.

Take a continuous statistical/random variable **X**. Assume its frequency or probability distribution respectively is described by the *density function* $f(x)$:

$$X \sim \begin{cases} -\infty < x < +\infty & \leftarrow \text{ possible values} \\ f(x) & \leftarrow \text{ density function} \end{cases}$$

We call *distribution function* of **X**:

$$F(x) = P[X \leq x]$$

Its concrete interpretation is that of percentage/probability in which **X** does not exceed x.

Here are two interpretations of $F(x)$.

Example 3.4.3 The possible values of **X** are between 0 and 1 and represent the possible political positions of individuals between 0 (extreme left wing) and 1 (extreme right wing). Assume that opinions are described by some density function $f(x)$:

$$X \sim \begin{cases} -\infty < x < +\infty & \leftarrow \text{ possible values} \\ \\ f(x) & \leftarrow \text{ density function} \end{cases}$$

Of course $f(x) \equiv 0$ for x non-belonging to $[0, 1]$. Choose any threshold x and ask yourselves "Which percentage of voters are not over the threshold x?". The answer is:

$$F(x) = \int_{-\infty}^{x} f(t)\,dt$$

Example 3.4.4 Assume that we are dealing with public investments. The investment project has a delivery date, but, as often occurs, the delivery date has a delay:

$$X \sim \begin{cases} -\infty < x < +\infty & \leftarrow \text{ possible values} \\ \\ f(x) & \leftarrow \text{ density function} \end{cases}$$

The AA. are Italian and consciously assign zero probability to negative delays[8]:

$$X \sim \begin{cases} -\infty < x < +\infty & \leftarrow \text{ possible values} \\ \\ 0 \text{ if } x \leq 0, \ \alpha e^{-\alpha x} \text{ if } x > 0 & \leftarrow \text{ density function} \end{cases}$$

The expected delay is:

$$E[X] = \int_{0}^{+\infty} x\alpha e^{-\alpha x} dx = \frac{1}{\alpha}$$

The distribution function is:

$$F(x) = \begin{cases} 0 & \text{for } x \leq 0 \\ \\ \int_{-\infty}^{x} f(t)\,dt = \int_{0}^{x} \alpha e^{-\alpha t} dt = 1 - e^{\alpha x} & \text{if } x > 0 \end{cases}$$

practical implication is evident. The expected delay in the completion of some project is 1 month. This means that $\alpha = 1$. The probability of a delay over the expected value is therefore:

$$\int_{1}^{+\infty} e^{-x} dt = e^{-1} \approx 36.8\%$$

May be an intriguing result for political science researchers. The Fig. 3.4 shows the graph of distribution function for **X**:

---

[8]The choice of an exponential density has a strong empirical background support. Its parameter α is obviously positive.

Fig. 3.4 Distribution of **X**

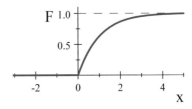

3.4.3   Discrete Distributions

We have seen above the case of discrete distributions we have met before. In that case there is no density[9], but the distribution function shines:

$$F(x) = P(\mathbf{X} \le x)$$

it is well defined.

While in the case of continuous distributions, the distribution function grows continuously, in this case of discrete distributions it jumps[10] at the points $\{x_1, x_2, \ldots, x_n\}$. An example should help.

Example 3.4.5 Think of a country where castes do prevail. Even if, for instance in India, they have been forbidden, their presence continues to be politically relevant. The castes are 1, 2, 3, 4, the population distribution among them is:

$$\mathbf{X} \sim \begin{cases} 1 \quad 2 \quad 3 \quad 4 \quad \leftarrow \text{ castes in growing order} \\[2mm] 0.4\ 0.3\ 0.2\ 0.1 \ \leftarrow \text{ caste frequency} \end{cases}$$

The variable x ranges from $-\infty$ to $+\infty$ and accumulates frequencies. The Fig. 3.5 shows that distribution function $F(x)$ jumps at 1, 2, 3, 4. The amount of each jump is equal to the frequency concentrated at that jump-point.

Fig. 3.5 Caste distribution

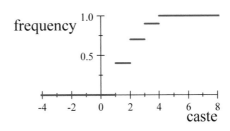

---

[9]True for normal beings, not for mathematicians, who are able to do that. We love our readers and we let them avoid this question.

[10]It's a good example of discontinuous functions, we met above.

3.5 Frequency/Probability Density Functions: Some Cases

The basic idea is that for some variable there is a range of possible values and that the description of the frequency/probability distribution on the set of possible values is defined through a function $f(x)$, we will call (frequency/probability) *density function* (sometimes *pdf* for short).

Let's start with a couple of examples.

Example 3.5.1 We are considering some electoral campaign. We do not know the random percentage **X** of the electors, who will vote some party. We have infos coming from polls, experts, etc. We want to describe our opinions about **X**. We call x any possible percentage. A reasonable description[11] could be provided by a function: $f : \mathbb{R} \to \mathbb{R}$ of the type:

$$f(x) = \begin{cases} 0 \text{ if } x < 0 \\ kx^{p-1}(1-x)^{q-1} & \text{where } p, q > 0; \\ 0 \text{ if } x > 1 \end{cases}$$

and $k = k(p, q) > 0$ is a constant such that

$$\int_{-\infty}^{+\infty} f(x)\, dx = \int_0^1 kx^{p-1}(1-x)^{q-1}\, dx = 1$$

The constant k is the reciprocal of a *special function B*, called "Beta function":

$$B(p, q) = \int_0^1 x^{p-1}(1-x)^{q-1}\, dx$$

Well, B is tabled and easily computable, and $k(p, q) = 1/B(p, q)$.

Here are some diagrams describing the shape of such density function for various values of the parameters. This distribution is named "Beta distribution with parameters p, q". The case of polarized preferences is well described by $p, q \in (0, 1)$. For instance, with $p = q = 0.9$ we would get the distribution plotted in Fig. 3.6a. With $p = q = 2$ we would characterize the situation in which a few people take extreme positions (Fig. 3.6b) whereas a more realistic case would be $p = 4, q = 7$ (Fig. 3.6c).

---

[11]Inherited from engineering-organization modeling.

Fig. 3.6 Beta distributions

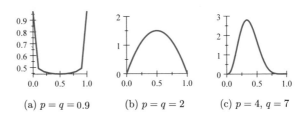

(a) $p = q = 0.9$ (b) $p = q = 2$ (c) $p = 4,\ q = 7$

If we want to compute the frequency/probability over some interval (a, b), we can compute:

$$\int_a^b f(x)\,dx$$

For instance, the frequency/probability over $(0.1, 0.3)$ is:

$$\frac{\int_{0.1}^{0.3} x^3 (1-x)^6\,dx}{\int_0^1 x^3 (1-x)^6\,dx} = 33.759\%$$

3.5.1 A Special Distribution

The importance of this distribution is a wonderful example of interactions between Mathematics and Politics.

Let's start from its name: in Germany it's *Gauss distribution*, in France *Laplace distribution*, in general in Europe, two names are used: "*Laplace–Gauss*" on the axis Paris-Berlin, but the general label, used almost all-over the world, is "*Gaussian*" or "*Normal*". The standard international way to describe this distribution is $N\,(\cdot, \cdot)$ the blanks will be filled quickly. For people with a physicist imprinting, the corresponding familiar vocabulary would suggest "*Law of errors*".

Politics tells us that when there are many people who pretend some object, it must have some value.

Where stays the reason of this importance?

We try to explain our intuitive point very simply.

Think of repeating some random experiment: the outcome of the experiment #*n* is $\mathbf{X}_n$. Consider the first two independent[12] trials and consider their sum:

$$\mathbf{Y}_2 = \mathbf{X}_1 + \mathbf{X}_2$$

The probability distribution of $\mathbf{Y}_2$ turns out to depend on the probability distributions of both $\mathbf{X}_1$ and $\mathbf{X}_2$.

Let's now consider the sum of n terms:

---

[12]Independent means that the outcome of a trial has no influence on the probability/frequency distribution of the other.

$$\mathbf{Y}_n = \mathbf{X}_1 + \mathbf{X}_2 + \cdots + \mathbf{X}_n$$

If the number of the $\mathbf{X}_s$ addenda is sufficiently large say $n = 50$, then, under broad conditions, the density of $\mathbf{Y}_n$ can be well approximated with the normal density function $N\left(m, \sigma^2\right)$ whose espression is:

$$f(x) = \frac{1}{\sigma\sqrt{2\pi}} e^{-\dfrac{(x-m)^2}{2\sigma^2}}$$

The normal density plot is bell–shaped. The Fig. 3.7 shows the case $m = 0$ and $\sigma = 1$.

Fig. 3.7 Normal density
$N(0, 1)$

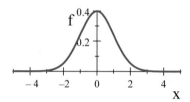

The corresponding frequency/probability (cumulative) distribution function is:

$$F(x) = \int_{-\infty}^{x} f(t)\, dt$$

In the case $m = 0$ and $\sigma = 1$, its graph is represented in Fig. 3.6.

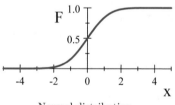

Normal distribution

This fact stems from a series of theorems, usually labelled as *Central Limit Theorems*.

Let's look at another interesting application of continuous distributions.

Example 3.5.2 **Income distribution, Vilfredo Pareto (1848–1923), Engineer, Economist and Sociologist.**—The income distribution is a crucial issue for economics, politics and sociology. There are more poor people than rich: almost obvious! But how? The historical model proposed by Vilfredo Pareto, in order to describe the income distribution turns out to work very well with empirical data also today. Pareto frequency density function is

$$p(x) = \frac{A}{(x+a)^\alpha}$$

We assume that the three parameters entering the expression of p are:

$$A = 1000000, \qquad a = 100, \qquad \alpha = 3$$

The total frequency implied is

$$\int_0^{+\infty} p(x)\mathrm{d}x = 50$$

Its normalized version (where observed frequencies are percentages) is

$$f(x) = \frac{1}{50}p(x)$$

The average income J is

$$J = \int_0^{+\infty} xf(x)\mathrm{d}x = 100$$

Consider now some income interval, say from 0 to 100. The fraction of people with income between 0 and 100 is

$$\int_0^{100} f(x)\mathrm{d}x = 0.75 \qquad (3.5.1)$$

If you change the value of α, the so-called Pareto index, you can see that as α increases, the fraction of people with income under some sufficient threshold turns out to increase. The estimation of the parameter α has been one of the earliest tools to measure possible tensions in a society: too high values of α can produce social instability. For instance, letting $\alpha = 4$, instead of 3, the fraction of population computed with (3.5.1) would increase to 0.875.

We sketch now how concretely the parameter α could be empirically estimated. We start from the distribution function:

$$F(x) = 1 - \frac{a^{\alpha-1}}{(a+x)^{\alpha-1}}$$

It tells us which % of the population has an income not exceeding x. Its complement to 1:

$$\Phi(x) = 1 - F(x)$$

It tells us the % of the population with income greater than x. It is interesting to use this function for the empirical estimation of the parameters a, α which characterize some income distribution. The analytic expression of the cumulative distribution function is

$$\Phi(x) = \frac{a^{\alpha-1}}{(a+x)^{\alpha-1}}$$

Taking logs:

$$\ln \Phi(x) = (\alpha - 1) \ln a - (\alpha - 1) \ln(a + x)$$

hence, starting from empirical data about the percentage of people with income beyond various thresholds, the method of Least Squares (see Example 15) could be fruitfully applied to evaluate both α and a.

3.6   People Survival

Important economic and social problems turn out to depend on survival of individuals. Life insurance and Pensions systems are bright examples.

Historically the Actuarial profession handles these problems. We will show the basics as a crucial application of separable differential equations.

People are born and die, but when? About the birthday of any author and any reader, no choice. It's given. About death, again, almost always no choice, but several possibilities.

In life insurance and in pension systems a key tool is represented by the *survival function*. We have some population of — say — 100, 000 (initially probably crying) newborns. Their age is 0. We would like to know how many of them $l(x)$ are expected[13] to survive at the age $x \geq 0$.

A naïve approach to the problem of estimating $l(x)$ would be that of observing for around one century the "crying population" in order to draw obsolete conclusions of the type: "this was true for our grandfathers/grandmothers, but today the world is completely different: hygiene, health-care, food,...". Fatal KD objection.

This remark suggests us that, for studying such long-term phenomena, which are crucial for Social and Political Sciences, a way-out could be to start from a locally observable behavior, in order to derive a global behavior from it.

Maths tells us that this is always possible numerically,[14] if we have full information, otherwise we can try to use the "old" analytic approach. In this field this bet is fully successful.

What Statistics is able to approximately provide us with is "the probability $q(x, h)$ that an individual of age x will die between the ages x and $x + h$.". The problem consists in finding a way to recover $l(x)$ from these "local" probs.

We can represent easily this terrible probability as:

---

[13] As we are working with large populations, the law of large numbers supports our rather naïve approach to the problem, thinking of it as if it was a deterministic one.

[14] As we will see later, the price to pay for that is far from being negligible.

$$q\left(x,h\right) = \frac{l\left(x\right) - l\left(x+h\right)}{l\left(x\right)} = -\frac{l\left(x+h\right) - l\left(x\right)}{l\left(x\right)}$$

If we assume that l is differentiable at x, we can write:

$$-\frac{l\left(x+h\right) - l\left(x\right)}{l\left(x\right)} = -\frac{l'\left(x\right)}{l\left(x\right)}h + o\left(h\right)$$

Dividing by h we get:

$$-\frac{l'\left(x\right)}{l\left(x\right)} + \frac{o\left(h\right)}{h}$$

which brings us to the *mortality rate*:

$$\mu\left(x\right) := -\frac{l'\left(x\right)}{l\left(x\right)} \tag{3.6.1}$$

This means that, for instance:

$$\mu\left(x\right)\frac{1}{365} = -\frac{l'\left(x\right)}{l\left(x\right)}\frac{1}{365} \tag{3.6.2}$$

approximates the probability for an individual of age x to die within 24 h.

Well, statistical evidence allows us to estimate $\mu\left(x\right)$, our problem consists in recovering $l\left(x\right)$.

Starting from (3.6.1) we get:

$$\frac{l'\left(x\right)}{l\left(x\right)} = -\mu\left(x\right)$$

Hence, given $l\left(0\right) = L$:

$$l\left(x\right) = L \cdot e^{-\int_0^x \mu(s)ds}$$

But, what's a credible functional form for $\mu\left(x\right)$? We present two models, respectively due to B. Gompertz (1779–1865) and W.M. Makeham (1826–1891). The first one works well with plants, cars, tumors, while the second one performs better with human beings. An important feature of these results is that the mechanics is the same: what changes across populations is the value of parameters, but not the functional form.[15]

Gompertz idea is that $\mu\left(x\right)$ increases exponentially with the age x.

This collapses into the equation:

---

[15]This point is of interest. In Nature sciences a common fact is that models do not depend on where you are, because they are universal in some sense. What can be observed in Social sciences is that some mathematical models do hold in general, but with location dependent parameters.

$$\frac{\mu'(x)}{\mu(x)} = \gamma > 0$$

Its solution is simply offensive:

$$D[\ln \mu(x)] = \gamma \Rightarrow \ln \mu(x) = \gamma x + c, c \text{ arbitrary}$$

hence:

$$\mu(x) = b \cdot e^{\gamma x} \text{ with } b = e^c > 0$$

From (3.6.1) we get:

$$l(x) = ke^{-\frac{b}{\gamma}e^{\gamma x}} = kg^{c^x} \tag{3.6.3}$$

where:

- k is an irrelevant positive scale constant;
- $g = e^{-\frac{b}{\gamma}}$ is the basis of some exponential function ($0 < g < 1$);
- $c > 1$,

With $g = 0.97$, $k = 100000$, $c : 1.05$ we get the Gompertz survival function

$$l(x) = 100000 \times 0.97^{1.05^x}$$

The Fig. 3.8 illustrate the behavior of $l(x)$.

The innovation proposed by Makeham was based on the idea that the mortality rate cannot depend only on the age, but must take into account that (unfortunately) there are death reasons independent of age: terrorism, transportation accidents, fires...

To model this Makeham proposes that:

$$\mu(x) = be^{\gamma x} + a$$

with $a > 0$ representing the accidental component of the mortality rate.

Fig. 3.8 Gompertz

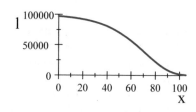

Computations quite similar to the ones that have brought us to (3.6.3), it is possible to reach a new survival function:

$$l(x) = g^{c^x} s^x$$

where $s = e^{-m}$ is a number between 0 and 1.

The Fig. 3.9 contains the "old" Gompertz survival function and the "new" Makeham model: we have assumed $m = 0.5\%$.

Fig. 3.9 Gompetz and Makeham

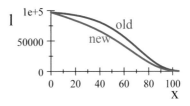

3.7 Exercises

Exercise 3.7.1 Show, using the antiderivative definition, that the following formulas

$$\text{(a)} \int (1 + x) e^x dx = xe^x + c \qquad \text{(b)} \int \frac{2x}{x^2 + 1} dx = \ln (x^2 + 1) + c$$

hold for any real number x (with $c \in \mathbb{R}$).

Exercise 3.7.2 Consider the function $f : (0, +\infty) \to \mathbb{R}$ defined by $f(x) = 1/x^2$.

(a) Calculate the set of antiderivatives G of f on $(0, +\infty)$.
(b) Find the antiderivative G^* of f on $(0, +\infty)$ passing through the point $(1/2, 1)$.

Exercise 3.7.3 Using a substitution of the type $ax + b = t$, with a, b real numbers, calculate the following definite integrals:

$$\text{(a)} \int_2^{e+1/2} \frac{1}{2x - 1} dx \quad \text{(b)} \int_{-1}^3 \left(\frac{1}{2x + 3} \right)^3 dx \quad \text{(c)} \int_0^2 \frac{4}{2x + 1} dx$$

$$\text{(d)} \int_{-2}^3 e^{3-2x} dx \qquad \text{(e)} \int_0^1 \frac{1}{\sqrt{5x + 4}} dx \qquad \text{(f)} \int_0^2 (2x - 1)^3 dx$$

Exercise 3.7.4 Consider the function $f : (0, +\infty) \to \mathbb{R}$ defined by

$$f(x) = \frac{1}{x} \sqrt[3]{\ln x}$$

(a) Using the substitution $\sqrt[3]{\ln x} = t$, find the set of all the antiderivatives G for f on the interval $(0, +\infty)$.

(b)   Find the antiderivative G^* of f on $(0, +\infty)$ whose graph passes through the point $(e, 2)$.

Exercise 3.7.5 Using the method of integration by parts, find the set of all the antiderivatives G for the functions $f : \mathbb{R} \to \mathbb{R}$ defined by:

$$
\begin{aligned}
&(a)\ f(x) = \ln(2x) \quad &(b)\ f(x) = \ln\sqrt{x} \quad &(c)\ f(x) = (\ln x)^2 \\
&(d)\ f(x) = \ln x^2 \quad &(e)\ f(x) = x\ln x \quad &(f)\ f(x) = xe^x
\end{aligned}
$$

where $A = (0, +\infty)$ from (a) to (e) and $A = \mathbb{R}$ for (f).

Exercise 3.7.6 Consider the function $f : (2, +\infty) \to \mathbb{R}$ defined by

$$f(x) = \frac{2x + 3}{(x - 2)^2}$$

(a)   Using the substitution $x - 2 = t$ $(t > 0)$, verify that

$$\int \frac{2x + 3}{(x - 2)^2}\,dx = 2\int \frac{1}{t}\,dt + 7\int \frac{1}{t^2}\,dt$$

(b)   Calculate the set of all antiderivatives G of f on $(2, +\infty)$ and verify that the result found is right.

(c)   Find the antiderivative G_1 of f on $(2, +\infty)$ whose graph passes through the point $(3, 1)$.

Exercise 3.7.7 Using the substitutions indicated, determine, the set of the antiderivatives G for the functions $f : A \subseteq \mathbb{R} \to \mathbb{R}$ defined by:

$$
\begin{aligned}
&(a)\ f(x) = \frac{1}{\sqrt{x} + 1}, \left[\sqrt{x} + 1 = t\right] \quad &(b)\ f(x) = \frac{1}{\sqrt{x}\left(\sqrt{x} + 1\right)}, \left[\sqrt{x} = t\right] \\
&(c)\ f(x) = e^{\sqrt{x}}, \left[\sqrt{x} = t\right] \quad &(d)\ f(x) = x\sqrt{x - 1}, \left[\sqrt{x - 1} = t\right]
\end{aligned}
$$

Exercise 3.7.8 Calculate the following definite integrals:

$$
\begin{aligned}
&(a)\ \int_1^2 3x\,dx \quad &(b)\ \int_1^4 \frac{\sqrt{x}}{3}\,dx \quad &(c)\ \int_{-1}^5 (2 - 3x)\,dx \\
&(d)\ \int_1^e \frac{x + 1}{x^2}\,dx \quad &(e)\ \int_{-1}^2 \left(2x - 3x^2\right)\,dx \quad &(f)\ \int_0^1 \frac{x - 1}{x + 2}\,dx
\end{aligned}
$$

Exercise 3.7.9 Calculate the following definite integrals:

$$
(a)\ \int_0^1 \left(e^{x/2} + e^{-x/2}\right)^2\,dx \quad (b)\ \int_1^e (\ln x)^2\,dx \quad (c)\ \int_{-1}^6 \left(x + \sqrt[3]{x + 2}\right)\,dx
$$

Exercise 3.7.10 Consider the function $f : \mathbb{R} \to \mathbb{R}$ defined by

$$f(x) = \frac{e^x}{e^{2x} + 3e^x + 2}$$

(a) Using the substitution $e^x = t$, calculate the set (or family) G, of antiderivatives of f on $\mathbb{R}$.
(b) Calculate the definite integral of f on $[-\ln 2, \ln 4]$.

Exercise 3.7.11 Consider the function $f : \mathbb{R} \to \mathbb{R}$ defined by

$$f(x) = (2x - 1)\, e^{2x-1}$$

(a) Using the following substitution $2x - 1 = t$, calculate the set (or family) of antiderivatives G of f on $\mathbb{R}$.
(b) Calculate the definite integral of f on $[-2, 2]$.
(c) Can you interprete the value found at (b) as an area? Justify your answer.

Exercise 3.7.12 A probability density function on the interval $[0, 5]$ has expression

$$f(x) = Ae^{-\frac{1}{5}x}$$

with $A > 0$.

(a) Compute A, so that the normalization condition is satisfied (the unitary area under the graph of f between 0 and 5).
(b) Compute the probability p concentrated on the interval $[0, 1]$.
(c) Compute the cumulative distribution function $F(x)$ over the interval $[0, 5]$.
(d) Integrating by parts, compute the expectation E.

Chapter 4
Dynamic Systems

4.1 Introduction

Dynamic Systems theory is a general mathematical tool that has been successfully used both in hard and in soft Sciences.

We will frequently denote them for short with DS:

DS stands for **Dynamic(al) Systems**

This chapter is directed to readers, interested in Social Sciences, with the aim to show that, even with relatively little mathematics, they could be able to construct possibly useful models in analyzing socio-political problems.

The set of cases we will consider is far from being exhausting, but it can provide a rough idea of how mathematics can help in studying this field.

A *dynamic system* is… a dynamic system.

Please, don't stop reading immediately.

The word **system** suggests that we are interested in studying some phenomena that could require several numbers to be described. Think, for instance, of the number of voters that four political parties p_1, p_2, p_3, p_4 are described by the following table:

Party	# of voters
p_1	120000
p_2	70000
p_3	85000
p_4	1200

Keeping fixed the listing order of the four parties, we could think to represent the voter distribution, at some given date, with the 4-vector:

© Springer Nature Switzerland AG 2018
L. Peccati et al., *Maths for Social Sciences*, UNITEXT - La Matematica
per il 3 + 2 113, https://doi.org/10.1007/978-3-030-02336-2_4

$$\mathbf{x} = \begin{bmatrix} 120000 \\ 70000 \\ 85000 \\ 1200 \end{bmatrix}$$

The adjective *dynamic* suggests that the system can evolve over time. For instance, the table presented above could describe the voters distribution today, at the date 0.

But we could be interested in the analogous situation 1 year later. We could write:

$$\mathbf{x}(0) = \begin{bmatrix} 120000 \\ 70000 \\ 85000 \\ 1200 \end{bmatrix}$$

to date that vector.

Perhaps 1 year later the new situation could be described by the vector[1]:

$$\mathbf{x}(1) = \begin{bmatrix} 100000 \\ 90000 \\ 80000 \\ 7200 \end{bmatrix}$$

Definition 4.1.1 This example suggests us that we can define a *dynamic system* as a *vector* $\mathbf{x}(t)$ depending on time t. Such a vector will be called[2] *state vector*. The number of its components will be called *dimension* of the DS.

In the introductory example, the dimension was $4 = \#$ of parties.

Remark 4.1.1 As scalars are special cases of vectors, we will continue to use *dynamic system* even if we work with a unique variable $x(t)$ depending on time. It will be called *state variable*. In this case the system dimension will be 1.

In dynamic systems, time can be modeled in two different ways.

It can be thought of as a *discrete* variable $t = 0, 1, 2, \ldots$ (today, tomorrow, the day after tomorrow, ...) or as a *continuous* variable $0 \leq t < +\infty$, which is not confined to take only integer values, but which can take any real value in a certain interval.

Definition 4.1.2 When time is discrete, we talk about *discrete systems*, while the label *continuous systems* is reserved to the ones in which t fills an interval.

Especially in the case of discrete systems, the notation $\mathbf{x}_t$ instead of $\mathbf{x}(t)$ enjoys some popularity. We don't like it because we prefer to use the index to denote the components of a vector, but that's life!

---

[1]Not necessarily the total number of voters will be the same.

[2]The word "state" has no political implication: it simply refers to the position (*state*) of the system.

The dynamics of a system can be represented naturally in Cartesian planes: always time on the horizontal axis, while we can use the vertical one for any state variable.

First let's consider a continuous system, describing the evolution of some population whose size increases by 10% per year.

Its evolution in continuous time is trivially given by (see, later, on p. 224):

$$x(t) = x(0) \cdot 1.1^t$$

Assume that its initial size is:

$$x(0) = 10000$$

The Fig. 4.1a describes its evolution.

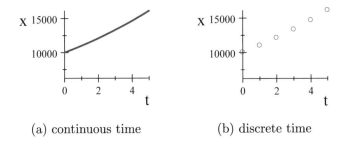

(a) continuous time   (b) discrete time

Fig. 4.1 Dynamical system evolution

In the discrete case, we will have a set of "points" representing the pictures of the system we take at 0, 1, 2, ... as the Fig. 4.1b shows.

If we know how $x(t)$, the state of the system at t, turns out to depend on t, this topic would be far from being exciting.

In reality, since the first appearance of DS in the 17th Century, often the only available information about them was not $x(t)$, but some local[3] evolution rule.

Substantially: we do not know $x(t)$, but we have information about how $x(t)$ evolves in time.

What we will see is that:

- in some (very) special cases we can obtain $x(t)$ using the information we have about its evolution rules;
- in general the reconstruction of $x(t)$ from its evolution rule is truly an impossible mission;
- in some relevant cases, the information about the evolution rule of the system will allow us to detect important features of its long-term behavior, even if $x(t)$ cannot be found.

---

[3]In continuous time, "local" is referred to what happens at epochs close to the current date t. In discrete time, for instance, which relationship must hold between $x(t)$ and $x(t+1)$. In continuous-time systems what counts is what happens between t and $t+h$, with h small.

4.2   Local Information: The Motion Law

Definition 4.2.1 We call *motion law* the set of rules that locally determines the evolution of a DS.

Later we will specify the structure of these rules.
We prefer to start from some simple examples.

4.2.1   Discrete Time

Example 4.2.1 A population, whose size is $x(t)$, is growing with an annual (constant) growth rate g:

$$\text{growth rate} = \frac{\text{population next year} - \text{population today}}{\text{population today}}$$

or, in formulas:

$$g = \frac{x(t+1) - x(t)}{x(t)}$$

this implies:

$$x(t+1) = (1+g)x(t)$$

If the growth rate varies (exogenously) over time, we would obtain:

$$x(t+1) = [1 + g(t)]x(t)$$

but we can consider more complex cases, in which, for instance, the growth rate also depends on the size itself of the population. We would find:

$$x(t+1) = [1 + g[x(t), t]]x(t)$$

Letting $x(0) = 10000$ and $g = 0.1$, we get the starting example above.

Example 4.2.2 Another very popular example, with several applications in Political Science too, is known as the *logistic model*. We can think of it as a model capturing two key features[4] of population growth:
(1) — The current size of the population pushes up its growth rate;
(2) — The closeness of the population size to its natural upper bound M (for food, space, ...) brakes its growth.
In formulas:

$$x(t+1) = x(t) + ax(t)[M - x(t)]$$

---

[4]They stand at the basis of the concrete current problem of immigration.

where $a > 0$ is a parameter. On the right-hand side (2nd addendum), apart from the modulating constant a, the first factor is the size of the population, while the second is the difference between the saturation level and the actual size at t. If, at some epoch $x(t)$ goes beyond M, then the second addendum will be negative and will bring the population size back down.

Example 4.2.3 This model is known as the *Colin Clark three sectors model*. The evolution of labor in an economic system is under scrutiny. This is clearly a relevant theme for politics. The economic system consists of three sectors (hence its name): (1) Agriculture, (2) Manufacturing, (3) Services. The model studies how the labor force, employed in these sectors, does evolve over time, potentially passing from any sector to any other. A transition matrix T is estimated:

$$T = \begin{bmatrix} 0.7 & 0.1 & 0.05 \\ 0.2 & 0.8 & 0.05 \\ 0.1 & 0.1 & 0.90 \end{bmatrix}$$

The sum of the elements in each column is 1. Its columns do inform us about the destiny at the date $t + 1$ of the workers distributed in the three sectors at t. Take, for instance, the first column, labeled "Agriculture column". It tells us that only $70\% = 0.7$ of people that at t were employed in the sector #1 (Agriculture), will remain in that sector, while 20% will migrate to Manufacturing and the remaining 10% will move directly to Services. An analogous interpretation is obvious for the other two columns. We maintain that the evolution of the occupation in the three sectors is governed by the motion law:

$$\mathbf{x}(t + 1) = T\mathbf{x}(t) \tag{4.2.1}$$

being

$$\mathbf{x}(t) = \begin{bmatrix} x_1(t) \\ x_2(t) \\ x_3(t) \end{bmatrix}$$

the state vector collecting the numbers of workers respectively employed in the three sectors at t. We check the correctness of this guess for Agriculture, but an analogous argument could be provided for the two other sectors too. If we write explicitly the Eq. (4.2.1), we get:

$$\begin{bmatrix} x_1(t + 1) \\ x_2(t + 1) \\ x_3(t + 1) \end{bmatrix} = \begin{bmatrix} 0.7 & 0.1 & 0.05 \\ 0.2 & 0.8 & 0.05 \\ 0.1 & 0.1 & 0.9 \end{bmatrix} \begin{bmatrix} x_1(t) \\ x_2(t) \\ x_3(t) \end{bmatrix}$$

Let us focus on $x_1(t + 1)$. The story we are told is that at $t + 1$, the total number of workers in the Agriculture sector is:

$$x_1 (t + 1) = 0.7x_1 (t) + 0.1x_2 (t) + 0.05x_3 (t)$$

which sounds very good. The first addendum $(0.7x_1 (t))$ simply counts those who were in the first sector at t and that did not move to another sector. The second addendum $(0.1x_2 (t))$ counts those that leave a factory to go to Agriculture, while the last one counts those who leave a bank to go and work in the countryside. An analogous interpretation can be provided for the rest of the vector $\mathbf{x} (t + 1)$.

4.2.2 Continuous Time

Example 4.2.4 A population grows with constant growth rate γ, instant after instant, over time. Its size $x (t)$ evolves according to this simple scheme:

$$x (t + h) = x (t) + \gamma x (t) h + o (h)$$

hence:

$$\frac{x (t + h) - x (t)}{h} = \gamma x (t) + \frac{o (h)}{h}$$

if we let h go to 0

$$\frac{x (t + h) - x (t)}{h} = \gamma x (t) + \frac{o (h)}{h}$$
$$\downarrow \qquad\qquad \downarrow \qquad\qquad \downarrow$$
$$x' (t) \qquad = \gamma x (t) + \qquad 0$$

The right-hand side approaches indefinitely $\gamma x (t)$ if h goes to 0. The left-hand side, being equal to the right one must approach[5] the *derivative* of x at t and therefore we have discovered that the population size moves according to the law:

$$x' (t) = \gamma x (t) \tag{4.2.2}$$

Example 4.2.5 It is easy to imagine that if we remove the simplifying assumption that the growth rate γ is constant, and more realistically we assume, for instance, that it depends both on the population size and on time:

---

[5]Assume that $x (t)$ is differentiable:

$$x (t + h) - x (t) = x' (t) h + o (h)$$

Divide both sides by h:

$$\frac{x (t + h) - x (t)}{h} = \frac{x' (t) h + o (h)}{h}$$

hence:

$$\frac{x (t + h) - x (t)}{h} = x' (t) + \frac{o (h)}{h}$$

When h approaches 0 the l.h.s. approaches $x' (t)$.

$$\gamma = \gamma\left[x\left(t\right), t\right] \tag{4.2.3}$$

then the motion law becomes:

$$x'\left(t\right) = \gamma\left[x\left(t\right), t\right] x\left(t\right)$$

something more realistic then (4.2.2).

Example 4.2.6 We will study some properties of an important growth model, due to Robert M. Solow [11]. An economic system evolves over time. The capital stock $K\left(t\right)$ and the number of workers $L\left(t\right)$ determine the amount of goods $Y\left(t\right) h$ the system produces between t and $t + h$:

$$Y\left(t\right) h = F\left[K\left(t\right), L\left(t\right)\right] h$$

The function F, transforming the volumes K, L of "production factors" into the product volume Y is called *production function*. There are serious reasons in Economic Theory about the assumption[6] that the product per capita:

$$\frac{Y\left(t\right)}{L\left(t\right)} h = \frac{F\left[K\left(t\right), L\left(t\right)\right]}{L\left(t\right)} h = f\left[k\left(t\right)\right] h$$

where

$$k\left(t\right) = \frac{K\left(t\right)}{L\left(t\right)}$$

is the stock of capital per capita. A part of the product (in percentage s) is saved and used to provide with appropriate capital new workers (that grow at the instantaneous rate g). It is not difficult to prove that, under these assumptions, the evolution of k is driven by the motion law:

$$k'\left(t\right) = sf\left[k\left(t\right)\right] - gk\left(t\right)$$

---

[6]Technically we assume that $F\left(K, L\right)$ is a *homogeneous function* of the first degree:

$$F\left(aK, aL\right) = aF\left(K, L\right); \qquad a > 0$$

Roughly speaking this means that if we increase by — say — 10% the amount of each production factor then the product amount turns out to be increased by *exactly* 10%. For instance, a popular example of a homogeneous function of the first degree:

$$F\left(K, L\right) = b\sqrt{KL}$$

In fact:

$$F\left(aK, aL\right) = b\sqrt{aKaL} = ab\sqrt{KL}$$

It is a special case of what the reader will learn in Economics is indicated in Economics as the Cobb–Douglas production function.

Nothing prohibits us assuming that s, g could vary over time, and the motion law would become:

$$k'(t) = s(t) f[k(t)] - g(t) k(t)$$

A particular interesting case (also for politics), we will consider later, is that of the dependence of g on k itself. In this case we would have:

$$k'(t) = s(t) f[k(t)] - g[k(t), t] k(t)$$

For an example, see below on p. 320.

4.2.3 Motion Law of a DS

After this battery of examples, it is natural to recognize common features, which bring us to define the motion law.

Definition 4.2.2 We call *motion law for a discrete DS* an equation providing us with the position of the DS at time $t + 1$, denoted with $\mathbf{x}(t + 1)$, as a function of $\mathbf{x}(t)$, the position at t, and (possibly) of time t itself:

$$\mathbf{x}(t + 1) = \mathbf{f}[\mathbf{x}(t), t]$$

Definition 4.2.3 We call *motion law for a continuous-time DS* an equation providing us with the evolution direction of the DS at time t, denoted with[7]

$$\mathbf{x}'(t) = \begin{bmatrix} x_1'(t) \\ x_2'(t) \\ \dots \\ x_n'(t) \end{bmatrix}$$

as a function of $\mathbf{x}(t)$, the position at t, and (possibly) of the time t itself

$$\mathbf{x}'(t) = \mathbf{f}[\mathbf{x}(t), t]$$

Example 4.2.7 This is a motion law for a discrete system of dimension 2:

$$\begin{cases} x_1(t + 1) = 3tx_1(t) - x_2(t) + t \\ x_2(t + 1) = 4x_1(t) - t^2 x_2(t) + e^t \end{cases}$$

The vector "tomorrow"

---

[7]It is nothing but the vector of the time derivatives of the state variables.

$$\begin{bmatrix} x_1 (t + 1) \\ x_2 (t + 1) \end{bmatrix} = \mathbf{x} (t + 1)$$

depends on the vector "today"

$$\begin{bmatrix} x_1 (t) \\ x_2 (t) \end{bmatrix} = \mathbf{x} (t)$$

and on today's date. Please, note that it turns out to be impossible to write the "history" of the first state variable $x_1 (t)$ alone, because it is intertwined with the one of the other state variable.

4.2.4 Autonomous Systems

The notion of motion law, which is crucial in the DS theory, is immediately fruitful in offering us an important distinction among DS.

Roughly speaking, it deals with the mere effect of time on the evolution rules of a DS. Compare these two discrete growth models for a population:

$$\begin{cases} \text{Model 1} - x (t + 1) = (1 + g) x (t) \\ \\ \text{Model 2} - x (t + 1) = [1 + g (t)] x (t) \end{cases}$$

In the first model the evolution rule (dictated by the constant g) is the same at any time: the population will grow with growth rate g, the same at any t. In the second one the growth rule changes over time, because the growth rate g can vary over time: $g = g (t)$.

Example 4.2.8 If the constant growth rate is $g = 0.05$, what we see is that the population will have a period increase of 5%. Starting from — say — $x (0) = 1000$, three periods later its size will be $x (3) = 1000 \times 1.05^3 = 1157.6$. Move now to the case of a variable growth rate, for instance $g (0) = 0.035$, $g (1) = 0.05$, $g (2) = 0.04$. In this case, always starting from $x (0) = 1000$, we would obtain: $x (3) = 1000 \times 1.035 \times 1.05 \times 1.04 = 1130.2$.

In the first model we can synthesize the dependence of future on the present within this scheme:

$$x (t) \longrightarrow x (t + 1)$$

where the arrow "$\longrightarrow$" means "determines".

In the second model the analogous scheme is slightly more complex because time t has two ways to influence the evolution of our system between t and $t + 1$:

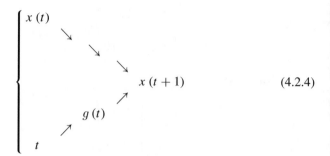

$$x(t+1) \qquad\qquad (4.2.4)$$

an indirect influence through $x(t)$ and a direct one through $g(t)$.

Let's think concretely, using this example.

Example 4.2.9 In the case of the first model, once the rules of growth (g) have been established, the population grows *autonomously*: $x(0)$ and g allows us to write the whole history. In the second model, together with the starting point $x(0)$, we need to know for each period t the appropriate growth rate $g(t)$. This means that the growth of the population is *non-autonomous* because the evolution rule (itself) changes over time.[8]

We are ready to distinguish among DS, two relevant categories. Such distinction is based on the fact that, in the motion law, the function $\mathbf{f}[\mathbf{x}(t),t]$ does depend directly on t or not.

Definition 4.2.4 Consider a DS, with motion law:

$$\left. \begin{array}{c} \mathbf{x}(t+1) \\[2mm] \mathbf{x}'(t) \end{array} \right\} = \mathbf{f}[\mathbf{x}(t),t]$$

We say that the system is *autonomous* if the function $\mathbf{f}$ is constant over time, in the sense that it varies only with the state of the system $\mathbf{x}(t)$, but not directly with time:

$$\mathbf{f}[\mathbf{x}(t),t] = \mathbf{f}[\mathbf{x}(t)]$$

When a system is not autonomous, we will call it *non-autonomous*.

Remark 4.2.1 The indication of the dependence on time of the state vector $\mathbf{x}$ or of its time derivative $\mathbf{x}'$ gives clarity to the exposition, but makes formulas a bit more complicate. Therefore in the case of continuous DSs the shorthand notation:

$$\mathbf{x}' = f(\mathbf{x}) \text{ instead of the more cumbersome } \mathbf{x}'(t) = \mathbf{f}[\mathbf{x}(t)]$$

is authorized.

---

[8]From a merely formal point of view, we could also reduce ourselves to the autonomous case. It would be sufficient to extend the state vector, in order to include time. This is mathematically correct but devastating for the applications, we will quickly see.

Motion laws of the types:

$$\begin{cases} x''(t) = f\left[x'(t), x(t), t\right] \\ x(t+2) = f\left[x(t+1), x(t), t\right] \end{cases}$$

describe *DSs of order* 2 as, in the continuous case they involve the second derivative, and in the discrete case, the maximum time distance between values of the state variable is 2 . The readers are sufficiently creative to imagine what is a DS of order $3, 4, \ldots, k$. An important fact is that a system of order $k > 1$, can be always transformed into a system of the 1st order, simply augmenting the system dimension to kn.

This implies that what we learn about DSs of the 1st order is all that we can learn about DSs of any order. The following example shows the way this trade-off order/dimension turns out to work.

Example 4.2.10 The motion law of a continuous DS is:

$$x'' = f(x)$$

Let $x_1(t) = x(t)$ and $x_2(t) = x'(t)$. We can rewrite the motion law as follows:

$$\begin{cases} x_1'(t) = x_2(t) \\ x_2''(t) = f[x_1(t)] \end{cases}$$

Miraculously a DS of dimension 1 and order 2 has been transformed into a DS of dimension 2, but of order 1.

To be concrete, here is a relevant example of an economic problem, which turns out to be interesting in Macroeconomics.

Example 4.2.11 We consider a very simple economic system. Let us look at what happens between two dates, close to each other, t and $t + h$, ignoring $o(h)$. The GNP (see p. 167) is: $Y(t)h$. The consumption is $C(t)h = cY(t)h$ (being $0 < c < 1$, the consumption propensity). Public investments are $G(t)h$, and they are given. As far as the amount of private investments is $J(t)h$ we assume that Companies invest, taking into account the consumption evolution, more precisely its acceleration $C''(t)$. Therefore $J(t)h = kC''(t)h = kcY''(t)h$ with $k > 0$, called the acceleration coefficient. A trivial balance equation must hold. It tells us that the total product must coincide with the sum of what is privately invested, publicly invested or consumed:

$$Y(t)h = J(t)h + G(t)h + C(t)h$$

Divide by h and represent all the quantities involved in terms of the state variable $Y(t)$. You get:

$$Y(t) = cY(t) + G(t) + kcY''(t) \tag{4.2.5}$$

hence:

$$Y''(t) = \frac{(1-c)Y(t) - G(t)}{kc}$$

We have a motion law of the type:

$$x''(t) = f\left[x'(t), x(t); t\right]$$

which is reasonably called of order 2. This model is very important because the trajectories it generates can be oscillating. The political relevance of such oscillations (the so called *economic cycle*) should be evident for all. The technical reason for such oscillations lays in the assumption that private investments depend on the acceleration of consumptions.

Remark 4.2.2 (*Vocabulary*) — It is important to be aware that sometimes DS models are differently named, essentially because of historical reasons.

Precisely, DS are identified with their motion law, which is described by an *equation* or by a *system of equations*. Such equations are often called *difference equations* in the discrete case and *differential equations* in the continuous one. The label *differential system* stands for "continuous DS of dimension >1".

4.3   Extracting Info from a Motion Law

An enormous literature about DSs is available today. The most general problem boils down to a question:

"If you know the motion law, what can you do with it?".

Usually three types of answers are proposed:

1. **Classic approach** — Find $x(t)$, possibly in closed form, so that, if you are given any initial position $x(0)$, you can compute directly the position of the system, at any time t, via some formula involving a function g you are able to find:

$$x(t) = g(t)$$

   In this case you must know precisely the motion function f.
2. **Numerical approach** — Given any initial position $x(0)$, and given the function f, compute (approximately) $x(t)$ for any t you could be interested in.
3. **Qualitative approach** — Even if you do not know completely f (for instance you know only that it is convex), draw some conclusions about the behavior of the system, on the basis of such a fuzzy information.

We can start from some well known facts, in order to understand the strengths and weaknesses of each approach.

1. When you succeed in deriving, in closed form, the time evolution law $\mathbf{x}(t)$ from the motion law, you must be very happy as, in general, this is not possible. The reason is that, in order to find $\mathbf{x}(t)$ in closed form, you have to compute integrals and you know that, in general, this is impossible in closed. For instance, given the following motion law:

$$x'(t) = e^{t^2}$$

it is possible to prove that it is impossible to find $x(t)$ in closed form. Of course, when the classic approach works, you get the best result you could expect. A closed expression for the state variable is generally useful in easily constructing theories.
2. The numerical approach works well always, but it requires you to know exactly $\mathbf{f}$. A numerical result is generally very useful in practice, but it is not always appropriate for constructing theories.
3. The qualitative approach works well with an autonomous system and sometimes it permits to draw conclusions also in the case of partial knowledge of $\mathbf{f}$.

The Table 4.1 synthesizes these facts.

Table 4.1 Approaches to dynamical systems

Approach	Strengths	Weaknesses
Classic	Elegant, good for theory	In general, it does not work
Numer.	It always works	Not good for theory
Qualit.	Only qualitative info, good for theory and practice	Only autonomous DS, partial results

The next sections are devoted to introduce these three approaches.
Applications of interest for Social sciences are briefly described.

4.4 Classic Approach

"Give me a motion law and a starting point and I'll provide you with $\mathbf{x}(t)$"

The sentence above deserves attention.[9]

In Social Sciences it is common to study jointly the interaction between demographic and sociological aspects of the evolution of a population. See, for example [12].

---

[9]Not Archimedes, but close to... It is attributed to P.S. Laplace (1749–1827), he was the the guy we already met when handling determinants or frequency/probability distributions.

For instance, demographic models, used in population projections, are rather complex and correctly consider the age-structure of the population under scrutiny. See, e.g., Example 4.4.11 at p. 246.

Sometimes such a complexity level is to high to allow for a proper combination of the demographic side with the sociological one and, like in the paper just cited, where a demographic model, without age structure, is introduced.

We will use repeatedly this scheme in the sequel in order to show how differential equations can be used in constructing demographic models oriented to sociological analysis.

4.4.1 Linear Discrete Systems

An important class of dynamic models, which are relevant for applications, is usually called "linear": we will confine here our attention only to these special motion laws. For a generic system of dimension n, the motion law is of the type we have seen above:

$$\mathbf{x}\,(t+1) = \mathbf{f}\,[\mathbf{x}\,(t)\,,t]$$

In principle, the structure of the function $\mathbf{f}$ can be very complicate, but, in the applications, very frequently we observe the following one:

$$\mathbf{x}\,(t+1) = A\,(t)\,\mathbf{x}\,(t) + \mathbf{b}\,(t) \tag{4.4.1}$$

in which the r.h.s. of the equation at any fixed time is a linear affine function of the state vector. More explicitly:

$$\begin{cases} x_1\,(t+1) = a_{11}\,(t)\,x_1\,(t) + a_{12}\,(t)\,x_2\,(t) + \cdots + a_{1n}\,(t)\,x_n\,(t) + b_1\,(t) \\ x_2\,(t+1) = a_{21}\,(t)\,x_1\,(t) + a_{22}\,(t)\,x_2\,(t) + \cdots + a_{2n}\,(t)\,x_n\,(t) + b_2\,(t) \\ \cdots \\ x_n\,(t+1) = a_{n1}\,(t)\,x_1\,(t) + a_{n2}\,(t)\,x_2\,(t) + \cdots + a_{nn}\,(t)\,x_n\,(t) + b_n\,(t) \end{cases} \tag{4.4.2}$$

where:

$$A\,(t) = \begin{bmatrix} a_{11}\,(t) & a_{12}\,(t) & \cdots & a_{1n}\,(t) \\ a_{21}\,(t) & a_{22}\,(t) & \cdots & a_{2n}\,(t) \\ & \vdots & & \\ a_{n1}\,(t) & a_{n2}\,(t) & \cdots & a_{nn}\,(t) \end{bmatrix} \quad \text{and } \mathbf{b}\,(t) = \begin{bmatrix} b_1\,(t) \\ b_2\,(t) \\ \vdots \\ b_n\,(t) \end{bmatrix}$$

Example 4.4.1 Consider this motion law:

$$\begin{aligned} x_1\,(t+1) &= 3t x_1\,(t)\,x_2\,(t) + t^2 \\ x_2\,(t+1) &= 4\,(t)\,x_1\,(t) - 5t^2 x_2\,(t) + t^3 \end{aligned}$$

Well, this specific motion law *does not characterize* a linear system because, in the first equation, the two state variables are multiplied with each other and this is incompatible with (4.4.2).

Example 4.4.2 Consider this motion law:

$$x_1 (t + 1) = 3tx_1 (t) - 2t^7 x_2 (t) + t^2$$
$$x_2 (t + 1) = 4tx_1 (t) - 5t^2 x_2 (t) + t^3$$

Well, this motion law *characterizes* a linear system because its structure is compatible with (4.4.2). Specifically:

$$\begin{cases} a_{11} (t) = 3t \\ a_{12} (t) = -2t^7 \\ a_{21} (t) = 4t \\ a_{22} (t) = 5t^2 \\ b_1 (t) = t^2 \\ b_2 (t) = t^3 \end{cases}$$

Remark 4.4.1 Note that what is required for the linearity of the system is not linearity of the r.h.s. w.r.t. time, but only w.r.t. the state variables.

It is easy to find an interpretation for the structure of linear dynamic systems, which is illuminating for the applications. What stays behind (4.4.1) or (4.4.2) is the idea that the future ($x (t + 1)$) is partially determined by the present of the system ($x (t)$) and partially by forces acting outside the system ($b (t)$). We can label the addendum $A (t) x (t)$ as the *endogenous part* of $x (t + 1)$ and $b (t)$ as the *exogenous* one:

$$x (t + 1) = \underbrace{A (t) x (t)}_{endog.} + \underbrace{b (t)}_{exog.}$$

Thanks to this interpretation, frequently, $b (t)$ is called *forcing term* in the motion law.

4.4.1.1 Types of Linear Systems

A special case of interest is the one in which the forcing term is identically null:

$$b (t) = 0 \text{ for } t = 0, 1, 2, \ldots \qquad (4.4.3)$$

Definition 4.4.1 When (4.4.3) occurs we say that the linear system is *homogeneous*, otherwise we say that it is *non-homogeneous*. In the homogenous case, the motion law turns out to be simply:

$$x (t + 1) = A (t) x (t)$$

Example 4.4.3 For instance, the system:

$$x_1 (t + 1) = 3tx_1 (t) - 2t^7 x_2 (t) + t^2$$
$$x_2 (t + 1) = 4tx_1 (t) - 5t^2 x_2 (t) + t^3$$

is non-homogeneous, while the following one:

$$x_1 (t + 1) = 3tx_1 (t) - 2t^7 x_2 (t)$$
$$x_2 (t + 1) = 4tx_1 (t) - 5t^2 x_2 (t)$$

is homogeneous.

Another relevant special case is characterized by the time independence of the coefficients $a_{r,s} (t)$.

Definition 4.4.2 If:

$$A (0) = A (1) = \cdots = A (t) = \cdots = A$$

we say that the system has *constant coefficients*.

Example 4.4.4 The system:

$$x_1 (t + 1) = 3tx_1 (t) - 2t^7 x_2 (t) + t^2$$
$$x_2 (t + 1) = 4tx_1 (t) - 5t^2 x_2 (t) + t^3$$

has not constant coefficients, because *at least one coefficient is not constant. In fact, all the coefficients are not constant. The system:*

$$x_1 (t + 1) = 3x_1 (t) - 2x_2 (t) + t^2$$
$$x_2 (t + 1) = 4x_1 (t) - 5x_2 (t) + t^3$$

is a system with constant coefficients. The constant matrix of coefficients is:

$$A = \begin{bmatrix} 3 & -2 \\ 4 & -5 \end{bmatrix}$$

Remark 4.4.2 A system with constant coefficients can have forcing term dependent on time. Constancy of coefficients implies concretely that the endogenous evolution rules are time-invariant. This does not imply that the exogenous influence is constant over time. In the case both A and **b** are constant:

$$\mathbf{x} (t + 1) = A\mathbf{x} (t) + \mathbf{b}$$

the linear system is also autonomous (see above the Definition 4.2.4 at p. 230).
The next question is 'How to solve linear systems?'.

Homogeneous Discrete Linear Systems

Let us start from the homogeneous case:

$$\mathbf{x}(t+1) = A(t)\mathbf{x}(t) \tag{4.4.4}$$

We assume given the coefficient matrix $A(t)$ and the starting point of the evolution $\mathbf{x}(0)$. It is not difficult to find a formula that provides us with $\mathbf{x}(t)$ in terms of $A(t)$ and of $\mathbf{x}(0)$. Letting $t = 0$ in the motion law $\mathbf{x}(0)$ (4.4.4), we get:

$$\mathbf{x}(1) = A(0)\mathbf{x}(0)$$

With $t = 1$, we get:

$$\mathbf{x}(2) = A(1)\mathbf{x}(1) = A(1)A(0)\mathbf{x}(0)$$

With $t = 2$, we get:

$$\mathbf{x}(3) = A(2)\mathbf{x}(2) = A(2)A(1)A(0)\mathbf{x}(0)$$

In general we would have:

$$\mathbf{x}(t) = A(t-1)A(t-2)\ldots A(2)A(1)A(0)\mathbf{x}(0) \tag{4.4.5}$$

or, more compactly:

$$\mathbf{x}(t) = \left[\prod_{s=1}^{t} A(t-s)\right]\mathbf{x}(0) \tag{4.4.6}$$

This expression can be written more simply, trough the introduction of an appropriate symbol for products of the type[10] appearing in (4.4.5) and in (4.4.6):

$$P(u,t) = A(t-1)A(t-2)\ldots A(2)A(1)A(u) = \prod_{s=1}^{t-u} A(t-s) \tag{4.4.7}$$

Note that $P(u,t)$, being a product of square matrices of the same order, is a square matrix of the same order. Please, note also that, if we let $u = 0$, we get exactly:

$$P(0,t) = A(t-1)A(t-2)\ldots A(2)A(1)A(0) = \prod_{s=1}^{t} A(t-s)$$

which are nothing but the two products appearing in (4.4.5) and in (4.4.6). Therefore these two equations become:

---

[10]We introduce a more general symbol for the needs we will find in handling the non-homogeneous case.

$$\mathbf{x}(t) = P(0, t)\,\mathbf{x}(0)$$

Remark 4.4.3 Rather frequently, at the exams, the AA. have observed students providing a *wrong version* of formula (4.4.5), namely:

$$\mathbf{x}(t) = A(0)\,A(1)\ldots A(t-2)\,A(t-1)\,\mathbf{x}(0) \tag{4.4.8}$$

The mistake stays in the fact that the order of factors in matrix multiplication is crucial.

Two special cases of (4.4.5) are practically useful. The first one concerns *constant coefficients*. If the motion law is:

$$\mathbf{x}(t+1) = A\mathbf{x}(t)$$

then the solution is:

$$\mathbf{x}(t) = A^t\mathbf{x}(0) \tag{4.4.9}$$

The second one concerns dimension 1. If $n = 1$, so that we have a unique state variable $x(t)$ and the matrix $A(t)$ shrinks to a scalar function of time. If so, (4.4.5) becomes[11]:

$$x(t) = x(0)\,a(0)\,a(1)\ldots a(t-1) = x(0)\prod_{s=0}^{t-1} a(s)$$

Example 4.4.5 Consider this motion law:

$$\begin{cases} x_1(t+1) = (1+t)\,x_1(t) + x_2(t) \\ x_2(t+1) = x_2(t) \end{cases}$$

or:

$$\begin{bmatrix} x_1(t+1) \\ x_2(t+1) \end{bmatrix} = \underbrace{\begin{bmatrix} 1+t & 1 \\ 0 & 1 \end{bmatrix}}_{A(t)} \begin{bmatrix} x_1(t) \\ x_2(t) \end{bmatrix}$$

The linear DS is homogeneous with non-constant coefficients as the coefficient of $x_1(t)$ in the first equation varies in time. The solution (4.4.5) is:

$$\begin{bmatrix} x_1(t) \\ x_2(t) \end{bmatrix} = \begin{bmatrix} t & 1 \\ 0 & 1 \end{bmatrix} \cdot \begin{bmatrix} t-1 & 1 \\ 0 & 1 \end{bmatrix} \cdots \cdots \begin{bmatrix} 2 & 1 \\ 0 & 1 \end{bmatrix} \cdot \begin{bmatrix} 1 & 1 \\ 0 & 1 \end{bmatrix} \cdot \begin{bmatrix} x_1(0) \\ x_2(0) \end{bmatrix}$$

---

[11] When handling scalars, the factor order is irrelevant.

It could be possible to represent the product of matrices $\prod\limits_{s=1}^{t} A(t-s)$ as follows:

$$P(0,t) = \prod_{s=1}^{t} A(t-s) = \begin{bmatrix} t! & t! \cdot [E(t) - E(u)] \\ 0 & 1 \end{bmatrix}$$

where $t!$ (with t non-negative integer is $1 \cdot 2 \cdot 3 \cdots \cdot t$, with $0! = 1! = 1$) and $E(t) = \sum_{s=0}^{t} \frac{1}{s!}$.

This result is far from being exciting. We will find something better later exploiting some special feature of $A(t)$ and what we will learn about the non-homogeneous case.

Example 4.4.6 Consider this motion law:

$$\begin{cases} x_1(t+1) = x_1(t) + x_2(t) \\ x_2(t+1) = x_2(t) \end{cases}$$

or:

$$\begin{bmatrix} x_1(t+1) \\ x_2(t+1) \end{bmatrix} = \underbrace{\begin{bmatrix} 1 & 1 \\ 0 & 1 \end{bmatrix}}_{A} \begin{bmatrix} x_1(t) \\ x_2(t) \end{bmatrix}$$

We know, that in the case of constant coefficients, we can write the solution according to formula (4.4.9) as:

$$\begin{bmatrix} x_1(t) \\ x_2(t) \end{bmatrix} = \begin{bmatrix} 1 & 1 \\ 0 & 1 \end{bmatrix}^t \cdot \begin{bmatrix} x_1(0) \\ x_2(0) \end{bmatrix}$$

Therefore all boils down to find an expression for A^t. It can be found intuitively. Let us compute A^2, A^3, A^4 and all should become clear:

$$A^2 = \begin{bmatrix} 1 & 1 \\ 0 & 1 \end{bmatrix} \cdot \begin{bmatrix} 1 & 1 \\ 0 & 1 \end{bmatrix} = \begin{bmatrix} 1 & 2 \\ 0 & 1 \end{bmatrix}$$

$$A^3 = A \cdot A^2 = \begin{bmatrix} 1 & 1 \\ 0 & 1 \end{bmatrix} \cdot \begin{bmatrix} 1 & 2 \\ 0 & 1 \end{bmatrix} = \begin{bmatrix} 1 & 3 \\ 0 & 1 \end{bmatrix}$$

$$A^4 = A \cdot A^3 = \begin{bmatrix} 1 & 1 \\ 0 & 1 \end{bmatrix} \cdot \begin{bmatrix} 1 & 3 \\ 0 & 1 \end{bmatrix} = \begin{bmatrix} 1 & 4 \\ 0 & 1 \end{bmatrix}$$

Every reader should have grasped that[12]:

$$A^t = \begin{bmatrix} 1 & t \\ 0 & 1 \end{bmatrix} \tag{4.4.10}$$

---

[12]It is interesting to note that those formula holds also for $t = 1, 0, -1, -2, \ldots$.

Therefore we are enabled to describe the evolution of the system, starting from any initial position:

$$\mathbf{x}(0) = \begin{bmatrix} x_1(0) \\ x_2(0) \end{bmatrix}$$

We get:

$$\mathbf{x}(t) = A^t \mathbf{x}(0) = \begin{bmatrix} 1 & t \\ 0 & 1 \end{bmatrix} \cdot \begin{bmatrix} x_1(0) \\ x_2(0) \end{bmatrix} = \begin{bmatrix} x_1(0) + tx_2(0) \\ x_2(0) \end{bmatrix}$$

The fact that we have succeeded in finding the simple expression (4.4.10) for A^t is rather exceptional, because A is "very special" and one could fail in trying to do the same with a generic matrix. What is true is that, in fact, there is a systematic way to represent the powers of a square matrix. We will not present it in this volume because of the mathematical burden it requires (complex numbers, trigonometric functions and spectral theory of matrices). It does not appear to us reasonable for this type of textbook.

Example 4.4.7 A Foundation has an endowment and part of it, of amount $x(0)$ will not be used for some years and it will be invested until a predictable need will arrive. The financial conditions at which this investment is made can vary in time. Let $i_0, i_1, i_2, \ldots, i_{t-1}$ be the yield rates of the investment between 0 and 1, between 1 and 2, ..., between $t-1$ and t. The motion law for the invested amount is:

$$x(t+1) = x(t)(1+i_t)$$

According to (4.4.8), we get the solution:

$$x(t) = x(0)(1+i_0)(1+i_1)\cdots(1+i_{t-1})$$

Note that, in the case the yield rate does not vary in time:

$$i_0 = i_1 = i_2 = \cdots = i_{t-1} = i$$

we would obtain:

$$x(t) = x(0)(1+i)^t$$

which is a well known formula from Finance, which allows for the computation the final value with compound interests.

Remark 4.4.4 **About means** — Something can be added, which is suggested by the previous example. A frequent problem in the applications requires to synthesize n numbers $x_1, x_2, \ldots, x_n$ into a single one $\bar{x}$. If these numbers are the interest rates $i_0, i_1, i_2, \ldots, i_{t-1}$ of the previous example the necessity could be that of asking: "Which is the mean yield rate at which $x(0)$ has been invested?" A naïve answer to this question suggests to compute the *arithmetic mean* of these numbers:

$$\overline{x} = \frac{x_1 + x_2 + \cdots + x_n}{n}$$

In fact, in Descriptive Statistics one encounters several different means and here was seriously born the problem of deciding which is the appropriate mean in a given circumstance. The ever growing use of statistical data in Social Sciences largely justifies what we are seeing now.

First of all, we show that the use of the arithmetic mean can be wrong. To do that, we recall the previous example. Assume that the investment of the endowment lasts two periods and that the yield rates for the two periods are 10 and 20% respectively. Let the amount invested be unitary $(x\,(0) = 1)$. The final value F is:

$$F = 1 \cdot (1 + 10\%)\,(1 + 20\%) = 1.1 \cdot 1.2 = 1.32$$

The arithmetic average of the two yield rates is:

$$\frac{10\% + 20\%}{2} = 15\%$$

If we compute the final value using this "representative yield rate" we do not get F:

$$1 \cdot (1 + 15\%)^2 = 1.15^2 = 1.3225 \neq 1.32$$

We show then how to face this problem in general, recalling the notion of *Chisini mean*.[13] The necessity of synthesis typically stems from some problem one wants to solve. In the light of this problem a quantity[14] I, depending from the data $x_1, x_2, \ldots, x_n$:

$$I\,(x_1, x_2, \ldots, x_n)$$

which is relevant for the problem (for instance the final value of the invested endowment) and which is *invariant* if the data are each replaced by their Chisini mean $\overline{x}$:

$$I\,(x_1, x_2, \ldots, x_n) = I\,(\overline{x}, \overline{x}, \ldots, \overline{x})$$

This is an equation with the only unknown $\overline{x}$. Solving for it we get the appropriate mean.

We apply now this point of view to the numerical example seen above and then we indicate the general solution to the problem of averaging yield rates, growth rates, decay rates…

The Chisini mean of the two interest rates above stems from the equation:

$$1 \cdot (1 + 10\%)\,(1 + 20\%) = 1 \cdot (1 + \overline{x})\,(1 + \overline{x})$$

---

[13] After the name of the Italian professor of Geometry Oscar Chisini (1889–1967), who proposed this innovative perspective in 1929.

[14] Nothing in common with an identity matrix.

which becomes:

$$1.32 = (1 + \overline{x})^2$$

hence:

$$\overline{x} = \sqrt{1.32} - 1 \approx 0.148\,91$$

or $\overline{x} \approx 14.891\%$.

In general, when required to average the rates $x_1, x_2, \ldots, x_n$ of some initial quantity C, Chisini equation results:

$$C\,(1 + x_1)\,(1 + x_2)\cdots(1 + x_n) = C\,(1 + \overline{x})^n$$

and hence:

$$\overline{x} = \sqrt[n]{\prod_{s=1}^{n} (1 + x_s)} - 1 = \sqrt[n]{(1 + x_1)\,(1 + x_2)\cdots(1 + x_n)} - 1$$

Non-homogeneous Linear systems

It is the time to extend what we have seen in the homogeneous case to the non-homogeneous one. We will use the same intuitive approach of p. 237. We start from the non-homogeneous linear motion law (4.4.1)

$$\mathbf{x}\,(t + 1) = A\,(t)\,\mathbf{x}\,(t) + \mathbf{b}\,(t)$$

With $t = 0$, it provides us with:

$$\mathbf{x}\,(1) = A\,(0)\,\mathbf{x}\,(0) + \mathbf{b}\,(0)$$

With $t = 1$ and taking into account the expression we have just obtained for $\mathbf{x}\,(1)$, we get:

$$\mathbf{x}\,(2) = A\,(1)\,[A\,(0)\,\mathbf{x}\,(0) + \mathbf{b}\,(0)] + \mathbf{b}\,(1) =$$
$$= A\,(1)\,A\,(0)\,\mathbf{x}\,(0) + A\,(1)\,\mathbf{b}\,(0) + \mathbf{b}\,(1) =$$
$$= \prod_{s=1}^{2} A\,(2 - s)\,\mathbf{x}\,(0) + A\,(1)\,\mathbf{b}\,(0) + \mathbf{b}\,(1)$$

With $t = 2$ we get:

$$\mathbf{x}\,(3) = \prod_{s=1}^{3} A\,(3 - s)\,\mathbf{x}\,(0) + \prod_{s=1}^{2} A\,(3 - s)\,\mathbf{b}\,(0) + A\,(2)\,\mathbf{b}\,(1) + \mathbf{b}\,(2)$$

With $t = 3$ one obtains:

$$\mathbf{x}(4) = \prod_{s=1}^{4} A(4-s)\mathbf{x}(0) + \prod_{s=1}^{3} A(4-s)\mathbf{b}(0) + \prod_{s=1}^{2} A(4-s)\mathbf{b}(1) + A(3)\mathbf{b}(2) + \mathbf{b}(3)$$

In general[15]:

$$\mathbf{x}(t) = \prod_{s=1}^{t} A(t-s)\mathbf{x}(0) + \sum_{u=0}^{t-1} \prod_{s=1}^{t-u-1} A(t-s)\mathbf{b}(u) \qquad (4.4.11)$$

Recalling the notation introduced in formula (4.4.7) above:

$$P(u,t) = \prod_{s=1}^{t-u} A(t-s)$$

we can rewrite (4.4.11) as follows:

$$\mathbf{x}(t) = P(0,t)\mathbf{x}(0) + \sum_{u=0}^{t-1} P(u+1,t)\mathbf{b}(u) \qquad (4.4.12)$$

Three special cases are particularly interesting:

- $A(t)$ independent of t: $A(t) \equiv A$;
- $A(t)$, $\mathbf{b}(t)$ independent of t: $A(t) \equiv A$ and $\mathbf{b}(t) \equiv \mathbf{b}$;
- only one state variable $x(t)$.

In the case of *time-invariance of the coefficient matrix* (4.4.11) reduces to:

$$\mathbf{x}(t) = A^{t}\mathbf{x}(0) + \sum_{u=0}^{t-1} A^{t-u-1}\mathbf{b}(u) \qquad (4.4.13)$$

If $\mathbf{b}(t)$ is time-invariant too, (4.4.11):

$$\mathbf{x}(t) = A^{t}\mathbf{x}(0) + \left(\sum_{u=0}^{t-1} A^{t-u-1}\right)\mathbf{b}$$

The sum of powers

$$\sum_{u=0}^{t-1} A^{t-u-1} = A^{t-1} + A^{t-2} + \cdots + A + I$$

---

[15]In general, a product $\prod_{p}^{q}$ with $p > q$ is assigned to have value 1. If $p = q$, the product has a single factor and the value of the product is that of its only factor.

can be simplified if $I - A$ is non-singular:

$$I + A + A^2 + \cdots + A^{t-1} = (I - A)^{-1} \left(I - A^t \right) = \left(I - A^t \right) (I - A)^{-1}$$

hence:

$$\mathbf{x}\,(t) = A^t \mathbf{x}\,(0) + (I - A)^{-1} \left(I - A^t \right) \mathbf{b}$$

Finally, in the scalar case:

$$x\,(t + 1) = a\,(t)\,x\,(t) + b\,(t)$$

we obtain:

$$x\,(t) = x\,(0) \prod_{s=0}^{t-1} a\,(s) + \sum_{u=0}^{t-1} b\,(u) \prod_{s=u+1}^{t-1} a\,(s)$$

Example 4.4.8 Consider the linear dynamic system, which is non-homogeneous and non-autonomous, similar to the one considered in the Example 4.4.5:

$$\begin{cases} x_1\,(t + 1) = (1 + t)\,x_1\,(t) + x_2\,(t) \\ x_2\,(t + 1) = x_2\,(t) + 1 \end{cases}$$

The motion law can be rewritten:

$$\begin{bmatrix} x_1\,(t + 1) \\ x_2\,(t + 1) \end{bmatrix} = \underbrace{\begin{bmatrix} 1 + t & 1 \\ 0 & 1 \end{bmatrix}}_{A(t)} \begin{bmatrix} x_1\,(t) \\ x_2\,(t) \end{bmatrix} + \begin{bmatrix} 0 \\ 1 \end{bmatrix}$$

The solution (4.4.12) is:

$$\begin{bmatrix} x_1\,(t) \\ x_2\,(t) \end{bmatrix} = P(0, t)\mathbf{x}\,(0) + \sum_{u=0}^{t-1} P\,(u + 1, t) \begin{bmatrix} 0 \\ 1 \end{bmatrix}$$

The expressions for P are the same computed above in the cited example.

Example 4.4.9 Retake into consideration the Example 4.4.6 and modify slightly the second equation

$$\begin{cases} x_1\,(t + 1) = x_1\,(t) + x_2\,(t) \\ x_2\,(t + 1) = x_2\,(t) + 1 \end{cases}$$

or:

$$\begin{bmatrix} x_1\,(t + 1) \\ x_2\,(t + 1) \end{bmatrix} = \underbrace{\begin{bmatrix} 1 & 1 \\ 0 & 1 \end{bmatrix}}_{A} \begin{bmatrix} x_1\,(t) \\ x_2\,(t) \end{bmatrix} + \begin{bmatrix} 0 \\ 1 \end{bmatrix}$$

Applying formula (4.4.13), and assuming that $I - A$ is non-singular, we get:

$$\mathbf{x}(t) = \begin{bmatrix} 1 & 1 \\ 0 & 1 \end{bmatrix}^t \mathbf{x}(0) + \sum_{u=0}^{t-1} \begin{bmatrix} 1 & 1 \\ 0 & 1 \end{bmatrix}^{t-u-1} \begin{bmatrix} 0 \\ 1 \end{bmatrix} =$$

$$= \begin{bmatrix} 1 & t \\ 0 & 1 \end{bmatrix} \mathbf{x}(0) + \left(A^{t-1} + A^{t-2} + \cdots + A^0 \right) \begin{bmatrix} 0 \\ 1 \end{bmatrix} =$$

$$= \begin{bmatrix} 1 & t \\ 0 & 1 \end{bmatrix} \mathbf{x}(0) + \left(\begin{bmatrix} 1 & t-1 \\ 0 & 1 \end{bmatrix} + \begin{bmatrix} 1 & t-1 \\ 0 & 1 \end{bmatrix} + \cdots \right.$$

$$\left. \cdots + \begin{bmatrix} 1 & 0 \\ 0 & 1 \end{bmatrix} \right) \begin{bmatrix} 0 \\ 1 \end{bmatrix} =$$

$$= \begin{bmatrix} 1 & t \\ 0 & 1 \end{bmatrix} \mathbf{x}(0) + \begin{bmatrix} t & t(t-1)/2 \\ 0 & t \end{bmatrix} \begin{bmatrix} 0 \\ 1 \end{bmatrix} =$$

$$= \begin{bmatrix} 1 & t \\ 0 & 1 \end{bmatrix} \mathbf{x}(0) + \begin{bmatrix} t(t-1)/2 \\ t \end{bmatrix}$$

hence:

$$\begin{cases} x_1(t) = x_1(0) + tx_2(0) + t(t-1)/2 \\ x_2(0) + t \end{cases}$$

Example 4.4.10 Reconsider the Example 4.4.7, we met before. Assume now that the final value to be computed does not stem only from the endowment $x(0)$, but that at the epochs $1, 2, 3, \ldots$ a constant amount b will increase the invested amount. A Foundation has an endowment and part of it, of amount $x(0)$ will not be used for some years and it will be invested until a predictable need will arrive. The financial conditions at which this investment is made can vary in time. Let $i_0, i_1, i_2, \ldots, i_{t-1}$ be the yield rates of the investment between 0 and 1, between 1 and 2, ..., between $t - 1$ and t. The motion law for the invested amount is:

$$x(t+1) = x(t)(1 + i_t) + b$$

According to (4.4.8), we get the solution:

$$x(t) = x(0) F(0, t) + b [F(1, t) + F(2, t) + \cdots + F(t, t)]$$

Where:

$$F(s, t) = (1 + i_s)(1 + i_{s+1}) \cdots (1 + i_{t-1}) \text{ and } F(t, t) = 1 \text{ for every } t$$

Note that, in the case the yield rate does not vary in time:

$$i_0 = i_1 = i_2 = \cdots = i_{t-1} = i$$

we would obtain:

$$x(t) = x(0)(1+i)^t + b \cdot \frac{(1+i)^t - 1}{i}$$

which is the well known formula from Finance for the computation of final values.

4.4.2   About Some Special Discrete Systems

We will work now with a (very) special discrete system. The motion law we will focus on is:

$$\mathbf{x}(t+1) = A\mathbf{x}(t) + \mathbf{b} \qquad (4.4.14)$$

where[16] $\mathbf{x}(t+1)$, $\mathbf{x}(t)$, $\mathbf{b}$, are n-vectors, A is a square matrix of order n.

The motivation of our interest can be explained via the following hyper-simplified:

Example 4.4.11 The demographic composition of a population is absolutely relevant for a politician. We simplify a lot. In the population we are targeting, there are three categories:

(1) young;
(2) middle aged;
(3) old people.

This population evolves according to very simple rules:

(a) - young people partially become middle aged and produce new young people;
(b) - middle aged partially become old people and produce (not so many) young people;
(c) - old people[17] do not produce young people and partially continue to be old[18]. The state vector decomposes the population into young (y), middle aged (m) and old (o) people:

$$\begin{bmatrix} y(t) \\ m(t) \\ o(t) \end{bmatrix}$$

The following matrix defines the evolution assumptions of our population:

$$A = \begin{bmatrix} 0.55 & 0.03 & 0 \\ 0.45 & 0.48 & 0 \\ 0 & 0.25 & 0.8 \end{bmatrix}$$

---

[16]We could allow for $A = A(t)$ and $\mathbf{b} = \mathbf{b}(t)$. The formulas are perfectly analogous. We will provide later the results, also in this more general case.

[17]Like one of the Authors.

[18]A nice way to say that they do not pass away.

To be interpreted (row by row) as follows:

[A]: (y) — people who are young at $t + 1$ are 50% of those who were young at t to be added to newborn people generated by young people (birth rate $0.05 = 5\%$) (total 55%) at t and to young people generated by middle aged people (birth rate $0.03 = 3\%$).

[B] (m) — who are middle aged at $t + 1$ are 45% of those who were young at t, to which 48% of the current middle aged, which remain so, must be added.

[C] The third group (o) grows as 25% of the middle age group becoming old, does not contribute to the other categories, but at $20\% = 1 - 80\%$ fades away.

Up to here the model could be classified as a *closed model* as there is no external influence. An obvious fact is that, while models can be closed, reality is *open*. We can think of this population that varies not only for endogenous reasons, but also because of migration/immigration. We can model this assuming, for instance, that during each period there are new net entries in the three age categories:

$$\mathbf{b} = \begin{bmatrix} 1000 \\ 1000 \\ 200 \end{bmatrix}$$

At this point, the motion law would be:

$$\mathbf{x}(t+1) = \begin{bmatrix} 0.55 & 0.03 & 0 \\ 0.45 & 0.48 & 0 \\ 0 & 0.25 & 0.8 \end{bmatrix} \mathbf{x}(t) + \begin{bmatrix} 1000 \\ 1000 \\ 200 \end{bmatrix}$$

The classic answer is natural and based on elementary linear algebra. At $t = 0$ the motion law provides us with:

$$\mathbf{x}(1) = A\mathbf{x}(0) + \mathbf{b}$$

At $t = 1$:

$$\mathbf{x}(2) = A\mathbf{x}(1) + \mathbf{b} = A[A\mathbf{x}(0) + \mathbf{b}] + \mathbf{b} = A^2\mathbf{x}(0) + (A + A^0)\mathbf{b}$$

Iterating we obtain:

$$\mathbf{x}(t) = A^t\mathbf{x}(0) + \sum_{s=0}^{t-1} A^s\mathbf{b} = A^t\mathbf{x}(0) + \left(\sum_{s=0}^{t-1} A^s \right)\mathbf{b} \qquad (4.4.15)$$

Formula (4.4.15) is valid in general, in the sense that, if we have a discrete system with motion law (4.4.14), then all its solutions are provided for us by (4.4.15).

Remark 4.4.5 These are special cases of the general one (4.4.15).

The set of all such solutions is called *general solution*. Choosing the (so called) *boundary condition*, i.e.; choosing **x** (0), we single out from the general solution the *particular solution* starting from that point.

It is interesting to note that if $I - A$ is non-singular it can be rewritten in a more simple form.

Take

$$\sum_{s=0}^{t-1} A^s = I + A + A^2 + \cdots + A^{t-1}$$

Call this sum S:

$$S = I + A + A^2 + \cdots + A^{t-1}$$

Left-multiply each side by A, getting:

$$AS = A + A^2 + \cdots + A^{t-1} + A^t$$

Subtract the two equalities side by side:

$$S - AS = I - A^t \Rightarrow (I - A) S = I - A^t$$

The non-singularity of $I - A$ allows to write:

$$S = (I - A)^{-1} \left(I - A^t\right)$$

Hence formula (4.4.15) can be explicitly rewritten as:

$$\mathbf{x}(t) = A^t \mathbf{x}(0) + (I - A)^{-1} \left(I - A^t\right) \mathbf{b}$$

A few lines above we have left-multiplied the sides of an equation. We would proud that our students would ask us: "But, what happens if we right-multiply? You know, for a Political Science student left and right are not the same". Our answer would be: "Thank you for the useful question. Keep calm on the political side because we would get for S the expression:

$$\left(I - A^t\right) (I - A)^{-1}$$

which seems different, but as the two factors do commute, the corresponding trajectory:

$$\mathbf{x}(t) = A^t \mathbf{x}(0) + \left(I - A^t\right) (I - A)^{-1} \mathbf{b}$$

would be exactly the same".

4.4.3 Continuous-Time Systems

Let's look at:

Example 4.4.12 The size $x(t)$ of a population evolves according to the motion law:

$$x'(t) = kx(t)$$

k being a positive constant. Note first that the null function:

$$x(t) \equiv 0 \qquad\qquad\qquad (4.4.16)$$

trivially satisfies the motion equation, both sides being null. If $x(t) \neq 0$, we can divide both sides of the equation by $x(t)$, getting

$$\frac{x'(t)}{x(t)} = k$$

The left hand side is the log-derivative of $x(t)$:

$$D \ln x(t) = \frac{x'(t)}{x(t)}$$

With an indefinite integration, from:

$$D \ln x(t) = k$$

we obtain:

$$\ln |x(t)| = kt + c$$

being c any real constant, hence, taking exponentials and letting $|C| = e^c$, we get the *general solution* (or *general integral*):

$$x(t) = Ce^t$$

C being any real constant. With $C = 0$, the solution (4.4.14) is recovered. Choosing C, we single out a *particular solution* (or *particular integral*).

Unlike the discrete case, in which a formula has been found, which solves the problem at large, in the continuous case the example has provided us with a solution for a very special case and it is worthwhile devoting the next two subsections to finding general formulas for more general settings. The preceding model is politically relevant under a number of respects. Two examples:

- In the case some state has a debt and does not repay anything, the model describes the evolution over time of such a debt. The parameter k is nothing but the debt rate of interest.
- Assuming k is negative the model describes the decay of materials over time, with k different being from material to material (e.g., wood, plastic, atomic residues…). We will see later a model about.

4.4.4   Continuous Systems: Separable Equations

Consider a continuous system of dimension $n = 1$, with motion law:

$$x' = f(x, t) \qquad (4.4.17)$$

and assume that f depends on the state x and on time t in a very special way, the only that allows to use the forthcoming method of *separation of variables*.

Definition 4.4.3   Given a function of two variables $f(x, y)$ we say that f is *product decomposable* if it can be written as a product of two functions, one only of x, the other only of y:

$$f(x, y) = a(x) b(y)$$

For instance, the function:

$$f(x, y) = x + xy$$

is product decomposable as:

$$x + xy = x \cdot (1 + y)$$

while the simpler function:

$$f(x, y) = x + y$$

is not.

We are going to generalize the procedure, introduced in the example above on p. 249, to this more general case:

$$x' = a(x) b(t)$$

usually called *separable equations*, because their r.h.s. is product decomposable.

The method replicates what we have seen in the introductory example, but in a more general setting.

First of all, let us note that any solution x^* of the equation:

$$a(x) = 0$$

generates a constant solution:

$$x\,(t) \equiv x^*$$ (4.4.18)

which annihilates both sides of the equation.

Once we rewrite the equation using Leibnitz notation for the derivative:

$$x'\,(t) = \frac{dx}{dt}$$

assuming $a\,(x) \neq 0$, if we divide both sides of the equation:

$$\frac{dx}{dt} = a\,(x)\,b\,(t)$$

by $a\,(x)$ and we multiply both sides by dt (recalling the definition of differential, we get (miraculously):

$$\frac{dx}{a\,(x)} = b\,(t)\,dt$$ (4.4.19)

We have "separated" the two variables: x appears only in the l.h.s. of the equation and t only in the r.h.s.

Denote with $A\,(x)$ any primitive of $1/a\,(x)$, with $B\,(t)$ any primitive of $b\,(t)$ and with c an arbitrary real constant. Integrating (4.4.19) side by side you get:

$$A\,(x) = B\,(t) + c$$

which, once solved for x will provide you an infinity of solutions (that can cater for special values of c or letting c go to ∞ also for the constant solutions found initially (see (4.4.18)).

Consider this simple:

Example 4.4.13 Let:

$$x' = -x^2 t$$

A first obvious constant solution is:

$$x\,(t) \equiv 0$$ (4.4.20)

Assuming $x \neq 0$ and separating variables we get:

$$-\frac{dx}{x^2} = t\,dt$$

hence:

$$\int -\frac{dx}{x^2} = \int t\,dt + c$$

or

$$\frac{1}{x} = \frac{t^2}{2} + c \implies x(t) = \frac{1}{t^2/2 + c}$$

if we let c grow indefinitely we recover the constant solution (4.4.20). The value of c is usually chosen with the aid of a further (boundary) condition, telling us that at a given time t_0 the state of the system is, say, x_0. If at 0 the state variable takes a unitary value:

$$x(0) = 1$$

we can find the corresponding value for the constant c:

$$1 = \frac{1}{0^2/2 + c} \implies c = 1$$

This version of the method works well if:

- the factors $a(x)$, $b(t)$ are analytically specified;
- we succeed in finding $A(x)$, $B(t)$.

Otherwise it is... wise to use a variant of the procedure, which singles out directly one particular solution satisfying a boundary condition of the type:

$$x(t_0) = x_0$$

This variant uses *definite integrals* instead of indefinite ones. This fact allows also for using efficient numerical methods to estimate the evolution of the state variable.

After having identified possible constant solutions, stemming from $a(x) = 0$, first of all, for esthetic reasons it is convenient to change the names of the two variables. For instance:

$$\begin{cases} t \text{ becomes } \tau \\ \\ x \text{ becomes } \xi \end{cases}$$

getting:

$$\frac{d\xi}{d\tau} = a(\xi) b(\tau)$$

We separate variables:

$$\frac{d\xi}{a(\xi)} = b(\tau) d\tau$$

and we integrate both sides: the left one from x_0 to x, the right one from t_0 to t:

$$\int_{x_0}^{x} \frac{d\xi}{a(\xi)} = \int_{t_0}^{t} b(\tau) d\tau$$

This equation in x, t defines the particular solution such that $x(t_0) = x_0$.

Let us reconsider the previous example:

Example 4.4.14 Let:

$$x' = -x^2 t$$

Assume also that we have the boundary condition:

$$x(0) = 1$$

A first obvious constant solution is:

$$x(t) \equiv 0 \tag{4.4.21}$$

but it is useless as it does not satisfy the boundary condition. Separating variables (after their translation to Greek) we get:

$$-\frac{d\xi}{\xi^2} = \tau d\tau$$

hence:

$$\int_1^x -\frac{d\xi}{\xi^2} = \int_0^t \tau d\tau$$

or

$$\frac{1}{x} - 1 = \frac{t^2}{2} \implies x(t) = \frac{1}{t^2/2 + 1}$$

The value of c, implied by the boundary condition, turns out to be $c = 1$, as we already found above, using indefinite integrals.

4.4.5 Continuous Systems: Linear Differential Equations of the First Order

As usual, we start from a real world problem. We have already investigated the problem in a simplified setting above at p. 234.

Example 4.4.15 The population of some country evolves over time according to endogenous variables (natality and mortality) and to exogenous ones (immigration/ emigration). The size of such population at t is $x(t)$. The endogenous component is of the type:

$$x(t+h) - x(t) = x(t) \cdot a(t) \cdot h + o(h)$$

where $a(t)$ is the demographic variation rate and the exogenous component is:

$$b(t) h + o(h)$$

being $b(t)$ the instantaneous migration rate. This brings to the evolution equation:

$$x(t+h) - x(t) = x(t) \cdot a(t) \cdot h + b(t) h + o(h)$$

If we assume that x is differentiable:

$$x(t+h) - x(t) = x'(t) h + o(h)$$

we get:

$$x'(t) h + o(h) = x(t) \cdot a(t) + b(t) + \frac{o(h)}{h}$$

or, finally, letting h go to 0:

$$x'(t) = a(t) x(t) + b(t)$$

Definition 4.4.4 A *linear differential equation of the first order* is an equation of the type

$$x'(t) = a(t) x(t) + b(t)$$

where a, b are given functions.

Remark 4.4.6 Note that if $b(t) \equiv 0$ then the method of separation of variables can be used.

4.4.6 An Interesting Socio-demographic Model/1

Consider a population, with (only) two genders: male (labelled 1) and female (labelled 2). Time $t \geq 0$ is a continuous non-negative variable.

The state vector of the system:

$$\mathbf{x}(t) = \begin{bmatrix} x_1(t) \\ x_2(t) \end{bmatrix}$$

informs us about the number of males and females constituting the population at the date t. The initial composition of the population is denoted with:

$$\mathbf{x}(0) = \begin{bmatrix} x_1(0) \\ x_2(0) \end{bmatrix}$$

When using the model to understand the population dynamics it is natural to think of $\mathbf{x}(0)$ as given.

We will construct a system of equations, describing the motion law[19] in continuous-time of our system.

Let us start from males. The state variable $x_1(t)$ is assumed to evolve in time because of:

- natality: the number of male newborns between t and $t + h$ is

$$\nu_1(t)\, x_2(t)\, h + o(h)$$

- mortality; the number males dying between t and $t + h$ is

$$\mu_1(t)\, x_1(t)\, h + o(h)$$

- migrations: the net number of males entering $(+)$ or exiting $(-)$ from the population between t and $t + h$ is

$$\lambda_1(t)\, x_1(t)\, h + o(h)$$

Let us briefly comment on these assumptions.

The small time interval between t and $t + h$ is considered. The increments of relevant variables over such time interval are assumed to be approximately proportional to the length h of the interval and to other specific quantities:

- for newborns specific quantity is the current number of females in the population and the proportionality constant is the male-fertility rate[20] $\nu_1(t)$;
- for male deaths the relevant quantity is the current number of males, with proportionality constant the male mortality rate $\mu_1(t)$;
- for the male migration balance the relevant quantity is the current number of males with proportionality constant the net migration rate $\lambda_1(t)$.

Note that the rates, we have introduced, turn out to be time-dependent. This makes harder the analysis, but more realistic the scheme.

The assumptions above bring to the equation[21]:

$$x_1(t + h) - x_1(t) = \nu_1(t)\, x_2(t)\, h - \mu_1(t)\, x_1(t)\, h + \lambda_1(t)\, x_1(t)\, h + o(h)$$

With the same arguments we used above, formula (4.2.2), we get:

$$x_1'(t) = \nu_1(t)\, x_2(t) - \mu_1(t)\, x_1(t) + \lambda_1(t)\, x_1(t)$$

Remark 4.4.7 — The fact that $x_1'(t)$, the evolution rate of x_1, depends on x_2, implies that we cannot study separately the behavior of the two state variables (males and females).

---

[19] See Definition 4.2.3 at p. 228.

[20] # of male newborns per woman.

[21] Never forget that $o(h) + o(h) + o(h) = o(h)$.

We can obtain a similar equation for females:

$$x_2'(t) = \nu_2(t) x_2(t) - \mu_2(t) x_2(t) + \lambda_2(t) x_2(t)$$

with obvious interpretation of the symbols.

The motion law can be described by the system of differential equations:

$$\begin{cases} x_1'(t) = \nu_1(t) x_2(t) - \mu_1(t) x_1(t) + \lambda_1(t) x_1(t) \\ x_2'(t) = \nu_2(t) x_2(t) - \mu_2(t) x_2(t) + \lambda_2(t) x_2(t) \end{cases} \tag{4.4.22}$$

which can be rewritten, with the language of linear algebra, as:

$$\begin{bmatrix} x_1'(t) \\ x_2'(t) \end{bmatrix} = \underbrace{\begin{bmatrix} -\mu_1(t) + \lambda_1(t) & \nu_1(t) \\ 0 & \nu_2(t) - \mu_2(t) + \lambda_2(t) \end{bmatrix}}_{A(t)} \begin{bmatrix} x_1(t) \\ x_2(t) \end{bmatrix} \tag{4.4.23}$$

or, more compactly:

$$\mathbf{x}'(t) = A(t)\,\mathbf{x}'(t)$$

Such a motion law is said *linear* and *homogeneous*, similarly to the scalar case we will consider at p. 254. In this model, the dynamics of the female components does not depend on the one of the male component. This is not standard for most males in everyday's life, but fully ascertained in Demography.

There we have learned how to handle the scalar case (only 1 state variable), reducing the computation of its general solution to the computation of integrals[22]. When the dimension of the system is >1, there is no general possibility to treat the problem analogously to the scalar case. However we will see that the special structure of the matrix $A(t)$ allows us to get an analogous result.

What is beneficial in such a structure is the entry 0 in $A(t)$ at the position $(2, 1)$. In fact, this implies that the dynamics of x_2, the number of females, is not influenced by the values x_1 takes. The second equation of the system (4.4.23):

$$x_2'(t) = [\nu_2(t) - \mu_2(t) + \lambda_2(t)] x_2(t)$$

can be treated as a separable equation (see p. 250). From:

$$\frac{dx_2}{x_2} = [\nu_2(t) - \mu_2(t) + \lambda_2(t)]\,dt$$

we get easily:

$$\int_{x_2(0)}^{x_2(t)} \frac{d\xi_2}{\xi_2} = \int_0^t [\nu_2(\tau) - \mu_2(\tau) + \lambda_2(\tau)]\,d\tau$$

---

[22]This task is easily treatable with numerical methods.

hence[23]:

$$\ln x_2(t) - \ln x_2(0) = \int_0^t [\nu_2(\tau) - \mu_2(\tau) + \lambda_2(\tau)]\,d\tau$$

and, finally:

$$x_2(t) = x_2(0)\,e^{\int_0^t [\nu_2(\tau) - \mu_2(\tau) + \lambda_2(\tau)]\,d\tau} \qquad (4.4.24)$$

Remark 4.4.8 — The expression (4.4.24) we have obtained for the number of females at t can be rewritten as follows:

$$x_2(t) = x_2(0)\,e^{\int_0^t \nu_2(\tau)\,d\tau}\,e^{\int_0^t -\mu_2(\tau)\,d\tau}\,e^{\int_0^t \lambda_2(\tau)\,d\tau}$$

which tells us something interesting about the female number. At time t it is equal to the initial value, multiplied by three independent factors, respectively determined by natality, mortality and net migration rate.

Remark 4.4.9 — If we introduce the demographic variation rate $V_2(t) = \nu_2(t) - \mu_2(t) + \lambda_2(t)$, the expression (4.4.24), can be shortly rewritten as:

$$x_2(t) = x_2(0)\,e^{\int_0^t V_2(\tau)\,d\tau} \qquad (4.4.25)$$

We will use this version immediately.

Now quickly back to the male equation in (4.4.22):

$$x_1'(t) = \nu_1(t)\,x_2(t) - \mu_1(t)\,x_1(t) + \lambda_1(t)\,x_1(t)$$

which trivially can be rewritten as:

$$x_1'(t) = [-\mu_1(t) + \lambda_1(t)]\,x_1(t) + \nu_1(t)\,x_2(t) \qquad (4.4.26)$$

as $x_2(t)$ now is known (remember (4.4.25)), we see that it is a linear non-homogeneous differential equation of the first order, we learned to handle above.

First of all let us introduce

- a (gross[24]) variation rate for males:

$$V_1(t) = -\mu_1(t) + \lambda_1(t)$$

---

[23] In our setting x_2 cannot take negative values and we omit absolute values.

[24] We call it "gross" because, unlike the female case, here the natality rate is absent as it enters the forcing term, owing to the fact that it must interact with the other state variable $x_2(t)$.

- a the symbol $N(t)$ for the approximate number of male newborns between t and $t+1$:

$$N_1(t) = \nu_1(t) \cdot x_2(t) \cdot 1 = \nu_1(t) x_2(t)$$

With these new symbols (4.4.26) should look prettier:

$$x_1'(t) = V_1(t) x_1(t) + N_1(t)$$

The equation can be rewritten as:

$$x_1'(\tau) e^{-\int_0^\tau V_1(u)\,du} - V_1(\tau) x_1(\tau) e^{-\int_0^\tau V_1(u)\,du} = N_1(\tau) e^{-\int_0^\tau V_1(u)\,du}$$

or:

$$D\left[x_1(\tau) e^{-\int_0^\tau V_1(u)\,du} \right] = N_1(\tau) e^{-\int_0^\tau V_1(u)\,du}$$

Integrating both sides of the equation from 0 to t, we obtain:

$$x_1(t) e^{-\int_0^t V_1(u)\,du} - x_1(0) = \int_0^t N_1(\tau) e^{-\int_0^\tau V_1(u)\,du}\, d\tau$$

or:

$$x_1(t) = x_1(0) e^{\int_0^t V_1(u)\,du} + \int_0^t N_1(\tau) e^{-\int_\tau^t V_1(u)\,du}\, d\tau$$

Remark 4.4.10 — This expression for the number of males deserves attention for the story it is telling us. The first addendum in the r.h.s. tells us the story of the initial stock of males in time: some of them die or determine migration phenomena according to the assumptions implicit in the first equation of the system (4.4.22). The structure of the evolution of the initial population is not too dissimilar from the evolution law of the stock of females. The second addendum tells us another story. It is the story of male newborns from 0 to t. The "number" of newborns at τ turns out to be multiplied by a factor, analogous to the one concerning the initial stock, but acting from their birthdate τ to t.

The fact that for our state vector $\mathbf{x}(t)$ we have obtained explicit expressions in terms of definite integrals is reassuring: if we cannot compute analytically these integrals, in practice we can successfully attack them with numerical methods.

However there is an interesting case, in which we can confine ourselves to the analytical treatment. It is the interesting special case in which the rates $\nu_s(t)$, $\mu_s(t)$, $\lambda_s(t)$ (with $s = 1, 2$) turn out to be constant:

$$\nu_s(t) \equiv \nu_s, \qquad \mu_s(t) \equiv \mu_s, \qquad \lambda_s(t) \equiv \lambda_s$$

With a slight notation abuse, we will use for these constants the same letters chosen above to denote the corresponding time functions.

The system (4.4.22) becomes:

$$\begin{cases} x_1'(t) = \nu_1 x_2(t) - \mu_1 x_1(t) + \lambda_1 x_1(t) \\ x_2'(t) = \nu_2 x_2(t) - \mu_2 x_2(t) + \lambda_2 x_2(t) \end{cases}$$

or, in the vector matrix language:

$$\begin{bmatrix} x_1'(t) \\ x_2'(t) \end{bmatrix} = \underbrace{\begin{bmatrix} -\mu_1 + \lambda_1 & \nu_1 \\ 0 & \nu_2 - \mu_2 + \lambda_2 \end{bmatrix}}_{A} \begin{bmatrix} x_1(t) \\ x_2(t) \end{bmatrix} \qquad (4.4.27)$$

From the second equation we get immediately:

$$x_2(t) = x_2(0) e^{(\nu_2 - \mu_2 + \lambda_2)t}$$

or:

$$x_2(t) = x_2(0) e^{V_2 t}$$

where $V_2 = \nu_2 - \mu_2 + \lambda_2$, in perfect analogy with the general case. We will also write V_1 for the difference $\lambda_1 - \mu_1$[25].

The first equation has solution:

$$x_1(t) = x_1(0) e^{V_1 t} + \nu_1 \left[\frac{e^{V_2 t} - e^{V_1 t}}{V_2 - V_1} \right] \qquad (4.4.29)$$

It seems reasonable to assume that $V_2 > V_1$. We list some empirical reasons in favor of this assumption.

• The two quantities to be compared are:

$$V_1 = \lambda_1 - \mu_1$$

$$V_2 = \nu_2 + \lambda_2 - \mu_2$$

---

[25]With these notation simplifications the coefficient matrix A in (4.4.27) becomes:

$$A = \begin{bmatrix} V_1 & \nu_1 \\ 0 & V_2 \end{bmatrix} \qquad (4.4.28)$$

- We can assume that $\lambda_1 \approx \lambda_2$, so that the migration rates should have no substantial influence on the comparison.
- The mortality rates of females are slightly higher than the ones of males, therefore $\mu_1 < \mu_2$ pushes V_1 above w.r.t. V_2.
- However this effect is largely compensated by the structural fact, we have already observed, that the natality rate ν_2 is present in V_2, while ν_1 does not contribute to V_1. In Demography the ratio ν_1/ν_2, commonly called *primary sex ratio* is stable in time and in space at a well known special level[26]:

$$\frac{\nu_1}{\nu_2} \approx 1.03 \text{ forever and everywhere}$$

Therefore the absence of ν_1 in V_1 should over-compensate the inequality between mortality rates and bring us to the conclusion that $V_2 > V_1$ is a largely acceptable assumption.

Now let us try to dig information in Eq. (4.4.29). The first addendum in the r.h.s.:

$$x_1(0)\, e^{V_1 t}$$

is a standard exponential-like formula. The qualitative behavior of this quantity is not influenced by the *positive* multiplicative constant $x_1(0)$, and is decided by the sign of V_1. We can provide some elementary conclusions:

(i)   if $V_1 > 0$ or $\lambda_1 > \mu_1 \Longrightarrow$ the first addendum increases indefinitely when t goes to $+\infty$;

(ii) if $V_1 = 0$ or $\lambda_1 = \mu_1 \Longrightarrow$ the first addendum is constant;

(iii) if $V_1 > 0$ or $\lambda_1 > \mu_1 \Longrightarrow$ the first addendum vanishes as t goes to $+\infty$.

Of course, the second case ($V_1 = 0$) is empirically irrelevant.

It should be evident that the sign of V_1 is determined by the (non)-capability of migrations to compensate deaths.

4.4.7   Linear Continuous Systems

What we have seen for linear discrete dynamic system can be proposed also for continuous ones. For a generic linear system of dimension n, the motion law is of the type we have seen at p. 234:

$$\mathbf{x}'(t) = \mathbf{f}\,[\mathbf{x}(t), t]$$

---

[26]The fact that the number of male newborns slightly exceeds that of females ($\approx +3\%$) is usually interpreted as a compensation of the fact that $\mu_1 > \mu_2$. This natural parameter can be altered in social systems where females are not welcome.

In principle, the structure of the function $\mathbf{f}$ can be very complicate, but, in the applications, very frequently we observe the following one:

$$\mathbf{x}'(t) = A(t)\,\mathbf{x}(t) + \mathbf{b}(t) \tag{4.4.30}$$

in which, at any fixed time, the r.h.s. of the equation is a linear affine function of the state vector. More explicitly:

$$\begin{cases} x_1'(t) = a_{11}(t)\,x_1(t) + a_{12}(t)\,x_2(t) + \cdots + a_{1n}(t)\,x_n(t) + b_1(t) \\ x_2'(t) = a_{21}(t)\,x_1(t) + a_{22}(t)\,x_2(t) + \cdots + a_{2n}(t)\,x_n(t) + b_2(t) \\ \qquad\qquad\qquad\qquad\qquad \vdots \\ x_n'(t) = a_{n1}(t)\,x_1(t) + a_{n2}(t)\,x_2(t) + \cdots + a_{nn}(t)\,x_n(t) + b_n(t) \end{cases}$$

$$\tag{4.4.31}$$

where:

$$A(t) = \begin{bmatrix} a_{11}(t) & a_{12}(t) & \cdots & a_{1n}(t) \\ a_{21}(t) & a_{22}(t) & \cdots & a_{2n}(t) \\ \vdots & & & \\ a_{n1}(t) & a_{n2}(t) & \cdots & a_{nn}(t) \end{bmatrix} \quad \text{and } \mathbf{b}(t) = \begin{bmatrix} b_1(t) \\ b_2(t) \\ \vdots \\ b_n(t) \end{bmatrix}$$

Example 4.4.16 — Consider this motion law:

$$x_1'(t) = 3tx_1(t)\,x_2(t) + t^2$$
$$x_2\grave{i}\,(t) = 4(t)\,x_1(t) - 5t^2x_2(t) + t^3$$

Well, this motion law *does not characterize* a linear system because, in the first equation, the two state variables are multiplied with each other and this is incompatible with (4.4.31).

Example 4.4.17 — Consider this motion law:

$$x_1(t+1) = 3tx_1(t) - 2t^7x_2(t) + t^2$$
$$x_2(t+1) = 4tx_1(t) - 5t^2x_2(t) + t^3$$

Well, this motion law *characterizes* a linear system because its structure is compatible with (4.4.31). Specifically:

$$\begin{cases} a_{11}(t) = 3t \\ a_{12}(t) = -2t^7 \\ a_{21}(t) = 4t \\ a_{22}(t) = 5t^2 \\ b_1(t) = t^2 \\ b_2(t) = t^3 \end{cases}$$

Remark 4.4.11 — Note that what is required for the linearity of the system is not linearity of the r.h.s. w.r.t. time, but only w.r.t. the state variables.

It is easy to find an interpretation for the structure of linear dynamic systems, which is illuminating for the applications. What stays behind (4.4.30) or (4.4.31) is the idea that the movement direction ($\mathbf{x}'(t)$) is partially determined by the present of the system ($\mathbf{x}(t)$) and partially by forces outside the system ($\mathbf{b}(t)$). We can label the addendum $A(t)\mathbf{x}(t)$ as the *endogenous part* of $\mathbf{x}'(t)$ and $\mathbf{b}(t)$ as the *exogenous* one (see above, for the discrete case at p. 235):

$$\mathbf{x}'(t) = \underbrace{A(t)\mathbf{x}(t)}_{\text{endog.}} + \underbrace{\mathbf{b}(t)}_{\text{exog.}}$$

Thanks to this interpretation, frequently, $\mathbf{b}(t)$ is called *forcing term* in the motion law.

4.4.7.1 Types of Linear Systems

A special case of interest is the one in which the forcing term is identically null:

$$\mathbf{b}(t) = 0 \text{ for every } t \geq 0 \text{ or } \mathbf{b}(t) \equiv 0 \qquad (4.4.32)$$

Definition 4.4.5 — When (4.4.32) occurs we say that the linear system is *homogeneous*, otherwise we say that it is *non-homogeneous*. In the homogenous case, the motion law turns out to be simply:

$$\mathbf{x}'(t) = A(t)\mathbf{x}(t)$$

Example 4.4.18 – For instance, the system:

$$x_1'(t) = 3tx_1(t) - 2t^7x_2(t) + t^2$$
$$x_2'(t) = 4tx_1(t) - 5t^2x_2(t) + t^3$$

is non-homogeneous, while the following one:

$$x_1'(t) = 3tx_1(t) - 2t^7x_2(t)$$
$$x_2'(t) = 4tx_1(t) - 5t^2x_2(t)$$

is homogeneous.

Another relevant special case is characterized by the time independence of the coefficients $a_{r,s}(t)$.

Definition 4.4.6 — If:

$$A(t) \equiv A$$

we say that the system has *constant coefficients*.

Example 4.4.19 — The system:

$$\begin{cases} x_1' \, (t) = 3tx_1 \, (t) - 2t^7 x_2 \, (t) + t^2 \\ x_2' \, (t) = 4tx_1 \, (t) - 5t^2 x_2 \, (t) + t^3 \end{cases}$$

is not a system with constant coefficients, because *at least one coefficient is not constant.* In this case abundantly: all the coefficients are not constant. The system:

$$\begin{cases} x_1' \, (t) = 3x_1 \, (t) - 2x_2 \, (t) + t^2 \\ x_2' \, (t) = 4x_1 \, (t) - 5x_2 \, (t) + t^3 \end{cases}$$

is a system with constant coefficients. The constant matrix of coefficients is:

$$A = \begin{bmatrix} 3 & -2 \\ 4 & -5 \end{bmatrix}$$

A system with constant coefficients can have forcing term dependent on time. Constancy of coefficients implies concretely that the endogenous evolution rules are time-invariant. This does not imply that the exogenous influence is constant. In the case $A, \mathbf{b}$ are both constant:

$$\mathbf{x} \, (t + 1) = A\mathbf{x} \, (t) + \mathbf{b}$$

the linear system is also autonomous.

4.4.7.2 How to Solve Linear Systems

Bad news, in general, w.r.t. the discrete case we have seen before.
  In the case the dimension n of the linear dynamic system is unitary:

$$x' \, (t) = a \, (t) \, x \, (t) + b \, (t) \tag{4.4.33}$$

it is possible to write an expression for the solution of the differential equation (4.4.33). When $n > 1$, there is no general explicit expression for the solution.
  In the case of homogeneous systems with constant coefficients:

$$\mathbf{x}' \, (t) = A\mathbf{x} \, (t) \tag{4.4.34}$$

there is an 'expensive' possibility to write down an expression for $\mathbf{x} \, (t)$. It is expensive because it requires some more pure mathematics (trigonometric functions, complex numbers, spectral theory of matrices), which would unbalance this textbook.
  For non-homogeneous linear dynamic systems with constant coefficients:

$$\mathbf{x}' \, (t) = A\mathbf{x} \, (t) + \mathbf{b} \, (t) \tag{4.4.35}$$

there are useful techniques, which allow, once the homogeneous equation (4.4.34) has been solved, to find the solution for the non-homogeneous (4.4.35) through some attempts suggested by the analytic structure of the forcing term $\mathbf{b}(t)$. However, this approach works pretty well for a rather large class of forcing terms, but not in general.

Example 4.4.20 An example of equation of the type (4.4.33) could be:

$$x'(t) = \frac{x(t)}{t} + t^2$$

In this case:

$$a(t) = 1/t \text{ and } b(t) = t^2$$

A procedure for solving the equation suggests first to put all the terms involving x, x', the state variable and its time derivative, in the l.h.s. of the equation:

$$x'(t) - a(t) x(t) = b(t)$$

Calling $A(t)$ any primitive of $a(t)$, we then multiply both sides of the equation by the so-called *integrating factor*:

$$e^{-A(t)}$$

We get:

$$x'(t) e^{-A(t)} - e^{-A(t)} a(t) x(t) = e^{-A(t)} b(t)$$

It is evident that the l.h.s. of the equation is the derivative[27] of $x(t) e^{-A(t)}$:

$$D\left[x(t) e^{-A(t)}\right] = e^{-A(t)} b(t)$$

hence:

$$x(t) e^{-A(t)} = \int e^{A(t)} b(t) \, dt + c$$

being c an arbitrary constant, and the general solution:

$$x(t) = e^{A(t)} \left[c + \int e^{-A(t)} b(t) \, dt\right]$$

Back to the preceding example, in order to see how the trick works:

---

[27] Remember the "hairdresser theorem":

$$D[\alpha(t) \beta(t)] = \alpha'(t) \beta(t) + \alpha(t) \beta'(t)$$

and that, thanks to the chain rule:

$$D\left[e^{\alpha(t)}\right] = e^{\alpha(t)} \alpha'(t)$$

Example 4.4.21 We have

$$x'(t) = \frac{x(t)}{t} + t^2 \text{ or } x'(t) - \frac{x(t)}{t} = t^2 \qquad (4.4.36)$$

For simplicity, let us restrict ourselves to $t > 0$. In this case:

$$a(t) = \frac{1}{t} \text{ and } b(t) = t^2$$

We also have:

$$A(t) = \ln|t| \Rightarrow e^{A(t)} = t \Longrightarrow e^{-A(t)} = \frac{1}{t}$$

Multiplying by the integrating factor $\frac{1}{t}$ we get:

$$x'(t)\frac{1}{t} - x(t)\frac{1}{t^2} = t$$

The l.h.s. is the derivative of:

$$x(t) \cdot \frac{1}{t}$$

and hence:

$$x(t) \cdot \frac{1}{t} = \int t \, dt + c$$

hence the general solution:

$$x(t) = t\left(\frac{t^2}{2} + c\right) = \frac{t^3}{2} + ct \qquad (4.4.37)$$

Two remarks.

1. There is, also in this case, a variant of the procedure, we saw on p. 252, that via definite integrals singles out the solution satisfying an appropriate boundary condition. However we do not present it in detail.
2. It is easy to check that each one of the (infinity of) functions (4.4.37) satisfies the motion law (4.4.36):

$$x'(t) = \frac{x(t)}{t} + t^2$$

In fact, differentiating (4.4.37) we get:

$$x'(t) = \frac{3}{2}t^2 + c$$

and substituting:

$$\frac{x\,(t)}{t} + t^2 = \frac{\dfrac{t^3}{2} + ct}{t} + t^2 = \frac{3}{2}t^2 + c$$

4.4.7.3 An Interesting Socio-demographic Model/2

Later we will introduce new concepts and new techniques in order to study the long term behavior of a dynamic system. It is worthwhile to meet informally some of these facts in advance, using a model of some interest like the one we have already considered. This anticipation should help the reader to grasp immediately the concrete relevance of these notions and the solutions of some natural problems.

Take above (4.4.27) and consider the following question 'Is it possible that the size and the composition of the population does not vary in time even if it is pushed to variation by the motion law (4.4.27)?'.

A partial answer to this question is rather intuitive: if $x_1\,(0) = x_2\,(0) = 0$, then it is obvious that the state vector $\mathbf{x}\,(t)$ will stay at $\mathbf{o}$ for every $t \geq 0$ (i.e.; forever, a matter of language).

However a reasonable further question is: 'Is this the only possibility to "freeze" the system?'.

We can answer asking us whether there are vectors $\mathbf{x}^* = \begin{bmatrix} x_1^* \\ x_2^* \end{bmatrix}$ such that:

$$\mathbf{x}\,(t) \equiv \mathbf{x}^* \text{ for every } t \geq 0$$

We will learn later that such points are called "equilibrium points" or more simply "equilibria" for the dynamic system.

After having observed that if $\mathbf{x}\,(t) \equiv \mathbf{x}^*$ for every $t \geq 0$, obviously $\mathbf{x}'\,(t) \equiv \mathbf{o}$ for every $t \geq 0$, we can derive from (4.4.27) the "equilibrium system":

$$\mathbf{o} = A\mathbf{x}^*$$

or, more explicitly:

$$\begin{bmatrix} 0 \\ 0 \end{bmatrix} = \underbrace{\begin{bmatrix} V_1 & \nu_1 \\ 0 & V_2 \end{bmatrix}}_{A} \begin{bmatrix} x_1^* \\ x_2^* \end{bmatrix}$$

Assuming that $V_1 V_2 \neq 0$, we have to face a Cramer system (see Chap. 1), which has a unique solution. Therefore the origin $\mathbf{o}$ is the only "freezing point".

Later, should we move the system from the "freezing point": for instance we construct a new population with — say — one male and one female, what would happen? Which will be the future of this population under the rules of the motion law: will it be brought to extinction or not? These questions will be studied later introducing the notions of "stability/instability" of an equilibrium.

4.4.7.4 An Interesting Socio-demographic Model/3

Let us reconsider the population model without age-structure we have introduced above at p. 234 and let us introduce a variant with respect to the case (4.4.27) of constant demographic rates ν_s, μ_s, λ_s, with $s = 1, 2$. The motion law in that model assumed that the migration balance, both for males and females, was proportional to the current size of each of the two sub-populations.

The corresponding addenda in the r.h.s. of the motion law (4.4.27) were:

$$\begin{cases} \lambda_1 x_1 (t) & \text{for males} \\ \lambda_2 x_2 (t) & \text{for females} \end{cases}$$

We explore here a different case, which will allow us to introduce some novelty of interest.

The difference stays in the fact that instead of *linear dependence* of migration balances on the corresponding sub-population sizes, we will assume *linear affine dependence*:

$$\begin{cases} \lambda_1 x_1 (t) + \kappa_1 & \text{for males} \\ \lambda_2 x_2 (t) + \kappa_2 & \text{for females} \end{cases}$$

This concretely means that there are (constant) migration phenomena, independent of the sub-populations sizes.[28]

The motion law (4.4.27) becomes:

$$\begin{bmatrix} x_1' (t) \\ x_2' (t) \end{bmatrix} = \underbrace{\begin{bmatrix} V_1 & \nu_1 \\ 0 & V_2 \end{bmatrix}}_{A} \begin{bmatrix} x_1 (t) \\ x_2 (t) \end{bmatrix} + \underbrace{\begin{bmatrix} \kappa_1 \\ \kappa_2 \end{bmatrix}}_{\kappa}$$

In the case the vector κ is not null ($\kappa \neq \mathbf{o}$), such a dynamic system will be called "non-homogeneous", because of the presence of a non-null forcing term. An economist would qualify this addendum as "exogenous".

The equilibrium system is:

$$\begin{bmatrix} 0 \\ 0 \end{bmatrix} = \underbrace{\begin{bmatrix} V_1 & \nu_1 \\ 0 & V_2 \end{bmatrix}}_{A} \begin{bmatrix} x_1^* \\ x_2^* \end{bmatrix} + \underbrace{\begin{bmatrix} \kappa_1 \\ \kappa_2 \end{bmatrix}}_{\kappa}$$

Under the natural assumption that $V_1 V_2 \neq 0$, we have again a Cramer's system. From:

$$\begin{bmatrix} x_1^* \\ x_2^* \end{bmatrix} = \begin{bmatrix} V_1 & \nu_1 \\ 0 & V_2 \end{bmatrix}^{-1} \begin{bmatrix} -\kappa_1 \\ -\kappa_2 \end{bmatrix}$$

we get:

---

[28]Recent facts (end of the Twenties) in Southern Europe should make this variant of some interest.

$$\begin{bmatrix} x_1^* \\ x_2^* \end{bmatrix} = \begin{bmatrix} \dfrac{\nu_1 \kappa_2 - \nu_2 \kappa_1}{V_1 V_2} \\[2ex] -\dfrac{\kappa_2}{V_2} \end{bmatrix} \tag{4.4.38}$$

In order that this stationary population can concretely make sense, it is essential that $\mathbf{x}^*$ has non-negative components. As concerns x_2^*, this requirement boils down to the following condition system:

$$\text{(a)} \ \begin{cases} \kappa_2 < 0 \\ V_2 > 0 \end{cases} \ \text{or (b)} \ \begin{cases} \kappa_2 > 0 \\ V_2 < 0 \end{cases} \tag{4.4.39}$$

The interpretation is straightforward. Recalling that $V_2 = \nu_2 - \mu_2 + \lambda_2$ is the variation rate of the female sub-population, acting on its current size, its positivity (case (a)) would guarantee an exponential increase of the number of females, which is incompatible with stationarity, so, necessarily the exogenous movement must be negative. The argument for the case (b) is symmetric. If $V_2 < 0$, in the absence of exogenous immigration ($\kappa_2 > 0$), the population would be brought to extinction.

The analogous interpretation concerning males is slightly more involved, because, as we know that females have their own history, the one of males is simply intertwined with the one of females.

Let us first rewrite appropriately the expression of x_1^* in the positivity condition of the stock of males in the population:

$$\frac{\nu_1}{V_1} \cdot \underbrace{\frac{\kappa_2}{V_2}}_{<0} > \frac{\kappa_1}{V_1} \tag{4.4.40}$$

The negativity of κ_2/V_2 stems from the sign analysis of x_2, we have just seen.

We must distinguish the two cases:

$$\text{(a) } V_1 > 0 \text{ and (b) } V_1 < 0$$

Recalling that $V_1 = \lambda_1 - \mu_1$, the case (a) corresponds to the case that endogenous immigration overcompensate male deaths. in this case, as $\nu_1/V_1 > 0$, the l.h.s. of (4.4.40) turns out to be negative. Therefore, the positivity condition for x_1^* requires that $\kappa_1/V_1 < 0$, hence $\kappa_1 < 0$. in this case stationarity requires that the exogenous migration must have a negative balance. In the case (b), when $V_1 < 0$, the l.h.s. of (4.4.40) is positive and therefore the positivity condition is of the type:

$$\frac{\kappa_1}{V_1} < \underbrace{\frac{\nu_1 \kappa_2}{V_1 V_2}}_{>0}$$

hence, multiplying both sides by $V_1 < 0$:

$$\kappa_1 > \nu_1 \underbrace{\frac{\kappa_2}{V_2}}_{<0}$$

requiring that the exogenous migration κ_1 must be non-negative or, if negative, not "too negative".

We conclude this (long) application, showing how these results can be used in Social Sciences.

The problem we will face is that of understanding the long term effects of some social aspect on the size of a population. It will be also an occasion for using jointly various tools presented in this textbook.

Focus your attention on the education level of females: a typical effect of a higher education level is the reduction of natality rates (ν_1 and ν_2).

Assuming that the population is at the equilibrium $\mathbf{x}^*$ (see (4.4.38) above) we study how the equilibrium sizes x_1^*, x_2^* do move when ν_2 varies. From:

$$x_2^* = \frac{-\kappa_2}{\nu_2 + \lambda_2 - \mu_2}$$

we can compute $\partial x_2^*/\partial \nu_2$. Let us assume that $V_2 = \nu_2 + \lambda_2 - \mu_2 > 0$, in which case necessarily $\kappa_2 < 0$ (see the case (a) in (4.4.39) above). We have:

$$\frac{\partial x_2^*}{\partial \nu_2} = \frac{\kappa_2}{(\nu_2 + \lambda_2 - \mu_2)^2} < 0$$

This means that when the natality rate for females declines, the equilibrium size of the female subpopulation decreases.

In order to understand what happens to the size of the male subpopulation, it is not necessary to use differential calculus. In fact, from the equation defining x_2^* in (4.4.38), we get[29]: $\kappa_2 = -V_2 x_2^*$. Substituting κ_2 in the equation defining x_1^*, again in (4.4.38), we get:

$$x_1^* = \frac{\nu_1 \left(-V_2 x_2^*\right) - V_2 \kappa_1}{V_1 V_2} = \left(-\frac{\nu_1}{V_1}\right) x_2^* - \frac{\kappa_1}{V_1}$$

This equation tells us another part of the story: at the equilibrium the number of males is a linear affine function of the number of females. To understand the direction of variation of x_1^* it is sufficient to ascertain the sign of the slope $(-\nu_1/V_1)$. As we have seen above, it is likely that $V_1 < 0$, hence the slope is positive: this means that if the number of females goes down, the number of males does the same.

Example 4.4.22 — In the preceding analysis, we have assumed a reduction in the female birth rate ν_2 and assumed that the birth rate for males ν_1 does not change.

---

[29]The concrete interpretation of this equality is evident: at the equilibrium, the net number of females leaving exogenously the population must balance the net endogenous number of new females.

This is not totally realistic, because, in general, these birth rates are proportional: $\nu_1 = 1.03\nu_2$. The reader is invited to repeat the analysis under this assumption and to show that the qualitative conclusions are the same.

Example 4.4.23 — The reader is invited to examine another connection between social aspects and demography of a population: health and equilibrium size. A better health system should reduce the mortality rates. Assuming that

$$\mu_1 = a\mu_2 \tag{4.4.41}$$

being $a > 1$ constant, find the effects on $\mathbf{x}^*$ of a reduction of μ_2 (and consequently of μ_1 according to the proportionality assumption (4.4.41)).

A Pair of Application to Economic Policy

Here are two applications of linear differential equations to a problem of Economic Policy: how to handle dynamically public debt.

Example 4.4.24 Take a Country, with public debt: this is not a difficult task. Continuous time is $t \geq 0$. The debt evolves in time because of accrued interests and repayment policies. Let $D(t)$ be the public debt at t. Let also $\delta(t)$ be the interest rate at t and $R(t)$ the repayment intensity at t. The basic model of the evolution of the debt is locally described:

$$D(t+h) = D(t) + D(t)\,\delta(t)\,h - R(t)\,h + o(h)$$

hence, assuming that D is differentiable:

$$D'(t) = \delta(t)\,D(t) - R(t) \tag{4.4.42}$$

The evolution of public debt can be described, solving the differential equation (4.4.42). Let $\Delta(t)$ be any antiderivative of $\delta(t)$, The integrating factor is:

$$e^{-\int \delta(t)dt} = e^{-\Delta(t)}$$

Multiplying both sides of (4.4.42), we get:

$$D'(t)\,e^{-\Delta(t)} - D(t)\,e^{-\Delta(t)}\delta(t) = R(t)\,e^{-\Delta(t)}$$

As the l.h.s. is nothing but:

$$\frac{d}{dt}\left[D(t)\,e^{-\Delta(t)}\right]$$

we get:

$$\frac{d}{dt}\left[D(t)\,e^{-\Delta(t)}\right] = R(t)\,e^{-\Delta(t)}$$

Hence:

$$D(t)\,e^{-\Delta(t)} = \int R(t)\,e^{-\Delta(t)}dt + c, \text{ with } c \in \mathbb{R}, \text{ arbitrary}$$

Now, we want to focus our attention on a very particular case:

$$\delta(t) \equiv \delta \text{ constant and } R(t) \equiv R \text{ constant}$$

In this case $\Delta(t) = \delta t$. The integrating factor turns out to be $e^{-\delta t}$ and, consequently:

$$D(t) = e^{\delta t}\left(c + \int Re^{-\delta t}dt\right) = -\frac{R}{\delta} + ce^{\delta t} \qquad (4.4.43)$$

The value of c strongly depends on $D(0)$, the initial amount of the debt. The connection between c, the initial debt can be found easily: it suffices to let $t = 0$ in (4.4.43) and solve for c:

$$-\frac{R}{\delta} + ce^{\delta \cdot 0} = D(0)$$

hence:

$$c = D(0) - \frac{R}{\delta}$$

Therefore:

$$D(t) = \frac{R}{\delta} + \left[D(0) - \frac{R}{\delta}\right]e^{\delta t}$$

The behavior of public debt under such assumptions is clear-cut. If the initial debt $D(0)$ is equal at the crucial level R/δ, $D(t) \equiv R/\delta$: the Government pays due interests and the principal component of the debt turns out to be constant. Such a repayment policy is terrible, if the starting point of the debt:

$$D(0) > \frac{R}{\delta}$$

The debt increases exponentially and explodes. The political reason is that the repayment rate R is even not able to pay the interests. The last inequality tells us, in fact, that:

$$R < \delta D(0)$$

If, as an alternative:

$$R > \delta D(0) \Rightarrow D(0) < \frac{R}{\delta} \qquad (4.4.44)$$

then the public debt will go down to 0. The equation in t:

$$\frac{R}{\delta} + \left[D(0) - \frac{R}{\delta}\right]e^{\delta t} = 0$$

provides us with the extinction date t^* of the debt. We get:

$$t^* = \frac{1}{\delta}\left[\ln\frac{R}{R - \delta D\,(0)}\right]$$

A numerical case is useful. Assume that $\delta = 1\%$, $D\,(0) = 1000$, $R = 12$. The condition (4.4.44) is satisfied as:

$$12 > 1000 \times 0.01 = 10$$

Therefore:

$$t^* = \frac{1}{0.01}\ln\left(\frac{12}{12 - 0.01 \times 1000}\right) = 179.18$$

If $\delta = 0.01$ and $R = 12$ then $R/\delta = 1200$ and the public debt evolution is

$$D\,(t) = \frac{R}{\delta} + \left[D\,(0) - \frac{R}{\delta}\right]e^{\delta t}$$

The Fig. 4.2 shows the trajectories starting from $D\,(0) = 1100$, 1200 and 1300 respectively.

The political lecture of this example is rather intuitive: if the cost of debt is constant over time and if you — Government — repay a constant amount, greater than due interests, you'll be able to extinguish your debt. If what you pay are exactly the interests, your debt will remain constant in time. If your repayment is not sufficient to pay at least interests, your debt will increase exponentially.

Fig. 4.2 Trajectories of public debt

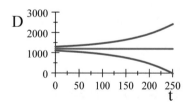

Example 4.4.25 Take the preceding example and consider this variant of the general model. It has some concrete interest in the (common) case of Countries with a relevant debt $D\,(0)$ at some starting date 0. We will preserve the assumption of constant cost δ of public debt, but we want to explore what happens if the reimbursement intensity $R\,(t)$ varies in time. A realistic assumption we are considering assumes that the initial reimbursement rate is not sufficient to repay the debt, but that the repayment intensity will increase exponentially in time:

$$R\,(t) = R\,(0)\,e^{\alpha t} \text{ with } \alpha > 0$$

We take the motion law equation:

$$D'(t) = \delta D(t) - R(0) e^{\alpha t}$$

we multiply by the integrating factor $e^{-\delta t}$ both sides of the new version of (4.4.42), getting:

$$D'(t) e^{-\delta t} - D(t) e^{-\delta t} \delta = -R(0) e^{(\alpha-\delta)t}$$

As already ascertained, the l.h.s. is nothing but:

$$\frac{d}{dt} \left[D(t) e^{-\delta t} \right] = -R(0) e^{(\alpha-\delta)t}$$

Hence:

$$D(t) e^{-\delta t} = -R(0) \frac{e^{(\alpha-\delta)t}}{\alpha - \delta} + c$$

with $c \in \mathbb{R}$, arbitrary, or, finally:

$$D(t) = c e^{\delta t} - R(0) \frac{e^{\alpha t}}{\alpha - \delta} \qquad (4.4.45)$$

In order to make the example interesting we introduce two additional assumptions:

(1) $D(0) > \dfrac{R(0)}{\delta}$;

(2) $\alpha > \delta$. Taking into account what we have seen in the previous example, the first condition tells us that, if the Country would repay the debt with constant intensity $R(0)$, in the long term, the public debt would explode. The second assumption tells us that the growth rate α of the repayment intensity is higher the the one at which the public debt grows because of accrued interests. We want to get a confirmation of what intuition should suggest to every reader: at the beginning the reimbursement intensity is too low for being able to brake the growth of debt, but after a sufficient time this should happen and the total repayment of the debt should occur.

Let us first determine the constant c as a function of the initial level of the debt $D(0)$. We take (4.4.45) above. Letting $t = 0$ we obtain:

$$D(0) = c - \frac{R(0)}{\alpha - \delta} \quad \text{and consequently} \quad c = D(0) + \frac{R(0)}{\alpha - \delta}$$

Therefore the debt trajectory starting at 0 by some given level $D(0)$ has equation:

$$D(t) = \left(D(0) + \frac{R(0)}{\alpha - \delta} \right) e^{\delta t} - R(0) \frac{e^{\alpha t}}{\alpha - \delta} \qquad (4.4.46)$$

Differentiating we get:

$$D'(t) = \left(D(0) + \frac{R(0)}{\alpha - \delta}\right)\delta e^{\delta t} - \frac{R(0)\,\alpha}{\alpha - \delta}e^{\alpha t}$$

Some tedious computations tell us that $D'(t) > 0$ for sufficiently small values of $t > 0$. The derivative of D is null at:

$$t^* = \frac{1}{\alpha - \delta}\ln\frac{\delta\,(\alpha - \delta)\,D(0) + R(0)}{R(0)\,\alpha}$$

and is negative for $t > t^*$. If we want to compute at what time the debt will be completely repaid, using (4.4.46) we have to find: t^{**} such that $D(t^{**}) = 0$:

$$\left(D(0) + \frac{R(0)}{\alpha - \delta}\right)e^{\delta t} - R(0)\frac{e^{\alpha t}}{\alpha - \delta} = 0$$

We find easily:

$$t^{**} = \frac{1}{\alpha - \delta}\ln\frac{(\alpha - \delta)\,D(0) + R(0)}{R(0)} = \frac{1}{\alpha - \delta}\ln\left[1 + (\alpha - \delta)\frac{D(0)}{R(0)}\right]$$

4.4.7.5   Linear Autonomous Systems

In several applications a frequent scheme, used by social researchers, boils down to:

$$\left.\begin{array}{c}\mathbf{x}(t+1)\\\mathbf{x}'(t)\end{array}\right\} = A\mathbf{x}(t) + \mathbf{b}$$

In this scheme, A is a square nth order matrix and $\mathbf{b}$ is an n-vector. We have already anticipated this scheme above.

Example 4.4.26 — For instance:

$$\left.\begin{array}{c}\mathbf{x}(t+1)\\\mathbf{x}'(t)\end{array}\right\} = \begin{bmatrix} 3 & -1 \\ 2 & 4 \end{bmatrix}\begin{bmatrix} x_1(t) \\ x_2(t) \end{bmatrix} + \begin{bmatrix} 3 \\ -4 \end{bmatrix} \qquad (4.4.47)$$

is such a system, while:

$$\left.\begin{array}{c}\mathbf{x}(t+1)\\\mathbf{x}'(t)\end{array}\right\} = \begin{bmatrix} 3 & -1 \\ 2 & 4 \end{bmatrix}\begin{bmatrix} x_1(t) \\ x_2(t) \end{bmatrix} + \begin{bmatrix} 3t \\ -4t \end{bmatrix}$$

is not, because, even if the coefficient matrix $\begin{bmatrix} 3 & -1 \\ 2 & 4 \end{bmatrix}$ is constant, the forcing term $\mathbf{b}(t) = \begin{bmatrix} 3t \\ -4t \end{bmatrix}$ depends on time t. For systems like (4.4.47), we are in the presence of autonomous linear dynamic models.

A natural question about such systems, in which both the endogenous evolution and the exogenous one do not vary in time, concerns equilibria, together with their stability.

The social interpretation of such question is rather obvious. Think of some population, think it evolves naturally under rather constant conditions, think that there are migration phenomena, independent of the population composition, which are constant in time. Under these conditions, a relevant question is whether there exists a long term equilibrium for such a population.

All of us are aware about the importance of such questions, together with the difficulty to let people understand them.

Let us start with the continuous time case.

4.4.7.6 Continuous Time Linear Autonomous Systems: Existence of Equilibria

Here is the motion law:
$$\mathbf{x}'(t) = A\mathbf{x}(t) + \mathbf{b}$$

Let us first look for the equilibrium condition (see at p. 286) or the idea that the dynamic system is... not dynamic:
$$\mathbf{o} = A\mathbf{x}^* + \mathbf{b}$$

hence, in the normal case that A is non-singular[30] we get:
$$\mathbf{x}^* = -A^{-1}\mathbf{b}$$

Remark 4.4.12 — If the system starts exactly at $\mathbf{x}^*$, it will continue to stay there forever.

Example 4.4.27 — Take (4.4.47), in the continuous time version:
$$\begin{bmatrix} x_1'(t) \\ x_2'(t) \end{bmatrix} = \begin{bmatrix} 3 & -1 \\ 2 & 4 \end{bmatrix} \begin{bmatrix} x_1(t) \\ x_2(t) \end{bmatrix} + \begin{bmatrix} 3 \\ -4 \end{bmatrix} \tag{4.4.48}$$

The equilibrium equation is:
$$\begin{bmatrix} 0 \\ 0 \end{bmatrix} = \begin{bmatrix} 3 & -1 \\ 2 & 4 \end{bmatrix} \begin{bmatrix} x_1^* \\ x_2^* \end{bmatrix} + \begin{bmatrix} 3 \\ -4 \end{bmatrix}$$

hence:
$$\begin{bmatrix} x_1^* \\ x_2^* \end{bmatrix} = -\begin{bmatrix} 3 & -1 \\ 2 & 4 \end{bmatrix}^{-1} \begin{bmatrix} 3 \\ -4 \end{bmatrix} = \begin{bmatrix} -4/7 \\ 9/7 \end{bmatrix}$$

---

[30]The case of singularity of A is mathematically relevant, but, empirically, irrelevant.

4.4.7.7 Discrete Time Linear Autonomous Systems: Existence of Equilibria

Example 4.4.28 — For instance:

$$\mathbf{x}(t+1) = \begin{bmatrix} 3 & -1 \\ 2 & 4 \end{bmatrix} \begin{bmatrix} x_1(t) \\ x_2(t) \end{bmatrix} + \begin{bmatrix} 3 \\ -4 \end{bmatrix} \tag{4.4.49}$$

is such a system, while:

$$\mathbf{x}(t+1) = \begin{bmatrix} 3 & -1 \\ 2 & 4 \end{bmatrix} \begin{bmatrix} x_1(t) \\ x_2(t) \end{bmatrix} + \begin{bmatrix} 3t \\ -4t \end{bmatrix}$$

is not, because, even if the coefficient matrix $\begin{bmatrix} 3 & -1 \\ 2 & 4 \end{bmatrix}$ is constant, the forcing term $\mathbf{b}(t) = \begin{bmatrix} 3t \\ -4t \end{bmatrix}$ turns out to depend on time t.

We consider the same problem we have met above in the continuous time case.

$$\mathbf{x}(t+1) = A\mathbf{x}(t) + \mathbf{b}$$

Let us first look for the equilibrium condition (see at p. 286):

$$I\mathbf{x}^* = A\mathbf{x}^* + \mathbf{b}$$

hence, in the case $I - A$ is non-singular[31]:

$$\mathbf{x}^* = (I - A)^{-1}\mathbf{b}$$

Example 4.4.29 — Take (4.4.49):

$$\begin{bmatrix} x_1(t+1) \\ x_2(t+1) \end{bmatrix} = \begin{bmatrix} 3 & -1 \\ 2 & 4 \end{bmatrix} \begin{bmatrix} x_1(t) \\ x_2(t) \end{bmatrix} + \begin{bmatrix} 3 \\ -4 \end{bmatrix} \tag{4.4.50}$$

The equilibrium equation is:

$$\begin{bmatrix} x_1^* \\ x_2^* \end{bmatrix} = \begin{bmatrix} 3 & -1 \\ 2 & 4 \end{bmatrix} \begin{bmatrix} x_1^* \\ x_2^* \end{bmatrix} + \begin{bmatrix} 3 \\ -4 \end{bmatrix}$$

hence:

$$\begin{bmatrix} x_1^* \\ x_2^* \end{bmatrix} = \left(\begin{bmatrix} 1 & 0 \\ 0 & 1 \end{bmatrix} - \begin{bmatrix} 3 & -1 \\ 2 & 4 \end{bmatrix} \right)^{-1} \begin{bmatrix} 3 \\ -4 \end{bmatrix} = \begin{bmatrix} -5/8 \\ 7/4 \end{bmatrix}$$

---

[31] The case of singularity of $I - A$ is mathematically relevant. Empirically, it is irrelevant.

4.4.7.8 Stability of the Equilibrium for Continuous Time Systems

In the case of continuous time systems there are reasonable conditions which are necessary and sufficient for the convergence of the state vector $\mathbf{x}\,(t)$ to the equilibrium $-A^{-1}\mathbf{b}$. The reader should have understood that in the toolbox of mathematicians there are several recurring tools. One of them has already played an important role in Chap. 2, when dealing with optimization problems.

The use of the NW principal minors of a square matrix, that could have produced nightmares when reading Chap. 3, is coming back!

Take a linear continuous time autonomous system, with motion law:

$$\mathbf{x}'\,(t) = A\mathbf{x}\,(t) + \mathbf{b}$$

If A is non-singular it admits the equilibrium:

$$\mathbf{x}^* = -A^{-1}\mathbf{b}$$

How to ascertain that the trajectories of the system $\mathbf{x}\,(t)$ will approach the equilibrium $\mathbf{x}^*$?

The first piece of the answer stays above, in some complements about square matrices, that have been introduced in the first chapter, starting at p. 87.

What we need first is the *characteristic polynomial* of A:

$$P\,(\lambda) = \det\left[\lambda I - A\right] = \lambda^n + \alpha_1\lambda^{n-1} + \alpha_2\lambda^{n-2} + \cdots \alpha_1\lambda + \alpha_n$$

We have then to derive a square matrix H of order n, usually entitled to Hurwitz, which uses the coefficients of the powers of λ in the characteristic polynomial $P\,(\lambda)$.

Here are the simple rules allowing us to construct the *Hurwitz matrix H*:

- the first row of H is provided by the polynomial $P\,(\lambda)$, through its odd index coefficients

$$\alpha_1, \alpha_3, \alpha_5, \ldots$$

and it is filled with 0's until the nth position;
- the second row starts first with the even index coefficients[32] in $P\,(\lambda)$:

$$\alpha_0 = 1, \alpha_3, \alpha_5, \ldots$$

etc. and it is filled with 0's until the nth position;
- the third row is like the first one, but its first entry is 0 and then the others repeat the first and of course there is one 0 less in filling the row;
- the fourth row replicates the second one, but with a null entry at the beginning and loss of one 0 on the right;

---

[32]The first entry in this row is the coefficient α_0 of λ^n, which is always unitary.

- the procedure continues with increasing shifts to the right until the nth row has been constructed

Solemnly:

Definition 4.4.7 — Given the characteristic polynomial of A:

$$P(\lambda) = \lambda^n + \alpha_1\lambda^{n-1} + \alpha_2\lambda^{n-2} + \cdots + \alpha_n$$

we call *Hurwitz matrix*, associated to $P(\lambda)$ the nth order square matrix:

$$\underset{n\times n}{H} = \begin{bmatrix} \alpha_1 & \alpha_3 & \alpha_5 & \alpha_7 & \cdots & \cdots & 0 & \cdots & 0 \\ 1 & \alpha_2 & \alpha_4 & \alpha_6 & \cdots & \cdots & 0 & \cdots & 0 \\ 0 & \alpha_1 & \alpha_3 & \alpha_5 & \alpha_7 & \cdots & 0 & \cdots & 0 \\ 0 & 1 & \alpha_2 & \alpha_4 & \alpha_6 & \cdots & 0 & \cdots & 0 \\ \cdots & \cdots & \cdots & \cdots & \cdots & \cdots & \cdots & \cdots & \cdots \\ 0 & 0 & \cdots & \cdots & \cdots & \cdots & \cdots & \cdots & \cdots \end{bmatrix}$$

In particular, if $n = 2$, the characteristic polynomial has this structure:

$$P(\lambda) = \lambda^2 + \alpha_1\lambda + \alpha_2$$

The associated Hurwitz matrix is:

$$H = \begin{bmatrix} \alpha_1 & 0 \\ 1 & \alpha_2 \end{bmatrix}$$

Recalling that such polynomial can be rewritten as (see the last section of Chap. 1):

$$\lambda^2 - \mathrm{tr}(A) + \det(A)$$

we have:

$$H = \begin{bmatrix} -\mathrm{tr}(A) & 0 \\ 1 & \det(A) \end{bmatrix} \tag{4.4.51}$$

if $n = 3$, the structure of the characteristic polynomial is:

$$P(\lambda) = \lambda^3 + \alpha_1\lambda^2 + \alpha_2\lambda + \alpha_3$$

and the associated Hurwitz matrix becomes:

$$H = \begin{bmatrix} \alpha_1 & \alpha_3 & 0 \\ 1 & \alpha_2 & 0 \\ 0 & \alpha_1 & \alpha_3 \end{bmatrix}$$

Now we are ready for a sensational result. There is a simple necessary and sufficient condition (usually called *Routh–Hurwitz condition*) that guarantees the (global) stability of the equilibrium $\mathbf{x}^* = -A\mathbf{b}$. This condition requires simply to ascertain the signs of the NW principal minors of H.

Theorem 4.4.1 (Routh–Hurwitz) — *The positivity of all the NW principal minors of H is necessary and sufficient for the global stability of $\mathbf{x}^*$.*

Remark 4.4.13 Recalling (4.4.51), in the case $n = 2$, the conditions of Theorem 4.4.1 boil down to:

$$\begin{cases} \operatorname{tr}(A) < 0 \\ \det(A) > 0 \end{cases} \tag{4.4.52}$$

Let us see some examples.

Example 4.4.30 Consider the dynamic system with motion law:

$$\mathbf{x}'(t) = \underbrace{\begin{bmatrix} -1/2 & 1/4 \\ 1/4 & -1/3 \end{bmatrix}}_{A} \mathbf{x}(t)$$

This system admits a unique equilibrium: the origin $\mathbf{o}$. The characteristic polynomial of the coefficient matrix is:

$$\det(\lambda I - A) = \lambda^2 + \frac{5}{6}\lambda + \frac{5}{48}$$

The Hurwitz matrix is:

$$H = \begin{bmatrix} 5/6 & 0 \\ 1 & 5/48 \end{bmatrix}$$

Its NW principal minors are:

$$H_1 = h_{11} = 5/6 \text{ and } H_2 = \det H = \frac{5}{6} \cdot \frac{5}{48} = \frac{25}{288}$$

Both are positive, hence the stability condition is satisfied. Note that we can reach the same conclusion through the conditions (4.4.52) noting that:

$$\begin{cases} \operatorname{tr}(A) = -1/2 - 1/3 = -5/6 < 0 \\ \det(A) = -1/2 \cdot (-1/3) - (1/4)^2 = 5/48 > 0 \end{cases}$$

Example 4.4.31 Consider the continuous-time dynamic system with motion law:

$$\mathbf{x}'(t) = \underbrace{\begin{bmatrix} 2 & 0 \\ 0 & -1 \end{bmatrix}}_{A} \mathbf{x}(t)$$

The characteristic polynomial of its coefficient matrix is:

$$\det \left(\lambda \begin{bmatrix} 1 & 0 \\ 0 & 1 \end{bmatrix} - \begin{bmatrix} 2 & 0 \\ 0 & -1 \end{bmatrix} \right) = \lambda^2 - \lambda - 2$$

Let us check for the stability of its only equilibrium through the Hurwitz matrix:

$$H = \begin{bmatrix} -1 & 0 \\ 1 & -2 \end{bmatrix}$$

Its NW principal minors are:

$$H_1 = h_{11} = -1 \text{ and } H_2 = \det H = 2$$

The first minor excludes immediately stability. The same diagnosis stems from trace the trace of the coefficient matrix: $tr(A) = 2 - 1 = 1 > 0$ and the conclusion is at hand on the basis of (4.4.52).

Example 4.4.32 Consider the characteristic polynomial of a matrix A of order 3.

$$P(\lambda) = \lambda^3 + \frac{9}{2}\lambda^2 + 5\lambda + \frac{3}{2}$$

The associated Hurwitz matrix is:

$$H = \begin{bmatrix} 9/2 & 3/2 & 0 \\ 1 & 5 & 0 \\ 0 & 9/2 & 3/2 \end{bmatrix}$$

Its NW principal minors are

$$H_1 = h_{11} = 9/2$$
$$H_2 = \det \begin{bmatrix} 9/2 & 3/2 \\ 1 & 5 \end{bmatrix} = \frac{45}{2} - \frac{3}{2} = \frac{42}{2}$$
$$H_3 = \det \begin{bmatrix} 9/2 & 3/2 & 0 \\ 1 & 5 & 0 \\ 0 & 9/2 & 3/2 \end{bmatrix} = \frac{63}{2}$$

For a dynamic system with motion law:

$$\mathbf{x}'(t) = A\mathbf{x}(t) + \mathbf{b}$$

the equilibrium $\mathbf{x}^* = -A^{-1}\mathbf{b}$ would be globally stable.

Example 4.4.33 Let us look at the following characteristic polynomial:

$$P(\lambda) = \lambda^4 + \frac{9}{2}\lambda^3 + 8\lambda^2 + 7\lambda + 2$$

The associated Hurwitz matrix is:

$$H = \begin{bmatrix} 9/2 & 7 & 0 & 0 \\ 1 & 8 & 2 & 0 \\ 0 & 9/2 & 7 & 0 \\ 0 & 1 & 8 & 2 \end{bmatrix}$$

Its NW principal minors are:

$$H_1 = h_{11} = 9/2$$
$$H_2 = \det \begin{bmatrix} 9/2 & 7 \\ 1 & 8 \end{bmatrix} = 29$$
$$H_3 = \det \begin{bmatrix} 9/2 & 7 & 0 \\ 1 & 8 & 2 \\ 0 & 9/2 & 7 \end{bmatrix} = \frac{325}{2}$$
$$H_4 = \det H = 325$$

As they are all positive, we can derive the global stability of the equilibrium for a continuous-time dynamic system with this characteristic polynomial for its coefficient matrix.

Example 4.4.34 Consider this characteristic polynomial:

$$P(\lambda) = \lambda^4 + \lambda^3 - 2\lambda^2 - 6\lambda - 4$$

The corresponding Hurwitz matrix turns out to be:

$$H = \begin{bmatrix} 1 & -6 & 0 & 0 \\ 1 & -2 & -4 & 0 \\ 0 & 1 & -6 & 0 \\ 0 & 1 & -2 & -4 \end{bmatrix}$$

Its first three NW principal minors are:

$$H_1 = h_{11} = 1$$
$$H_2 = \det \begin{bmatrix} 1 & -6 \\ 1 & -2 \end{bmatrix} = 4$$
$$H_3 = \det \begin{bmatrix} 1 & -6 & 0 \\ 1 & -2 & -4 \\ 0 & 1 & -6 \end{bmatrix} = -20$$

In order to have guarantees for the global stability of the equilibrium it is even non-necessary to compute H_4, as the negativity of H_3 precludes the equilibrium possibility.

4.5 Numerical Approach

"Give me a starting point, a motion law and a computer and I'll evaluate $\mathbf{x}(t)$"

4.5.1 Discrete Systems

This approach is trivial. You are given the motion law:

$$\mathbf{x}(t+1) = \mathbf{f}[\mathbf{x}(t), t]$$

and a starting point:

$$\mathbf{x}(0) = \mathbf{x}^0$$

From such a point, using $\mathbf{f}$ you can compute:

$$\mathbf{x}(1) = \mathbf{f}[\mathbf{x}(0), 0]$$

then:

$$\mathbf{x}(2) = \mathbf{f}[\mathbf{x}(1), 1]$$

etc.

Example 4.5.1 Let:

$$\begin{bmatrix} x_1(t+1) \\ x_2(t+1) \end{bmatrix} = \begin{bmatrix} x_1(t) + x_2(t) - t x_1(t) x_2(t) \\ x_1(t) + x_2(t) \end{bmatrix}$$

be the motion law and $\mathbf{x}(0) = \begin{bmatrix} 1 \\ 2 \end{bmatrix}$. We get:

$$\mathbf{x}(1) = \begin{bmatrix} 1 + 2 - 0 \cdot 1 \cdot 2 \\ 1 + 2 \end{bmatrix} = \begin{bmatrix} 3 \\ 3 \end{bmatrix}$$

and then:

$$\mathbf{x}(2) = \begin{bmatrix} 3 + 3 - 1 \cdot 3 \cdot 3 \\ 3 + 3 \end{bmatrix} = \begin{bmatrix} -3 \\ 6 \end{bmatrix}$$

the next system position will be:

$$\mathbf{x}(3) = \begin{bmatrix} -3 + 6 - 2 \cdot (-3) \cdot 6 \\ -3 + 6 \end{bmatrix} = \begin{bmatrix} 39 \\ 3 \end{bmatrix}$$

etc.

4.5.2 Continuous Systems

We illustrate the strategy in dimension 1, but nothing changes in higher dimensions.
First of all we approximate the continuous model with a discrete one.
In the approximation the epochs we take into consideration are:

$$t = 0; h, 2h, 3h, \ldots, nh, \ldots$$

where $h > 0$ is the step of the *discretization* process.
The continuous-time motion law:

$$x'(t) = f[x(t), t]$$

is approximated with:

$$x(t + h) - x(t) \approx f[x(t), t]h \Longrightarrow x(t + h) \approx x(t) + f[x(t), t]h$$

At this point we can work as in the discrete case. $x(0) = x_0$ is given. We can approximate

$$x(h) \approx x(0) + f[x(0), 0]h$$

and then:

$$x(2h) \approx x(h) + f[x(h), h]h$$

$$x(3h) \approx x(2h) + f[x(2h), 2h]h$$

etc.

Example 4.5.2 Let:

$$x'(t) = x(t) - t \tag{4.5.1}$$

with starting point $x(0) = 1$. The discrete approximation of the motion law is:

$$x(t + h) \approx x(t) + [x(t) - t]h$$

We choose $h = 0.1$. We get:

$$x(0.1) \approx 1 + (1 - 0) \cdot 0.1 = 1.1$$

Iterating:

$$x(0.2) \approx 1.1 + (1.1 - 0.1) \cdot .1 = 1.2$$

etc. As we are able to solve the Eq. (4.5.1), we can compare the approximated solution
with the exact one. The equation we have at hand is linear. Using the technique seen
at p. 264 we find the general solution:

$$x(t) = ce^t + t + 1$$

The boundary condition $x(0) = 1$ provides us with the equation:

$$1 = ce^0 + 0 + 1 \Longrightarrow c = 0$$

therefore the exact solution is:

$$x(t) = t + 1$$

and, owing to the linearity of this solution, there is no approximation error.
In the case the initial condition be $x(0) = 0.5$, c must satisfy:

$$0.5 = ce^0 + 1 + 0 \Longrightarrow c = -0.5$$

hence the exact solution is:

$$x(t) = -0.5e^t + t + 1$$

The Table 4.2 provides us with the exact values $x(t)$, the approximate values $y(t)$ and the % approximation error.

Table 4.2 Trajectory starting from 0.5

Time	Exact x	Approx. y	% Error
0	0.5	0.5	0
0.1	0.547	0.55	0.548
0.2	0.589	0.595	1.018
0.3	0.625	0.635	1.6
0.4	0.654	0.668	2.141
0.5	0.676	0.695	2.811
0.6	0.689	0.714	3.628
0.7	0.693	0.726	4.762
0.8	0.687	0.728	5.968
0.9	0.67	0.721	7.612
1	0.641	0.703	9.672

As you can see the approximation is good at the beginning, but it worsens later. The choice of a smaller step h would help. The Fig. 4.3 illustrates the numerical results.

Fig. 4.3 Exact and
approximate

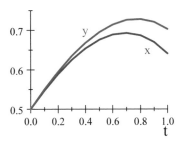

4.6 Qualitative Approach

4.6.1 *Equilibria: Notion and General Systems*

Let us consider the motion law of a DS:

$$\left.\begin{array}{c} \mathbf{x}\,(t+1) \\[2mm] \mathbf{x}'\,(t) \end{array}\right\} = \mathbf{f}\,[\mathbf{x}\,(t)\,,t] \qquad\qquad (4.6.1)$$

We start from the definition of some points in the state space of a dynamic system, which are interesting per se and which will turn out to be crucial for implementing this approach. In words, these points are such that if the system starts at one of them, it will stay there forever.

Definition 4.6.1 We say that a constant (vector) $\mathbf{x}^*$ is an *equilibrium point* for a DS if:

$$\mathbf{x}\,(t) \equiv \mathbf{x}^*$$

defines a solution.

To be clear, if $\mathbf{x}^*$ is an equilibrium point for a discrete system, then:

$$\mathbf{x}\,(0) = \mathbf{x}\,(1) = \mathbf{x}\,(2) = \cdots = \mathbf{x}\,(t) = \cdots = \mathbf{x}^*\ \text{for every}\ t = 0, 1, 2, \ldots$$

while, for a continuous system:

$$\mathbf{x}\,(t) = \mathbf{x}^*\ \text{for every}\ t \geq 0$$

When studying separable equations we have encountered several constant solutions: all of them were identifying equilibria for those DSs. Look, for instance, on p. 251.

Let us try to find a condition which characterizes equilibria. It is sufficient for excluding the interest of equilibria for non-autonomous systems.[33] Let us take (4.6.1). If the system is at some equilibrium $\mathbf{x}^*$, then, owing to (4.6.1), we must have:

$$\mathbf{x}^* \equiv \mathbf{f}\left[\mathbf{x}^*, t\right] \qquad (4.6.2)$$

In principle, this condition turns out to be a bit exotic, because the r.h.s. depends on time t, while its l.h.s. does not, with very special exceptions.

In the continuous case, in the case of a constant solution:

$$\mathbf{x}'(t) \equiv \mathbf{o}$$

hence:

$$\mathbf{o} \equiv \mathbf{f}\left[\mathbf{x}^*, t\right] \qquad (4.6.3)$$

which turns out to be once again a bit exotic too.

Therefore we decide to confine our attention to autonomous systems, where $\mathbf{f}$ is independent of time t:

$$\mathbf{f}\left[\mathbf{x}(t)\right] \text{ instead of } \mathbf{f}\left[\mathbf{x}(t), t\right]$$

Remark 4.6.1 In an applied perspective the autonomy of a system can be seen as structural endogeneity of its evolution: the future of a system depends *only* on where the system is and not on *when*.

We look at some examples, that should help in understanding this point.

Example 4.6.1 Let us start from a discrete DS:

$$x(t+1) = x(t) + t$$

If this system has an equilibrium x^*, the equilibrium condition:

$$x^* \equiv x^* + t$$

turns out to be impossible: the two sides are equal only at $t = 0$.

Example 4.6.2 We continue with the continuous DS with motion law:

$$x'(t) = x(t) + t$$

---

[33] We inform our readers that for non-autonomous systems there is a parallel notion we will not take into consideration.

the equilibrium condition becomes:

$$0 \equiv x^* + t \tag{4.6.4}$$

once again impossible, as above.

We provide also an exceptional example of non-autonomous system exhibiting an equilibrium.

Example 4.6.3 Consider a population, whose initial size is $x(0)$. The growth coefficient for this population is $g(t)$ from $t - 1$ to t:

$$\frac{x(t+1)}{x(t)} = g(t) \tag{4.6.5}$$

hence the motion law:

$$x(t+1) = g(t) x(t)$$

We can easily obtain that:

$$x(t) = x(0) g(0) g(1) \cdots g(t-1)$$

We maintain that $x^* = 0$ is an equilibrium:

$$x(t) = 0 \cdot g(0) g(1) \cdots g(t-1) = 0$$

It's true, but far from being interesting. At $t = 0$ we cannot use (4.6.5) because, at $t = 0$, we should divide by 0, but its version $x(t+1) = g(t) x(t)$ works.

4.6.2 How to Find Equilibria for Autonomous Systems?

It is easier than the reader could think. It is possible to write an equation or a system of equations, characterizing equilibria.

Everything works pretty well because, when we have stated the equilibrium properties above (see: (4.6.2) and (4.6.3)) we have met "strange" equations, in which the number of unknowns were greater than that of the equations. Solving such equations for the state variables what we obtained was the definition of the equilibrium as a (true) function of t, while, by definition, the equilibrium position of the system must not depend on t. Take, for instance, the Eq. (4.6.4) in the example above. If we solve it for x^* we get the contradictory answer:

$$x^* = -t$$

as x^* should not depend on t.

In this case we had:

- one equation;
- two unknowns.

In the case of any autonomous system such an asymmetry disappears: the numbers of the equations and of the unknowns are equal. This fact should sound familiar to the readers: "well posed" systems of equations have the same number of equations and unknowns.

In the autonomous case, the equilibrium condition for DSs:

$$\left.\begin{array}{c} \mathbf{x}\,(t+1) \\[2mm] \mathbf{x}'\,(t) \end{array}\right\} = \mathbf{f}\,[\mathbf{x}\,(t)]$$

produces these (systems of) equation(s):

$$\begin{cases} \text{in the discrete case: } \mathbf{x}^* = \mathbf{f}\,(\mathbf{x}^*) \\[2mm] \text{in the continuous case: } \mathbf{o} = \mathbf{f}\,(\mathbf{x}^*) \end{cases}$$

If n is the system dimension, they are systems of n equations in n unknowns.
   If $n = 1$, what we have is exactly 1 equation (in 1 unknown).
   Let us look at a couple of examples.

Example 4.6.4 Let:

$$\begin{cases} x_1\,(t+1) = 2x_1\,(t) + 3x_2\,(t) + 4 \\[2mm] x_2\,(t+1) = 5x_1\,(t) + 6x_2\,(t) + 7 \end{cases} \tag{4.6.6}$$

At an equilibrium:

$$\mathbf{x}^* = \begin{bmatrix} x_1^* \\[2mm] x_2^* \end{bmatrix}$$

the following conditions must hold:

$$\begin{cases} x_1^* = 2x_1^* + 3x_2^* + 4 \\[2mm] x_2^* = 5x_1^* + 6x_2^* + 7 \end{cases}$$

or:

$$\begin{cases} x_1^* + 3x_2^* = -4 \\[2mm] 5x_1^* + 5x_2^* = -7 \end{cases}$$

equivalent to:

$$\begin{bmatrix} 1 & 3 \\ 5 & 5 \end{bmatrix} \begin{bmatrix} x_1^* \\ x_2^* \end{bmatrix} = \begin{bmatrix} -4 \\ -7 \end{bmatrix}$$

hence:

$$\begin{bmatrix} x_1^* \\ x_2^* \end{bmatrix} = \begin{bmatrix} 1 & 3 \\ 5 & 5 \end{bmatrix}^{-1} \begin{bmatrix} -4 \\ -7 \end{bmatrix} = \begin{bmatrix} -1/10 \\ -13/10 \end{bmatrix}$$

The only equilibrium for this DS of dimension 2, has been obtained solving a system of 2 equations in 2 unknowns. If the system starts exactly at $\begin{bmatrix} -1/10 \\ -13/10 \end{bmatrix}$ and obeys the motion law, it will stay there forever.[34]

Example 4.6.5 Consider this DS of dimension 1:

$$x'(t) = 3x(t)[1 - x(t)] \tag{4.6.7}$$

Its equilibrium condition turns out to be:

$$0 = 3x(1 - x)$$

Trivially, the equation exhibits 2 solutions, hence two equilibria:

$$\begin{cases} x^* = 0 \\ x^{**} = 1 \end{cases}$$

If the system starts exactly from 0 or 1 and obeys the motion law, it will stay there forever.[35]

Remark 4.6.2 (*a bit philosophical*) — When looking for solutions to a difference/ differential equation, we are looking for something "difficult" (= a function), while, when looking for equilibria, we are looking for points (a number or a vector). The reader should perceive a clear downgrade in the problem difficulty.

4.6.3 Nature of an Equilibrium Point

We will shorten our language replacing "equilibrium point" with the (exciting) acronym EQ.

---

[34]Suggestion for the reader: using (4.6.6), try to check this fact!
[35]Suggestion for the reader: try to check that both:

$$\begin{cases} x(t) \equiv 0 \\ x(t) \equiv 1 \end{cases}$$

are constant solutions of the differential equation: they obey the motion law.

In the qualitative approach it turns out to be crucial to stress some substantial difference among EQs. We introduce three different positions, labelled as

- **UE** for *unstable equilibrium*;
- **LSE**[36] for *locally stable equilibrium*;
- **GSE** for *globally stable equilibrium.*

Here are their descriptions:

- **UE** — x^* is an EQ: if the DS starts at x^*, it remains there forever, but, if the system starts from some initial point $x(0)$, even slightly different from x^*, the behavior of the system could turn out to be very different. In practice, the concrete relevance of x^* is negligible, because it is a possible history of the system, but such a history should be completely rewritten if starting from a slightly different position. We will see some interesting politically relevant examples later.
- **LSE** — x^* is an EQ: if the DS starts at x^*, it remains there forever, but, if the system starts from some initial point $x(0)$ not too far from x^*, the ultimate behavior of the system will be close to x^*. The interest for x^* is very high, because we can think of a DS that, starting not too far from x^*, it behaves substantially as if it was starting from the EQ. We will label this case saying that x^* is a *local attractor*.
- **GSE** — x^* is an EQ: if the DS starts at x^*, it remains there forever, but, if the system starts from some initial point $x(0)$ different from x^*, the ultimate behavior of the system will be close to x^*. Its concrete relevance is absolutely very high, because we can think of a DS that, no matter how far from x^* it starts, it behaves substantially as if it was starting from the EQ. We will label this case saying that x^* is possibly a *global attractor*.

The reader will discover that the identification of the nature of any EQ in dimension 1 is very painless. The technical tool needed is named "phase diagram".

4.7 A Newcomer: The Phase Diagram

4.7.1 *Notion*

Decades of experience tell us that several problems in understanding these techniques are originated by the superposition of two different descriptions of the same dynamic phenomenon. For simplicity, we will try to clarify this point using examples of unitary dimension.

We are back to page 222, where the notion of dynamic system was introduced.

We were given a function $x(t)$ and we represented it in the (t, x) plane.

---

[36]Nothing to do with the prestigious London School of Economics and Political Science, usually labelled LSE.

When we know such a function, that is the most natural and efficient way to describe the behavior of the system in time.

However, in several cases, we are not given this function, we cannot even hope to find it, the only information we have is the motion law of the system

$$x'(t) = f[x(t)]$$

The function $x(t)$ cannot be found for a number of possible reasons:

- we do not know the analytic expression of $f[x(t)]$, because we only have qualitative information about it[37];
- we know $f[x(t)]$, but once separated the variables (see above on p. 250):

$$\frac{dx}{f(x)} = dt$$

it turns out that we are not able to compute $\int dx/f(x)$.

Luckily several properties of the dynamic system can be directly extracted from the geometric representation (in a phase diagram) of the motion law itself, in the (x, x') plane.

Please, pay attention to the fact that the geometric description of the system dynamics at p. 222 is an object living in the (t, x) plane, while the geometric description of the motion law is an object living in the (x, x') plane.

Definition 4.7.1 Given an autonomous dynamic system of dimension 1, with motion law:

$$\left.\begin{array}{c} x(t+1) \\ \\ x'(t) \end{array}\right\} = f[x(t)]$$

we call *phase diagram* of the system the diagram with $x(t)$ in abscissae and $x(t+1)$ or $x'(t)$ on the ordinate axis. The diagram is completed with a straight line, respectively:

$$\begin{cases} \text{the bisector of the 1st-3rd quadrant } x(t+1) = x(t) \\ \\ \text{the horizontal axis } x' = 0 \end{cases}$$

---

[37]This situation is common in Social Sciences and makes them different from Hard Sciences like Physics. When studying Solow's model, in the Example 4.2.6 on p. 227, we have handled a production function $F(K, L)$ asking that this function is homogeneous of degree 1 (constant returns to scale), that it is concave (decreasing marginal productivity of the factor) and we have examined a small piece of the neoclassical Economic Theory. If we specify F:

$$F(K, L) = aK^\alpha L^{1-\alpha}$$

according to the well known scheme by Cobb–Douglas, the quality of our analysis goes down, moving from theory to an exercise.

We will usually call *phase curve* the graph of f and *equilibria line*[38], respectively, the bisector or the abscissae axis.

Example 4.7.1 Take the discrete motion law:

$$x(t+1) = \underbrace{\frac{1}{2}x(t)}_{\text{\# of survivors}} + \underbrace{1}_{\text{new entry}}$$

It is the case of a population in which, during each period, one half of it survives and exactly one new individual arrives. The phase diagram is the following one:

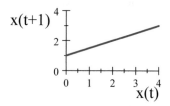

Example 4.7.2 Take the continuous motion law:

$$x'(t) = -\frac{1}{2}x(t) + 1$$

Its interpretation is straightforward if we multiply both sides of the equation by h, thought of as the length of a short time interval[39]:

$$dx(t) = -\frac{1}{2}x(t)h + h$$

This equation tells us the same story as the previous example: in a short time interval the population loses a part of its members (50% on a year basis ($h = 1$) and therefore the remaining part survives) and new entries are registered (1 per year as above):

$$\underbrace{dx}_{\text{population variation}} = \underbrace{-\frac{1}{2}xh}_{\text{lost population}} + \underbrace{h}_{\text{new entries}}$$

---

[38]The reason for this label will appear clear very soon.

[39]Recall that:

$$x'(t)h = dx$$

is the (time) differential of x, or an approximation of its increment $x(t+h) - x(t)$.

Here is the phase diagram portraying the motion law:

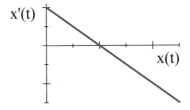

4.7.2 Equilibria in a Phase Diagram

Let's learn to read a phase diagram. The equilibria for an autonomous DS can be easily detected looking at the phase diagram. We must distinguish between the discrete and the continuous case.

4.7.2.1 Discrete Case

An equilibrium for the DS is a point x^* at which two conditions are satisfied:

- $x(t+1) \equiv x(t) = x^*$;
- $x(t+1)$ and $x(t)$ respect the motion law, or $x(t+1) = f[x(t)]$, or $x^* = f(x^*)$.

Geometrically, the first condition requires that:

- x^* belongs to the bisector (here's why we called it *equilibria line*);
- x^* belongs to the phase curve too.

If we look at the phase diagram in the Example 4.7.1 on p. 292, we see there is a unique equilibrium for the system: $x^* = 2$: it is the solution of the equation:

$$x = \frac{1}{2}x + 1 \Longrightarrow x^* = 2$$

Concluding: for discrete autonomous DSs the equilibria are nothing but the intersection(s) between the phase curve and the bisector of the 1st and 3rd quadrant in the phase diagram.

Let's look at another example:

Example 4.7.3 Reconsider the discrete logistic model, we met at p. 224:

$$x(t+1) = x(t) + ax(t)[M - x(t)] \text{ with } a > 0$$

The equilibrium equation is:

$$x = x + ax(M - x)$$

or:

$$x(M - x) = 0$$

and it exhibits the two solutions $x^* = 0$ and $x^* = M$. You find below the phase diagram: the phase curve is a parabola. It intersects the 1st–3rd quadrant bisector at two points. The Fig. 4.4 is the one with $a = 1$ and $M = 2$.

Fig. 4.4 Logistic model

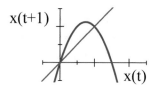

4.7.2.2   Continuous Case

An equilibrium for the DS is a point x^* at which two conditions are satisfied:

- $x'(t) \equiv 0$;
- $x(t)$ respects the motion law, either $x'(t) = f[x(t)]$, or $0 = f(x^*)$.

Geometrically, the first condition requires that:

- x^* belongs to the abscissae axis (here's why we call it *equilibria line*);
- x^* belongs to the phase curve.

If we look at the phase diagram in the Example 4.7.2 on p. 292, we see there is a unique equilibrium for the system: $x^* = 2$: it is the solution of the equation:

$$0 = -\frac{1}{2}x + 1 \Longrightarrow x^* = 2$$

In conclusion: for continuous autonomous DSs the equilibria are given by nothing but the intersection(s) between the phase curve and the abscissae axis in the phase diagram.

Another example.

Example 4.7.4 Consider the continuous-time version of the logistic model, we saw at p. 224:

$$x'(t) = ax(t)[M - x(t)] \text{ with } a > 0$$

The equilibrium equation is:
$$0 = ax\,(M - x)$$

or:
$$x\,(M - x) = 0$$

and it exhibits the two solutions $x^* = 0$ and $x^* = M$. In Fig. 4.5 you find the phase diagram, with $a = 1$ and $M = 10$. The phase curve is a parabola. It intersects the abscissae axis at 0 and at 10.

Fig. 4.5 Continuous logistic

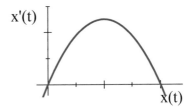

4.7.3 Behaviors Revealed by a Phase Diagram

We announced that, using phase diagrams, it is possible to gather ideas about the (long-term) behavior of a DS.

The practice of the corresponding techniques is different for discrete- and continuous-time DSs. This is the reason for two different subsubsections.

4.7.3.1 Discrete Case

Let us consider the motion law:
$$x\,(t + 1) = 3 + 0.5x\,(t)$$

and construct the phase diagram for the DS obeying it.

This system has a unique equilibrium x^*. It can be found solving:
$$x^* = 3 + 0.5x^*$$

hence:
$$x^* = 6$$

Let us choose a starting point, for instance, $x(0) = 1$. We can compute the following position of the system through the motion law:

$$x(1) = 3 + 0.5x(0) = 3 + 0.5 \times 1 = 3.5$$

and iterating:

$$\begin{cases} x(2) = 3 + 0.5x(1) = 3 + 0.5 \times 3.5 = 4.75 \\ x(3) = 3 + 0.5 \times 4.75 = 5.375 \\ x(4) = 3 + 0.5 \times 5.375 = 5.6875 \\ x(5) = 3 + 0.5 \times 5.6875 = 5.8438 \\ x(6) = 3 + 0.5 \times 5.843\,8 = 5.921\,9 \\ x(7) = 3 + 0.5 \times 5.921\,9 = 5.961\,0 \\ \dots \end{cases} \qquad (4.7.1)$$

The behavior of the system is pretty clear, when t increases indefinitely the state of the system becomes closer and closer to $x^* = 6$. Starting from any other point, we could reach the same conclusion. The price to pay for reaching such conclusion consists in iterative computations like the ones of (4.7.1).[40]

It should appear truly surprising that we can reach the same conclusion, without computations, simply using the phase diagram.

Construct it, choose the starting point $x(0) = 1$ and, using the phase curve, construct the next position of the system as shown in Fig. 4.6.

Fig. 4.6 Phase diagram: first step

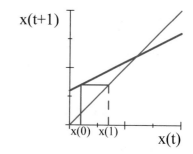

Looking now for a new position, it is sufficient lo look vertically at the phase curve. Through a sequence of segments (the arrows $\updownarrow$ means "vertical" move, while $\longleftrightarrow$ means "horizontal" move) we construct graphically the sequence of system positions and we reach the same conclusion of convergence to the equilibrium as above. The Table 4.3 summarizes the moves you have do to.

The Fig. 4.7 shows the first 4 steps above described.

---

[40]In this case some iterations gave us a rather clear perspective, but examples could be easily constructed, which require millions of iterations to reveal the end of the story.

Suggestion — Repeat the experience starting from different positions: you will reach the same conclusion: if you start from any $x(0) > x^* = 6$ you will get decreasing values for $x(t)$, indefinitely approaching the equilibrium from above, if you start from $x^* = 6$, the phase diagram "freezes" the system at $x^* = 6$.

The graphic method is a **winning** one because:

Table 4.3 How to plot a trajectory

Step	Move
1	Start from the point with abscissa $x(0)$ and ordinate 0
2	↕ go to the phase curve
3	⟷ go to the bisector
4	↕ go to the phase curve
5	⟷ go to the bisector
...	...

Fig. 4.7 Phase diagram: first 4 steps

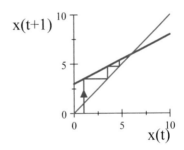

(1) it does not require computations;
(2) it can be sufficient to have only a qualitative graph of f;
(3) you can easily test the effect on the long-term behavior of the system of different choices of the initial condition.

Take this example:

Example 4.7.5

$$x(t+1) = \frac{1}{1+x(t)}$$

Assume we are interested only on a positive equilibrium $x^* > 0$. It should solve the equation:

$$x = \frac{1}{1+x} \Rightarrow x^2 + x - 1 = 0$$

its only positive root is[41]:

---

[41] It is one of the most important numbers of Mathematics and Art. It is known as *Golden Ratio* (or *Aurea Sectio* in Latin). For instance, it governs proportions in Ancient Greece architecture. See, for instance, the masterpiece [8].

$$x^* = \frac{-1 + \sqrt{1 + 4}}{2} = \frac{\sqrt{5} - 1}{2} \approx 0.618\,03$$

Let us construct its phase diagram and the steps we should have learn to do will generate a sort of cobweb[42] suggesting the convergence of the system toward its equilibrium…. The Fig. 4.8a illustrates this case.

The phase diagram also allows us to detect cases of divergence. Take a look at the next example.

Example 4.7.6 The motion law is:

$$x\,(t + 1) = 5 - 2x\,(t)$$

There is only one equilibrium point, solving:

$$x = 5 - 2x \Rightarrow x^* = \frac{5}{3}$$

The phase diagram in Fig. 4.8b allows us to conclude that the DS will escape to ∞. As a starting point we have chosen $x\,(0) = 1.5 < x^*$.

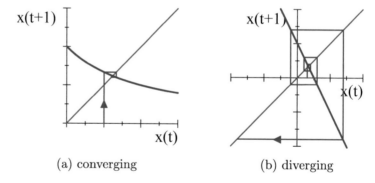

(a) converging   (b) diverging

Fig. 4.8 Cobweb trajectories

We have seen, via the examples, that an equilibrium can attract trajectories or repel them. Such properties of equilibria are labelled as *stability/instability* properties.

A DS can even have several equilibria, with possibly different stability/instability properties.

We present the case in the following example, which will turn out to be crucial for interesting applications, we will see later, in Political Science.

Example 4.7.7 Let us consider a motion law of the type:

$$x\,(t + 1) = f\,[x\,(t)]$$

---

[42] A vivid alternative name for this method is *cobweb method*.

The state variable is confined between 0 and 1:

$$0 \leq x(t) \leq 1$$

Let's think of it as representing the relevant fraction of some population, for instance the portion of voters, who plan to vote some candidate:

$$\begin{cases} x = 0 : \text{the relevant population fraction } x \text{ is null: } 0\% \\ 1 > x > 0 : \text{some positive, but not } 100\% \\ x = 1 : \text{the fraction covers the whole population: } 100\% \end{cases}$$

we do not reveal the analytic expression of f, in order to show you how powerful the qualitative approach is. Assume that the phase diagram be of type A as showed in Fig. 4.9.

Fig. 4.9 Type A

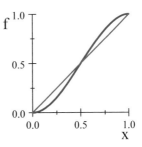

The diagram exhibits three equilibria:

$$\begin{cases} x^* = 0 \\ x^{**} = 1/2 \\ x^{***} = 1 \end{cases}$$

We will see that both $x^* = 0$ and $x^{***} = 1$ are attractors: if $x(0)$ is not too far from them, than our DS will approach them indefinitely. On the other side, x^{**} is not stable: even if the DS starts close to x^{**} it moves away.

The Fig. 4.10 shows that: (i) if we choose $x(0) = 0.4 < x^{**} = 1/2$ then the DS is attracted by $x^* = 0$; (ii) if we choose $x(0) = 0.6 > x^{**} = 0.5$ then the DS is attracted by $x^{***} = 1$.

This phenomenon is very interesting as it presents a *bifurcation case*: the starting point is crucial for the behavior of the system. If $x(0) = x_0$, the dynamic system is at an equilibrium, but if the starting point is not exactly x_0 it will escape to 0 or to 1.

In some sense, the next example is symmetrical with respect to the previous one. Once again we do not reveal the analytic expression of f, but its qualitative properties are sufficient to dig out the long-term behavior of the system.

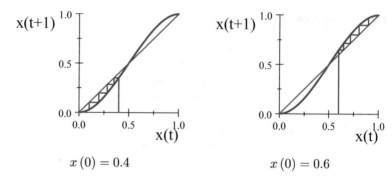

$$x(0) = 0.4 \qquad\qquad x(0) = 0.6$$

Fig. 4.10 Bifurcation

Example 4.7.8 Let us consider again a motion law of the type:

$$x\,(t+1) = f\,[x\,(t)]$$

the state variable is always confined to between 0 and 1:

$$0 \le x\,(t) \le 1$$

As announced, instead of specifying the analytic expression of f, we provide in Fig. 4.11 the phase diagram of type B.

Fig. 4.11 Type B

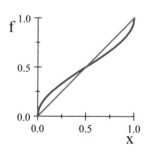

Example 4.7.9 The phase diagram exhibits three equilibria:

$$\begin{cases} x^* = 0 \\ x^{**} = 1/2 \\ x^{***} = 1 \end{cases}$$

We will see that both $x^* = 0$ and $x^{***} = 1$ repel trajectories: even if $x\,(0)$ is not too far from them (0 or 1), the DS will move away. On the other side, x^{**} is stable: even if the DS does not start close to x^{**} it will move toward it.

The Fig. 4.12 illustrates the two cases:

(i) if the DS starts from some $x(0) = 0.2 < x^{**} = 1/2$ then it approaches $x^{**} = 1/2$;
(ii) if the DS moves from an initial point $x(0) = 0.8 > x^{**} = 0.5$ then it is attracted by $x^* = 1/2$.

In this case the equilibrium x^{**} globally attracts trajectories (with the only extreme exceptions $x(0) = 0$ or 1). We will reflect later on the political implications of this fact.

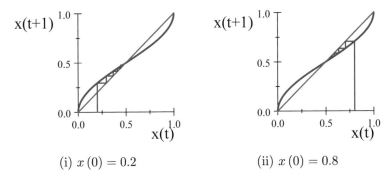

(i) $x(0) = 0.2$ $\qquad\qquad\qquad$ (ii) $x(0) = 0.8$

Fig. 4.12 Global attractor

A general result could be proved, which provides us with a sufficient condition for local stability:

Theorem 4.7.1 *Given a discrete autonomous DS, with motion law:*

$$x(t+1) = f[x(t)]$$

and an equilibrium point x^*:

$$x^* = f(x^*)$$

if the absolute value of the slope of the phase curve at x^* is smaller than 1:

$$|f'(x^*)| < 1 \ or \ -1 < f'(x^*) < 1 \qquad\qquad (4.7.2)$$

then x^* is (at least) a local attractor.

Remark 4.7.1 The previous theorem provides us with a *sufficient condition*, which is not necessary. However, note that in the previous examples, local attractivity was always accompanied by (4.7.2) and divergence by its contrary.

The next example is a companion of "Games and Grexit", where we have learned that, in a reasonable specification of the "game" between to countries (Lender $=$ L and Borrower $= B$), an equilibrium[43] does exist. We will refer to the companion pages above for the notation and the starting point. The basic idea consists in complementing the static framework we have constructed with a dynamic bargaining process between L and B. We will draw some conclusions, which turn out to be significant from a political point of view.

Example 4.7.10 Consider the model presented in the Example Grexit (C) at p. 190.

Starting point — In that model, the joint decision of two Countries was taken into consideration. Country L was providing money, while country B was expected to provide effort. Assumptions were introduced about the function f, putting together the ingredients and looking at the results 1 period later. In that Example, we assumed that the final wealth for the Borrowing country determined by its effort e and by u, the amount of money allowed by the Lending country. A possible specification follows:

$$f\,(e, u) = h\,(e)\,\phi\,(u) = ae^{\beta}u^{k} \text{ with } a > 0 \text{ and } 0 < \beta, k < 1$$

Such a function can be inserted in the general model (2.3.17).
Second step — The FOCs boil down to:

$$\begin{cases} \alpha vake^{\beta}u^{k-1} = 1 & \text{for L} \\ \\ (1 - \alpha)\,a\beta e^{\beta-1}u^{k} = c \text{ for B} \end{cases} \qquad (4.7.3)$$

These equations define the optimal policy for each country, conditional on the policy the other country will choose.
Third step — We explore such a dependence. We will extract from (4.7.3) two functions:

$$\begin{cases} e_L\,(u) \text{ for L} \\ \\ e_B\,(u) \text{ for B} \end{cases}$$

Their interpretation follows:

$$\begin{cases} e_L\,(u) \text{ is the optimal effort desired by L if L chooses } u \text{ for L} \\ \\ e_B\,(u) \text{ is the effort B thinks as optimal given } u \qquad \text{for B} \end{cases}$$

The two parameters have concrete interpretation is:

---

[43] Such an equilibrium is technically said to be a *Nash-equilibrium*, from the name of the Nobel Prize for Economics John Nash, made popular by the film *A beautiful mind*. A Nash-equilibrium is a point at which both the players are at an optimum. If one of them moves from that point looses something. If they are there and act rationally they do not move. A delicate question is "What happens if the two players are not at a Nash-equilibrium?" This is exactly what we are exploring here. This example is brand new.

$$\begin{cases} \beta \text{ is the elasticity of the effort of B w.r.t. how much L leaves to B} \\ \\ k \text{ is the productivity of B in transf. money at 0, into money at 1} \end{cases}$$

Their sum: $s = k + \beta$ can be smaller or greater than[44] 1. What we will see is that $s \lesseqgtr 1$ induces different portraits of the problem. They are different as far as mathematics is concerned, but we will see that the *political implications are exactly the same*. Here are the two announced functions:

$$\begin{cases} e_L(u) = \left(\dfrac{1+i}{\alpha a k}\right)^{1/\beta} u^{(1-k)/\beta} \\ \\ e_B(u) = \left[\dfrac{(1-\alpha)\,a\beta}{c}\right]^{1/(1-\beta)} u^{k/(1-\beta)} \end{cases}$$

Fourth step: the behavior of e_L and e_B. They have some relevant common features: both e_L and e_B are null at 0:

$$e_L(0) = e_B(0) = 0$$

Both e_L and e_B go to $+\infty$ if u goes to $+\infty$. Both e_L and e_B increase with u. As far as their concavity/convexity is concerned, they have opposite behaviors. Such a behavior is fully compatible with the existence of a unique relevant positive equilibrium (u^*, e^*), perfectly analogous to the one, we already met in GG. The Table 4.4 illustrates the possible cases.

Fifth Step: bargaining. At this point, a deeper analysis asks to be made. Next subsection is devoted.

Table 4.4 Lender and borrower efforts

Case	1	2
$k + \beta$	<1	>1
e_L	Convex	Concave
e_B	Concave	Convex
$e'_L(0)$	0	$+\infty$
$e'_B(0)$	$+\infty$	0
Plots		

----

[44]For obvious reasons we do not discuss the very, very... special case $k + \beta = 1$. It is of interest only for mathematicians.

4.7.3.2 Bargaining

We examine the dynamics of a simple bargaining process. The reference setting is described appropriately in the Cartesian plane (u, e).

The bargaining story needs three initial inputs:

- **Curve position**: we can meet two cases of convex versus concave or concave versus convex reaction curves for the pair B, L;
- **Initiative of the bargaining process**: L starts with an initial offer u_0 while B starts with an initial offer e_0.
- **Level of the initial offer w.r.t. the equilibrium value**: u_0, e_0 below or above the equilibrium levels u^*, e^*.

All of these aspects will turn out to be relevant. The analysis is simpler than its description as the technique is quite not new, but simply inherited from the well known phase diagram analysis. We hope that our expository choice will be rather understandable.

The single relevant[45] cases are $2^3 = 8$. As described in Table 4.4, there are 2 curve positions, to be multiplied by 2 possible startuppers, to be multiplied by 2 locations of the starting point w.r.t. the equilibrium.

As the technical machinery is the same, we will illustrate in detail one single case (out of eight) and provide the reader with simple suggestions about how to handle the other seven cases. Here is the *pilot* case:

- *curve positions*: B curve is convex, while L curve is the concave one (case 2 in Table 4.4).
- *initiative*: to L, which provides the allowance offer $u_0 = 0.9$.
- *position w.r.t. equilibrium*: the equilibrium is at $(1, 1)$ hence $u_0 < u^* = 1$.

The Fig. 4.13 illustrates the dynamics of the system in this pilot case.

Fig. 4.13 Pilot case

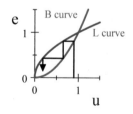

The bargaining process fails, even if the initial offer of L is close to the equilibrium value: 0.9 is close to 1, but $0.9 < 1$. DS theory tell us the story that, even if you are close to an equilibrium, it is not said that you will approach it indefinitely, concretely speaking, you will reach it. This should sound familiar in Political Science.

Here are the information/instruction for the reader, useful for the analysis the remaining seven cases.

1. the coordinate axes must be properly assigned: the abscissae axis is devoted to L proposals, the ordinate axis to B proposals;
2. the sequence of vertical/horizontal segments, constituting the broken line illustrating the bargaining dynamics, must start from the initiative axis (abscissae for L, ordinates for B);
3. if the starting point is L decision (like in the sample case), the segment must end when hitting the B curve and vice-versa;
4. consecutive segments are orthogonal.

4.7.3.3 Political Implications

We try to summarize our conclusions. We start from the two Tables 4.5 and 4.6 corresponding to the two possible reciprocal positions of the two curves.

The rows of each table correspond to the two possibilities of initiative (labelled IN):

- B starts, proposing an initial effort level e_0;
- L starts. proposing an initial allocation amount u_0.

The columns of each table correspond to the comparison between the initial proposal (labelled SP = Starting Point) and the equilibrium value e^*, u^* respectively.

The state variables (SV) are the levels of e, u proposed during the bargaining process.

Table 4.5 Case 1

IN\SP	$e_0 < e^*$ or $u_0 < u^*$	$e_0 > e^*$ or $u_0 > u^*$
B	SV $\to 0$	SV $\to +\infty$
L	SV $\to 0$	SV $\to e^*, u^*$

Table 4.6 Case 2

IN\SP	$e_0 < e^*$ or $u_0 < u^*$	$e_0 > e^*$ or $u_0 > u^*$
B	SV $\to 0$	SV $\to +\infty$
L	SV $\to e^*, u^*$	SV $\to e^*, u^*$

We try to translate in words the content Tables 4.5 and 4.6. This can be done according to two different perspectives. Precisely we can distinguish between different situations according to:

- the position of the starting point w.r.t. the equilibrium value;
- the country B, L which has the first move.

Starting Point Perspective

If the starting point is *above* the equilibrium value, the bargaining is *always successful*. The type of success depends on which country has the initiative. If the first mover is B (offering an effort $e_0 > e^*$) a virtuous circle starts and effort and allowance proposed increase significantly, while, if the first mover is L (offering an allowance $u_0 > u^*$) effort and allowance approach indefinitely the equilibrium values. These conclusions are the same both in case 1 and in case 2.

If the starting point is *under* the equilibrium value, the position of the curves, case (1 or 2) makes the difference. In the case 1, the bargaining process always fails, no matter which is the first mover. In the case 2, the first mover matters: failure if the initiative is up to B, equilibrium if the first mover is L.

First Mover Perspective

If the *first mover is B*, no matter which is the case (1 or 2), starting under the equilibrium ($e_0 < e^*$) brings always to failure. If the *first mover is L* there is convergence to the equilibrium, with the exception of a single combination, which brings to failure: L starts with an allowance $u_0 < u^*$.

A pair of remarks deserve some attention.

Remark 4.7.2 The bargaining process appears to be strongly dependent on its starting point: *first mover* and *position w.r.t the equilibrium*. Sometimes also the *reciprocal position of curves* turns out to be relevant.

Remark 4.7.3 The relevance of the position w.r.t. the equilibrium is an interesting explanation of what has been repeatedly observed in 2015 at the times of the possible Grexit. In the evening were frequently diffused announcements about a possible agreement between Germany and Greece and the day after failure of the bargaining turned out to be announced. A well known diplomatic habit in this type of bargaining activities consists in making offers, which are lower than the ones each party could afford. This attitude can bring to initial proposals *below* the equilibrium values, with the failure consequences we have found frequently in our analysis.

Mathematics provides crude lessons.

- if L offers too small allowances, the bargaining process will bring to nothing;
- if L offers allowances above the equilibrium value u^* two scenarios are possible:
  (1) unlimited growth in case 1, (2) acceptable growth in the case 2.

4.7.4 Continuous Systems

As usual, we start with a couple of examples.

Example 4.7.11 We are dealing with the exploitation of natural resources[46]: also a serious political problem. Consider a fish farm with size $x(t)$ and assume that the exploitation of this resource is made capturing a fixed annual number of fishes. The evolution of the size of the population obeys the motion law:

$$x(t+h) = x(t) + ax(t)h - ch + o(h)$$

where h is the length of a (short) observation time period, a is the natural evolution rate of the fish population (newborns − naturally dead between t and $t + h = ax(t)h + o(h)$) and c is the planned number of captures per year. Subtracting $x(t)$ from both sides, dividing by h and thinking of a vanishing h, we get the motion law:

$$x'(t) = ax(t) - c$$

What we know about sexuality and fertility of fishes[47] allows us to think that $a \gg 1$. This DS has a unique equilibrium:

$$0 = ax^* - c \Longrightarrow x^* = \frac{c}{a}$$

If $a = 2$ and $c = 10$, its phase diagram appears to be the following, with an equilibrium at $x^* = 5$

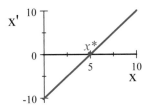

What we want to dig out from the phase diagram is what happens to any solution $x(t)$, starting at some point $x(0)$. The relevant information is provided by the phase curve. Assume that $x(0) = 4 < x^*$. As the graph of f at 4 is under the horizontal axis, the time derivative of the state variable at 4 (= ordinate at 4 of the point on the phase curve) is negative. This will imply that if the DS starts at 4, it will move to its left, until its extinction at 0. We can represent this fact putting arrows on the abscissae axis:

---

[46]Relevant for politics.
[47]Think of whitebait, to have an idea.

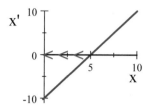

Imagine starting from an initial position $x(t) > x^*$, the positivity of the ordinate at these points for the phase curve implies that $x'(t) > 0$ and that therefore the system will move to its right. The Fig. 4.14 illustrates that the equilibrium we have detected repels trajectories therefore it is unstable.

Fig. 4.14 Unstable equilibrium

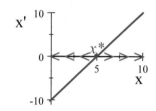

The policy implications should be pretty clear. If, for some reason, the capture moves an equilibrium population even slightly below the equilibrium level, the population will become extinct, while if the capture policy lets the population go beyond its equilibrium level, the population will grow (more and more quickly: think, for instance of Australian rabbits). Such an equilibrium deserves political monitoring.

Example 4.7.12 Again fish catching, but now the catching rule is proportional: fishers have an estimate of the size of the fish population at any time and they catch only some percentage of such a population. In this case, the parameter a is largely smaller and could even become negative: for simplicity we will assume that this happens and we will re-label a as:

$$a = -\alpha \text{ (with } \alpha > 0)$$

In the absence of any exogenous restocking strategy, the trivial effect of this catching policy would be the extinction of the population, if α is positive:

$$x'(t) = -\alpha x(t) \Rightarrow x(t) = x(0) e^{-\alpha t} \to 0 \text{ as } t \to +\infty$$

Call r the number of units added to the fish population per year. The ensuing motion law is:

$$x'(t) = -\alpha x(t) + r$$

The only equilibrium for this DS is[48]:

$$x^* = \frac{r}{\alpha}$$

Figure 4.15 provides the phase diagram with $\alpha = 2, r = 10$ and, consequently, $x^* = 5$. The arrows indicate that the system move towards 5 both from the left and from the right. Therefore the point x^* is a global attractor.

Fig. 4.15 Global attractor

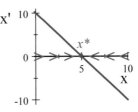

This equilibrium does not require political monitoring because the motion law "brings it back home" in the case of external shocks.

For continuous systems we can state a proposition, which is analogous to the one we met for discrete case.

Theorem 4.7.2 *Given a continuous autonomous DS, with motion law:*

$$x'(t) = f[x(t)]$$

and an equilibrium point x^*:

$$0 = f(x^*)$$

if the slope of the phase curve at x^* is negative:

$$f'(x^*) < 0 \qquad\qquad (4.7.4)$$

then x^* is a local attractor.

Remark 4.7.4 The thesis also follows if we assume that f decreases at x^*: no need for differentiability.

4.8 Some Politically Relevant Applications

This section is devoted to show how the tool of DSs can be fruitfully used in studying some political problems.

---

[48]It solves the equation:
$$0 = -\alpha x^* + r.$$

We do not maintain that Math is the main key tool for a political scientist, but we strongly affirm that Math can provide useful insights into some political problems.

The value of this assertion is related to the fact that some mathematical results, concerning social phenomena are counterintuitive.

If you learn something you were suspecting, the value added to your information is almost irrelevant, but if Math tells (and proves) something you did not expect, well: there is more value added.

Example 4.8.1 **Solow growth model** — The intensive production function we assume is:

$$f(x) = ax^\alpha$$

with $a = 10$, $\alpha = 0.3$. The assumed saving propensity is $s = 30\%$. Horizon time $T = 100$, $t = 1, \ldots, T$. Labor growth rate $g = 10\%$. Let k be the capital/worker ratio with $k_0 = 200$. We have

$$k(t+1) = \frac{k(t) + s\,f[k(t)]}{1+g}$$

The equilibrium value for capital per worker can be computed, solving the equation in k:

$$k = \frac{s}{g} f(k)$$

Back to our intensive production function:

$$f(k) = ak^\alpha$$

The equilibrium condition becomes:

$$k = \frac{sa}{g} k^\alpha$$

The trivial solution is $k = 0$ while the interesting one is:

$$k^* = \left(\frac{sa}{g}\right)^{1/(1-\alpha)}$$

4.8.1   *Consensus Diffusion*

We deal with a population, in which some novelty appears (see the seminal paper [3] by M. Granovetter[49]), but our analysis is significantly deeper:

---

[49]The mathematical analysis of the model is not due to the Author of the paper, who is a sociologist, but to Christopher Winship, who has had quite a different career.

- a new political leader;
- a new product;
- ....

Individuals in this population very realistically decide to approve the novelty, according to the opinion of others. The model we construct is a continuous-time one: it has to be considered as an approximation of reality. The driver of individual decisions is the prevailing opinion of others. We set this population at total size 1, therefore, when we refer to it, we are thinking in terms of %. Each individual has a personal threshold τ: if at least τ, out of the population, that individual approves. The threshold τ varies in the population. A *frequency density functions* $f(\tau)$ describes the statistical distribution of thresholds of the variable threshold **T**. As we learned in Chap. 3, we can describe the statistical distribution of **T** also through the cumulative distribution function:

$$F(\tau) = \int_{-\infty}^{\tau} f(s)\, ds$$

For instance, in the case these thresholds are uniformly distributed over [0, 1]: people with switching threshold between 0.2 and 0.5 are $0.5 - 0.2 = 0.3$. In this special case, we would have:

$$f(\tau) = \begin{cases} 0 & \text{for } \tau \leq 0 \\ 1 & \text{for } 0 \leq \tau \leq 1 \\ 0 & \tau > 1 \end{cases}$$

and:

$$F(\tau) = \begin{cases} 0 & \text{for } \tau \leq 0 \\ \tau & \text{for } 0 \leq \tau \leq 1 \\ 1 & \tau > 1 \end{cases}$$

The motion law of the consensus system is rather simple: $x(t+1)$, the consensus percentage at $t+1$ is nothing but the population percentage of people with threshold not beyond $x(t)$:

$$x(t+1) = F[x(t)] \tag{4.8.1}$$

We consider two types of population:

$$\begin{cases} (1) - unimodal\ distribution : \text{low frequencies for } t \text{ close to 0 and 1} \\ \\ (2) - bimodal\ distribution : \text{high frequencies for } t \text{ close to 0 and 1} \end{cases}$$

The Fig. 4.16 provides two examples of unimodal and bimodal density distributions. In the first one (1), few are ready to give their consensus on the basis of already few supporters, and few require a largely extended consensus to add their own. In the second one (2) a lot of (enthusiastic) people provide their consensus on the basis of a smaller acquired one, there is also a lot of (diffident) people who require a very extended consensus to give their own. Only a small part of the population gives his/her adhesion on the basis of a medium existing consensus.

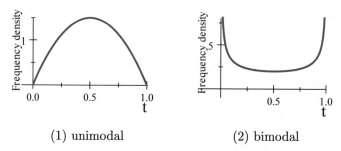

(1) unimodal   (2) bimodal

Fig. 4.16 Unimodal and bimodal distributions

It could appear intuitive that in a population of type (1) we can expect a stabilization of consensus around medium values, while, for type (2) we would expect the DS approaches an extreme position (consensus 0% or consensus 100%).

Well, Math says "No, you're wrong", exactly the opposite of what you expect is true.

In the case of the unimodal distribution (type (1)), the phase diagram is the one of Fig. 4.9, on p. 298 and, as we know very well both 0 and 1 are local attractors, while x^{**} is unstable. In the case of bimodal distribution (Type (2)), the phase diagram is the one of Fig. 4.11, and, as we know very well, x^{**} is a (politically) global attractor.

Note that in the type (2) distribution we have assumed that the most frequent threshold were the extreme ones. This is not very realistic and we could think of a more reasonable distribution of thresholds – say of type (3) – as Fig. 4.17 shows.

Fig. 4.17 Bimodal of type (3)

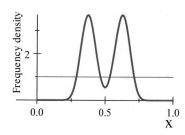

The population is a mixture of enthusiastic people and of skeptic people, but extreme positions are banned.

The Fig. 4.18 shows the phase diagram of the DS (4.8.1) for a bimodal distribution of type (3).

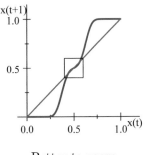

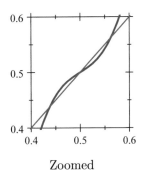

Better to zoom   Zoomed

Fig. 4.18 Phase diagram with density of type (3)

It exhibits 5 equilibria:

$x_0^* = 0$	attractor
$0 < x_1^* < x_2^* < x_3^* < 1$	the second out of 3 is locally stable, the others not
$x_4^* = 1$	attractor

The phase Fig. 4.18 brings us to depict three substantially relevant[50] scenarios:

(1) — The initial level of consensus satisfies:

$$x(0) < x_1^*$$

(2) — The initial level of consensus satisfies:

$$x_1^* < x(0) < x_3^*$$

(3) — The initial level of consensus satisfies:

$$x(0) > x_3^*$$

The political implications are rather evident:

(1) — starting too low implies default;
(2) — starting at a medium level can generate a mid-level stable consensus;
(3) — starting sufficiently high leads to full consensus.

Macron — The preceding model provides us with some insights in the electoral success of Emile Macron at the 2017 presidential elections in France. An almost

---

[50]The display of scenarios excludes the two unstable initial conditions:

$$x(0) = x_1^* \text{ and } x(0) = x_3^*.$$

unknown candidate was able to present himself as a brand new political leader, different from all the other candidates, and conquered very rapidly a vast majority. The Granovetter model we have seen above about consensus diffusion can suggest us a partial but credible insight. Macron profile is that of a new politician and he's against the rest of the candidates. The voter population had a strong desire of new politics, and was available to support any innovative candidate, without serious restrictions about the consensus the innovator had already obtained. This can be translated in terms of Granovetter's model saying that the frequency density f (u) of the adhesion thresholds was left-skew and concentrated a lot of frequency on low threshold levels:

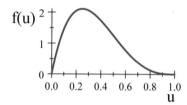

In the diagram you see above the mean value of the threshold is $1/3 \approx 33.33\%$. We already met the motion law for Granovetter's model:

$$x(t+1) = F[x(t)]$$

being F the cumulative distribution function of thresholds:

$$F(s) = \int_{-\infty}^{s} f(u)\,du$$

which defines the phase curve in the phase plane:

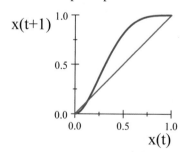

There are two locally stable equilibria (0 and 1). The other (unstable equilibrium) is x^*, which solves the equation:

$$s = F(s), \text{ with } 0 < s < 1$$

hence, numerically:
$$x^* \approx 15.8\%$$

If the initial consensus exceeds this level, the dynamic consensus grows very rapidly and approaches 100%. Note that the critical level is largely under the mean value of thresholds. What seems crucial to us is that an even small initial consensus can produce such spectacular results. we propose to the reader two points:

(1) — the role of Granovetter's rule for consensus evolution;
(2) — the ensuing consequences.

Let us change the shape parameters of f and, consequently, of F. We consider a new case, in which the mean threshold switches from $1/3$ to — say — 20%. The new density function turns out to be:

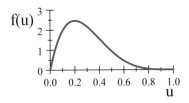

In the diagram you see above the mean value of the threshold now is $1/5 = 20\%$. With this frequency distribution, the unstable-threshold equilibrium will change. What we get is that now:
$$x^* = 9.6\%$$

This fact implies that what's relevant is the level of initial consensus. If the starting point is above the structural level x^*, success is expected.

4.8.2 Arab Springs

The nice paper [7] proposes a model for the interpretation of the so-called "Arab Springs" (2010–2011). In that paper, the treatment is essentially numerical, with all the weaknesses we know of this approach. We propose an improvement of the model, that can be treated with our toolbox. We preserves the continuous nature of that model and extracts interesting information using only qualitative tools.

4.8.2.1   Qualitative Analysis of Lang-Berck Model

We work with fractions of the population, whose size is standardized to the interval [0; 1], with the usual conventions:

- 0 means nobody;
- 1 means everybody;
- 0.75 — say — means 75% of the population.

The symbol $r = r(t)$ represents the state variable, i.e.; the fraction of *rioters* in the population. Its evolution over time is driven by two forces:

- its current size has an influence on its natural growth rate: a herd behavior is assumed;
- the current political regime tries to enforce a repression policy to stop the revolution starting.

The variation rate of r is modeled as follows:

$$r' = v(r)(1 - r) - rp(r)$$

This equation can be better understood as follows:

$$\underbrace{\frac{dr}{\Delta \text{riot. in } [t,\, t+dt]}}_{} = \underbrace{v(r)(1-r)\, dt}_{\text{new riot. in } [t,\, t+dt]} - \underbrace{rp(r)\, dt}_{\text{elim. riot. in } [t,\, t+dt]}$$

where both sides have been multiplied by dt.

Reasonable qualitative assumptions about v, p, assumed to be continuous and, if needed, differentiable, follow:

— $v(r) = $ instantaneous growth rate of the # of rioters): we can assume that $v(0) > 0$ and that v increases: this reflects the idea that revolution has an initial appeal and that such appeal increases with its own success.

— $p(r) = $ repression efficiency of the current regime: we can assume that $p(0) > 0$, that it is non-negative for every r, that p decreases as r increases.

In order to construct the phase diagram for this DS, whose motion law is:

$$r' = f(r)$$

being:

$$f(r) = v(r)(1 - r) - rp(r)$$

we start examining the behavior of f at the extrema of the relevant interval.

Value of f at 0 and 1 —   We obtain:

$$f(0) = v(0) \times (1 - 0) - 0 \times p(0) = v(0) > 0$$

Therefore the phase curve starts at 0 with a positive ordinate. At the beginning of the revolution, the movement has success because of the positivity of f near 0. We also get:

$$f(1) = v(1)(1-1) - 1 \times p(1) = -p(1) \leq 0$$

If $p(1) = 0$, than 1 is an equilibrium. If $p(1) < 0$, then there exists[51] an equilibrium $r^* < 1$, which is (at least) a local attractor. In principle the phase curve could cross the abscissae axis several times, but if we consider its last crossing, necessarily at that point, f decreases.

Value of f' at 0 and 1. — We have:

$$f'(r) = v'(r)(1-r) - v(r) - p(r) - rp'(r)$$

At 0:

$$f'(0) = v'(0)(1-0) - v(0) - p(0) - 0 \times p'(0) = v'(0) - v(0) - p(0)$$

The first addendum can be assumed to be positive, the second is negative, like the third one: the sign of $f'(0)$ turns out to be ambiguous, but this is irrelevant because near 0 we are only interested in the sign of f and not in the one of f'. At 1 we get:

$$f'(1) = v'(1)(1-1) - v(1) - p(1) - 1 \times p'(1) = -v(1) - p(1) - p'(1)$$

so, likely, $f'(1) < 0$. This means that the phase curve decreases near 1. This fact has two implications: (1) — If $f(1) = 0$, so that 1 is an equilibrium, it is an attractor: the revolution will have total success or (2) — If $f(1) < 0$, then the largest r^* such that $f(r^*) = 0$ is an attractor: the revolution has a partial success, in the sense that the rioter population becomes approximately r^*. In order for this to happen, we must distinguish two sub-cases: (2a) — r^* is the unique equilibrium, in which case, starting from every $r(0)$, the system approaches r^* (it is a global attractor); (2b) — there are other equilibria, smaller than r^*: if the system starts from $r(0)$ and there is no other equilibrium between the starting point and r^*, then r^* will attract the system.

The findings above stem from the analysis of a pair of phase diagrams represented in Fig. 4.19 even though they are only partially completed.

---

[51] The phase curve starts at 0 with positive ordinate, it is a continuous curve and it ends at 1 with a negative ordinate. Therefore there must be x^*, between o and 1 at which it crosses the abscissae axis.

The first one concerns the case $p(1) > 0$, while the second one the case $p(1) = 0$.

We can summarize the "political" conclusions coming from the qualitative analysis of the model as follows:

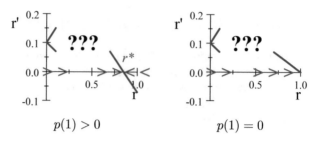

Fig. 4.19 Phase diagrams for Arab springs

- **Math ground $\Longrightarrow$ Political consequence**
- $f(0) > 0 \Longrightarrow$ When a revolution starts it has at least a temporary success.
- r^* is not unique $\Longrightarrow$ The revolution may stop and not continue without an external shock.
- $p(1) > 0 \Longrightarrow$ There exists a max level of success $r^* < 100\%$ which is an attractor. There are two subcases

  (i) r^* is unique $\Longrightarrow$ The % of rioters is always attracted by r^*.
  (ii) r^* is not unique $\Longrightarrow r^*$ is able to attract $r(t)$ if $r(0)$ is sufficiently large.

- $p(1) = 0 \Longrightarrow$ There exists a max level of success $r^* = 100\%$ which is an attractor. For sufficiently large $r(0)$ the revolution attains 100%.

Remark 4.8.1 We must make a crucial comment: this model is an *autonomous* DS. This means that the evolution rules of the DS do not change over time. What happens frequently in reality is that such rules can change. Two examples:

(1) — If a spreading revolution is not perceived $p(r) \equiv 0$, until the date t^*, at which the regime is alerted. In this case an appropriate model would be:

$$r' = \begin{cases} v(r)(1-r) & \text{for } t < t^* \\ v(r)(1-r) - rp(r) & \text{with } p(r) > 0, \text{ for } t > t^* \end{cases}$$

but the independence of the motion law of time would disappear.

(2) — The case of Egypt, for instance: a military *coup* occurs and the dynamics of the revolution is dramatically changed.

4.8.3 Growth and Demographic Trap

Consider the Solow model, already encountered in Example 4.2.6 at p. 227:

$$k' = sf(k) - gk \tag{4.8.2}$$

We will call $F(k)$ the r.h.s. of (4.8.2). A well known phenomenon consists in the fact that when an economic system grows, the fertility rate of its population decreases. Formally, this means that in (4.8.2) we should replace the constant fertility rate g with some decreasing function $g = g(k)$. Figure 4.20 provides an example.

Fig. 4.20 Fertility rate

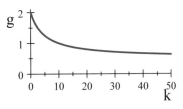

In the basic case (g constant), under broad conditions,[52] a unique equilibrium point exists for the system. Look at this

Example 4.8.2 The Table 4.7 reports the specification we assume in this example.

Table 4.7 Specification for Solow model

What	Assumption
Saving propensity	$s = 0.5$
Production function per capita	$f(k) = 10k^{1/2}$
Growth rate of the workforce	$g = 0.2$

The motion law is:

$$k' = F(k)$$

where:

$$F(k) = 0.5 \times 10k^{1/2} - 0.2k = 5\sqrt{k} - 0.2k$$

The phase diagram turns out to be:

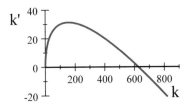

---

[52]It suffices that f satisfies the Inada conditions:

- $f'(0) = +\infty$;
- $f'(+\infty) = 0$;
- f increasing and concave.

and, together with the trivial equilibrium at 0, there is the relevant one at $k^* = 625$.

In the next example, we modify the assumptions about the growth rate of the labor force. Instead of assuming that g is constant, we replace it with a (linear decreasing) function of k:

$$g(k) = 3.5 - 0.2k$$

Example 4.8.3 Here are the specifications:

What	Assumption
saving propensity	$s = 0.5$
production function per capita	$f(k) = 10k^{1/2}$
growth rate of the workforce	$g(k) = 3.5 - 0.2k$

The new motion law is:

$$k' = F(k)$$

now being:

$$F(k) = 0.5 \times 10k^{1/2} - (3.5 - 0.2k)k = 5k^{1/2} - 3.5k + 0.2k^2$$

If we look for equilibria (via $\psi(k) = 0$), we find three equilibria:

$$\begin{cases} 0 & \text{trivial} \\ k^* \approx 2.95 & \psi'(k^*) < 0 \Longrightarrow k^* \text{ is a local attractor} \\ k^{**} \approx 9.30 & \psi'(k^{**}) > 0 \Longrightarrow \text{if } k > k^{**}, k \text{ will grow indefinitely} \end{cases}$$

A qualitative view of the phase diagram follows:

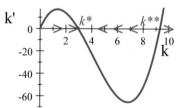

The relevant conclusions for politics are that, if the fertility goes down with growth, in a closed system there could be a *demographic trap at k^**, which prevents the possibility of growth of the economic system. How to avoid such a trap? There are at least two tools:

(1) — Immigration, in order to compensate the fertility decline;
(2) — A significant capital investment bringing the capital per capita above k^{**}. The task of politics is to choose between these two strategies or possible mixtures of them. See [3, 7].

4.9 Military Applications

We show how DS theory is useful in understanding some problems, that politics must unluckily face.

Two models are presented: the first one concerns war, while the second considers guerrilla.

4.9.1 War

There are two armies: Red and Blue. Let's consider a war between the two armies: Reds are firing Blues and vice versa. Let's indicate, respectively, with $r(t)$ and $b(t)$ the numbers of survivors, of Reds and Blues at t. The conflict model is based on:

- a time horizon T;
- a clock $t = 0, \ldots, T - 1$;
- two constant efficiency parameters, one for each army, indicated with $\rho > 0$ (for Reds), $\beta > 0$ (for Blues). They represent the expected number of enemies killed in a unitary period by each soldier of the two armies.
- the number of soldiers in each army at time 0:

$$\begin{bmatrix} r(0) \\ b(0) \end{bmatrix}$$

The discrete evolution rule is defined by the following simple system of equations

$$\begin{cases} r(t+1) = r(t) - \beta b(t) \\ b(t+1) = -\rho r(t) + b(t) \end{cases} \tag{4.9.1}$$

or

$$\begin{bmatrix} r(t+1) \\ b(t+1) \end{bmatrix} = A \cdot \begin{bmatrix} r(t) \\ b(t) \end{bmatrix}$$

where:

$$A = \begin{bmatrix} 1 & -\beta \\ -\rho & 1 \end{bmatrix}$$

The evolution rule (4.9.1) tells us that, for each army, the number of soldiers today is given by the number of soldiers yesterday, minus the number of fallen soldiers, which, in turn, is proportional to the number of enemies: the proportionality constants are nothing but the efficiency parameters (of the enemies).

Example 4.9.1 Let's assume that $\rho = 0.15$, $\beta = 0.20$. We choose the time horizon $T = 3$. The number of soldiers in each army today

$$\begin{bmatrix} r\,(0) \\ b\,(0) \end{bmatrix} = \begin{bmatrix} 1000 \\ 1000 \end{bmatrix}$$

The evolution rule is:

$$\begin{bmatrix} r\,(t+1) \\ b\,(t+1) \end{bmatrix} = \begin{bmatrix} 1 & -0.2 \\ -0.15 & 1 \end{bmatrix} \cdot \begin{bmatrix} r\,(t) \\ b\,(t) \end{bmatrix}$$

The number of soldiers 1 period later is:

$$\begin{bmatrix} r\,(1) \\ b\,(1) \end{bmatrix} = \begin{bmatrix} 1 & -0.2 \\ -0.15 & 1 \end{bmatrix} \cdot \begin{bmatrix} 1000 \\ 1000 \end{bmatrix} = \begin{bmatrix} 800 \\ 850 \end{bmatrix}$$

2 periods later we get:

$$\begin{bmatrix} r\,(2) \\ b\,(2) \end{bmatrix} = \begin{bmatrix} 1 & -0.2 \\ -0.15 & 1 \end{bmatrix} \cdot \begin{bmatrix} 800 \\ 850 \end{bmatrix} = \begin{bmatrix} 630 \\ 730 \end{bmatrix}$$

and at the end of the war we get:

$$\begin{bmatrix} r\,(3) \\ b\,(3) \end{bmatrix} = \begin{bmatrix} 1 & -0.2 \\ -0.15 & 1 \end{bmatrix} \begin{bmatrix} 630 \\ 730 \end{bmatrix} = \begin{bmatrix} 484 \\ 635.5 \end{bmatrix}$$

Exercise 4.9.1 It is of some interest to examine the dynamics, varying the ratios between the initial army sizes and the analogous ratios between the efficiency parameters: (a) $\beta/\rho < 1$; (b) $\beta/\rho = 1$; (c) $\beta/\rho > 1$. We suggest to explore numerically what happens with huge and poorly efficient armies versus small and very efficient ones. Some war stories can be well depicted in these terms.

4.9.2 Guerrilla

There are two actor groups: Defenders and Guerrillas. Defenders are moving through some area, looking for Guerrillas. Guerrillas create ambushes for Defenders. Let's indicate, respectively, with $d\,(t)$ and $g\,(t)$ the numbers of survivors in each group at time t. The simplest version of this conflict model is based on the following elements:

- a time horizon T (years);
- a unitary period $t = 1, 2, \ldots, T$ (one year);
- two efficiency parameters, one for each group: $\delta > 0$ (for Defenders), $\gamma > 0$ (for Guerrillas): the expected number of enemies killed in a unitary period by each soldier of the two groups; what will turn out to be different w.r.t. the preceding application is the condition set under which the two groups are acting;
- the initial numbers of Defenders and Guerrillas:

$$\begin{bmatrix} d\,(0) \\ g\,(0) \end{bmatrix}$$

The evolution rule cannot be modeled linearly. When the battle starts Guerrillas fire on Defenders, who are in full view. The loss for Defenders is reasonably proportional to the number of firing Guerrillas. These ones prepare an ambush for approaching Defenders, who fire blindly as they do not see Guerrillas. Therefore, the number of Guerrilla losses can be assumed proportional to both the number of firing Defenders and the number of Guerrillas in that area. The proportionality constant δ is the number of killed Defenders for every Guerrilla, in the case there is a single Defender.

Therefore, the evolution of the discrete system is regulated by the (vector) equation:

$$\begin{pmatrix} d\,(t+1) \\ g\,(t+1) \end{pmatrix} = \begin{pmatrix} d\,(t) - \gamma g\,(t) \\ g\,(t) - \delta d\,(t)\,g\,(t) \end{pmatrix}$$

The first equation tells us that, for each army, the number of regular soldiers today is given by the number of soldiers at t, minus the number of fallen soldiers, which, in turn, is proportional to the number of enemies at t: the proportionality constant is nothing but the efficiency parameters (of Guerrillas). The second equation tells us that the number of fallen Guerrillas is proportional to both the number of Defenders in the area and the number of Guerrillas itself.

Exercise 4.9.2 It is of some interest to examine the dynamics, varying the ratios between the initial army sizes and the analogous ratios between the efficiency parameters. We suggest to explore numerically what happens with huge and poorly efficient armies *vs.* small and very efficient ones.

Example 4.9.2 Let $T = 3$ be the time horizon measured in years. The numbers of Defenders and Guerrillas today are:

$$\begin{bmatrix} d\,(0) \\ g\,(0) \end{bmatrix} = \begin{bmatrix} 100 \\ 100 \end{bmatrix}$$

The two efficiency parameters are respectively $\delta = 0.1\%$ and $\gamma = 20\%$. The motion law for the discrete-time DS is

$$\begin{bmatrix} d\,(t+1) \\ g\,(t+1) \end{bmatrix} = \begin{bmatrix} d\,(t) - 0.2 \cdot g\,(t) \\ g\,(t) - 0.001 \cdot d\,(t)\,g\,(t) \end{bmatrix}$$

with $t = 0, 1, 2$. At time 1 we get:

$$\begin{bmatrix} d\,(1) \\ g\,(1) \end{bmatrix} = \begin{bmatrix} d\,(0) - 0.2g\,(0) \\ g\,(0) - 0.01d\,(0)\,g\,(0) \end{bmatrix} = \begin{bmatrix} 100 - 0.2 \cdot 100 \\ 100 - 0.001 \cdot 100 \cdot 100 \end{bmatrix} = \begin{bmatrix} 80 \\ 90 \end{bmatrix}$$

At time 2, we obtain:

$$\begin{bmatrix} d(2) \\ g(2) \end{bmatrix} = \begin{bmatrix} d(1) - 0.2g(1) \\ g(1) - 0.01d(1)g(1) \end{bmatrix} = \begin{bmatrix} 80 - 0.2 \cdot 90 \\ 90 - 0.001 \cdot 80 \cdot 90 \end{bmatrix} = \begin{bmatrix} 62.0 \\ 82.8 \end{bmatrix}$$

At time 3 the state of the system is:

$$\begin{bmatrix} d(3) \\ g(3) \end{bmatrix} = \begin{bmatrix} d(2) - 0.2g(2) \\ g(2) - 0.01d(2)g(2) \end{bmatrix} = \begin{bmatrix} 62 - 0.2 \cdot 82.8 \\ 82.8 - 0.001 \cdot 62 \cdot 82.8 \end{bmatrix} = \begin{bmatrix} 45.44 \\ 77.67 \end{bmatrix}$$

At the end of the war (3 years later) the death percentages for Defenders and Guerrillas are respectively:

$$\frac{100 - 45.44}{100} = 54.56\% \qquad \frac{100 - 77.67}{100} = 22.33\%$$

4.10 USA Against USSR: An Old Story

We start from the paper [14]. The paper is interesting, even if it contains mathematical flaws and some aspects of the model are not completely clarified. We present here an equivalent correct scheme which illustrates how to use linear discrete-time dynamic systems for this political problem.

The model focuses on the time evolution of decisions about the military expenditure of USA and USSR in the period before the fall of Berlin Wall (1989).

Time is discrete: $t = 0, 1, \ldots, T$.

The state variables in the model are the same for both countries and are of two types: *flows* and *stocks*. Flow variables represent the expenditure over a certain period, from $t - 1$ to t, whilst stock variables are the value of weapon stock at a certain date t.

The Table 4.8 collects these state variables, together with the notation we will use for them in the sequel.

Table 4.8 Model state variables

USA	State (endogenous) variable	USSR
$x(t)$	Military expenditure in the period $(t - 1, t)$	$y(t)$
$X(t)$	Value of the weapon stock at t	$Y(t)$

We add two exogenous variables, different from country to country.

- **USA** — A part of $x(t)$, denoted with $c(t)$, supports current wars. At the time the model was proposed, USA were involved both in Korean and in Vietnam wars. Obviously, this part of the annual expenditure does not increase the value of the weapon stock. It is the sum of the Korean component $k(t)$ and of the Vietnamese component $v(t)$. For the first ones we could think of an exponentially decreasing behavior:

$$k(t+1) = k(t)(1-r) \text{ with } 0 < r < 1$$

hence:

$$k(t) = k(0)(1-r)^t$$

while, for the continuous engagement of USA in the Vietnam war, we could take:

$$v(t) \equiv v(0)$$

At this point we gather:

$$c(t) = k(t) + v(t) = v(0) + k(0)(1-r)^t$$

- **USSR** — A part of the planned expenditure $y(t)$, denoted with $\gamma(t)$, was cancelled because of budget restrictions imposed by the Soviet planning policy. A simple assumption would be that $\gamma(t) \equiv \gamma(0)$.

We start now the construction of the motion law for our system. It concerns jointly the four state variables $x(t)$, $X(t)$, $y(t)$, $Y(t)$ we have introduced above together with the exogenous ones $c(t)$, $\gamma(t)$.

We list here the assumptions about the various state variables.

Period USA expenditure $x(t)$: it is driven by three forces.

1. An autonomous growth rate $a > 0$.
2. The comparison between the values of weapon stocks of USA and USSR, which pushes USA to update their investment, taking into account the parallel USSR position: we'll use the parameter b, to be interpreted as "extra-\$ for any gap between USSR and USA".
3. The current wars expenditure $c(t)$.

Period USSR expenditure $y(t)$: it is driven by two forces.

1. The alignment of the weapon stock of USSR to the one of USA, proportional through β (absolutely analogous to the parameter b above, to the perceived gap between the weapon stocks.
2. The expenditure restrictions discussed above.

Weapon stock of USA $X(t)$: it is driven by three forces.

1. Normal depreciation of the stock at some rate $\delta > 0$.
2. Period USA expenditure $x(t)$.
3. Period cost of current wars $c(t)$.

Weapon stock of USSR $Y(t)$: it is driven by two forces.

1. Normal depreciation of USSR to the one of USA, with depreciation rate $\varepsilon > 0$.
2. Period expenditure, under the restrictions discussed above.

We are ready to crystallize these assumptions into a system of difference equations, satisfied by the vector:

$$\mathbf{X}(t) = \begin{bmatrix} x(t) \\ y(t) \\ X(t) \\ Y(t) \end{bmatrix}$$

$$\mathbf{X}(t+1) = \begin{bmatrix} (1+a)\,x(t) + b\,[Y(t) - X(t)] \\ y(t) + \beta\,[X(t) - Y(t)] \\ (1-\delta)\,X(t) + x(t) - c(t) \\ (1-\varepsilon)\,Y(t) + y(t) - \gamma(t) \end{bmatrix}$$

or:

$$\mathbf{X}(t+1) = \begin{bmatrix} 1+a & 0 & -b & b \\ 0 & 1 & \beta & -\beta \\ 1 & 0 & 1-\delta & 0 \\ 0 & 1 & 0 & 1-\varepsilon \end{bmatrix} \mathbf{X}(t) + \begin{bmatrix} 0 \\ 0 \\ -c(t) \\ -\gamma(t) \end{bmatrix} \qquad (4.10.1)$$

Call A the coefficient matrix:

$$A = \begin{bmatrix} 1+a & 0 & -b & b \\ 0 & 1 & \beta & -\beta \\ 1 & 0 & 1-\delta & 0 \\ 0 & 1 & 0 & 1-\varepsilon \end{bmatrix}$$

and:

$$\mathbf{b}(t) = \begin{bmatrix} 0 \\ 0 \\ -c(t) \\ -\gamma(t) \end{bmatrix}$$

the exogenous forcing term.

The motion law of this system is then:

$$\mathbf{X}(t+1) = A\mathbf{X}(t) + \mathbf{b}(t)$$

Even if we could assume that the USSR budget reduction is constant:

$$\gamma(t) \equiv g$$

the forcing term, turns out to depend on time (*via* $c(t)$). This does not imply we are not able to trace the path of the system. It is sufficient we refer to the Sect. 4.4.2 "About some special discrete systems", where the case:

$$\mathbf{x}(t+1) = A\mathbf{x}(t) + \mathbf{b}(t)$$

is appropriately examined.

As discussed above we could assume that $\gamma(t) \equiv g$ (constant).

A further interesting simplifying assumption would consist in taking $r = 0$ or that $c(t) \equiv C$ (non-negative constant). We will use **b** to denote the vector

$$\mathbf{b} = \begin{bmatrix} 0 \\ 0 \\ -C \\ -g \end{bmatrix}$$

In this case the equilibrium vector $\mathbf{v}^* = \begin{bmatrix} x^* \\ y^* \\ X^* \\ Y^* \end{bmatrix}$ should satisfy:

$$\mathbf{v}^* = A\mathbf{v}^* + \mathbf{b}$$

or, more explicitly:

$$\begin{bmatrix} x^* \\ y^* \\ X^* \\ Y^* \end{bmatrix} = \begin{bmatrix} 1+a & 0 & -b & b \\ 0 & 1 & \beta & -\beta \\ 1 & 0 & 1-\delta & 0 \\ 0 & 1 & 0 & 1-\varepsilon \end{bmatrix} \begin{bmatrix} x^* \\ y^* \\ X^* \\ Y^* \end{bmatrix} + \begin{bmatrix} 0 \\ 0 \\ -C \\ -g \end{bmatrix} \qquad (4.10.2)$$

This system can be rewritten as:

$$\left(\begin{bmatrix} 1 & 0 & 0 & 0 \\ 0 & 1 & 0 & 0 \\ 0 & 0 & 1 & 0 \\ 0 & 0 & 0 & 1 \end{bmatrix} - \begin{bmatrix} 1+a & 0 & -b & b \\ 0 & 1 & b & -b \\ 1 & 0 & 1-\delta & 0 \\ 0 & 1 & 0 & 1-\varepsilon \end{bmatrix} \right) \begin{bmatrix} x^* \\ y^* \\ X^* \\ Y^* \end{bmatrix} = \begin{bmatrix} 0 \\ 0 \\ -C \\ -g \end{bmatrix}$$

the determinant of the coefficient matrix boils down to:

$$\Gamma = \det \left(\begin{bmatrix} 1 & 0 & 0 & 0 \\ 0 & 1 & 0 & 0 \\ 0 & 0 & 1 & 0 \\ 0 & 0 & 0 & 1 \end{bmatrix} - \begin{bmatrix} 1+a & 0 & -b & b \\ 0 & 1 & \beta & -\beta \\ 1 & 0 & 1-\delta & 0 \\ 0 & 1 & 0 & 1-\varepsilon \end{bmatrix} \right) =$$

$$= \det \begin{bmatrix} -a & 0 & b & -b \\ 0 & 0 & -\beta & \beta \\ -1 & 0 & \delta & 0 \\ 0 & -1 & 0 & \varepsilon \end{bmatrix} = -a\beta\delta$$

In the case $a\beta\delta \neq 0$, it is a Cramer's system with a unique solution:

$$\mathbf{v}^* = \begin{bmatrix} 0 \\ g - C\varepsilon/\delta \\ -C/\delta \\ -C/\delta \end{bmatrix}$$

provided by

$$\mathbf{v}^* = \left(\begin{bmatrix} 1 & 0 & 0 & 0 \\ 0 & 1 & 0 & 0 \\ 0 & 0 & 1 & 0 \\ 0 & 0 & 0 & 1 \end{bmatrix} - \begin{bmatrix} 1+a & 0 & -b & b \\ 0 & 1 & \beta & -\beta \\ 1 & 0 & 1-\delta & 0 \\ 0 & 1 & 0 & 1-\varepsilon \end{bmatrix} \right)^{-1} \begin{bmatrix} 0 \\ 0 \\ -C \\ -g \end{bmatrix} \qquad (4.10.3)$$

Let us now move from algebra to the political implications. The fact that a unique equilibrium for the system boils down to the requirement that $\Gamma \neq 0$ enjoys of some political interpretation.

The mathematical consequences follow:

- the USA want to vary their freely decided military expenditure (concretely): $a \neq 0$;
- the USA are alerted about the possible gap in the weapon stock (concretely): $b \neq 0$;
- the depletion rate of the USA stock of weapons is positive: $\delta > 0$.

All of these consequences are rather trivial. But, if we look at the equilibrium point, non-trivial remarks can be added.

What we have found above is:

$$\mathbf{v}^* = \begin{bmatrix} x^* \\ y^* \\ X^* \\ Y^* \end{bmatrix} = \begin{bmatrix} 0 \\ g - C\varepsilon/\delta \\ -C/\delta \\ -C/\delta \end{bmatrix}$$

Let us start from the third and fourth component: the common value of the weapon stock for both USA and USSR should be negative. the immediate consequence is that such an equilibrium has no sense: therefore there is no logic acceptable equilibrium.

That's is Science.

Intuitively you can find solutions for several social problems, but you can find to be deceived.

For a politician, it's normal to invent solutions and frequent to accept that they cannot have any concrete implementation.

We will explore the case in which the expenses for current wars of the USA are set to 0 ($\Rightarrow C = 0$). In this case the equilibrium vector $\mathbf{v}^*$ would become:

$$\mathbf{v}^* = \begin{bmatrix} x^* \\ y^* \\ X^* \\ Y^* \end{bmatrix} = \begin{bmatrix} 0 \\ g \\ 0 \\ 0 \end{bmatrix}$$

In such case, of course, forUSSR should be let $g = 0$, not for imitation but for un-necessity. The equilibrium point would turn out to be **o**. This fact is politically important as it tells us that aggressive/competing policies do not bring to any relevant equilibrium.

At this point a further step would be needed, concerning the stability of such pacifist equilibrium. But for such an analysis, the tools we present in this book are insufficient. What we could say after a deeper analysis is that it is not stable.

4.10.1 An Electoral Application

Consider hundred parties[53] during an electoral campaign. We refer to the standard political model based on the left/right continuum between 0 and 1 (e.g., see: [5]).

According to some political literature, candidates can play an electoral game in three different ways:

1. F: maintain their position **fixed** on the left/right continuum (that is, a left-wing party remains a left-wing one);
2. C: **change** their position on the left/right continuum (i.e.; for instance, a left-wing party becomes a center-left one);
3. V: play on **valence** issues (i.e.; their leader is presented as competent, honest …).

It is also widely accepted that leaders need to follow strict rules in order to make an efficient campaign:

1. if a party plays C (i.e.; it changes its position) at time t, it cannot play V at time $t + 1$ (i.e., it cannot base its campaign on valence issues);
2. if a party plays V (i.e., focuses the campaign on valence issues) at time t it cannot play C at time $t + 1$ (i.e., it cannot change its position in that period).

The following vector represents the absolute frequency of the three strategies (F, C, V) observed at time 0:

$$\mathbf{x}(0) = \begin{bmatrix} 10 \\ 80 \\ 10 \end{bmatrix} \begin{matrix} \leftarrow F \\ \leftarrow C \\ \leftarrow V \end{matrix}$$

This means that, at time 0, only 10 of parties do not change their positions on the left/right continuum, 80 decide to change their positions and only 10 play on valence issues.

The following transition matrix describes the evolution of party strategies:

$$T = \begin{matrix} & F & C & V & \\ & \begin{bmatrix} 30\% & 55\% & 40\% \\ 35\% & 45\% & 0\% \\ 35\% & 0\% & 60\% \end{bmatrix} & & & \begin{matrix} F \\ C \\ V \end{matrix} \end{matrix}$$

---

[53] We're Italian.

The first column of the matrix T tells us that: at time 1, 30% of the parties, that had a fixed position F, keep a fixed position, 35% change their position on the left/right continuum and 35% start using valence issues. The other columns of T can be interpreted analogously. The evolution over 10 periods ($t = 0, 1, 2, \ldots, 10$) is described by:

$$\mathbf{x}(t+1) = T \cdot \mathbf{x}(t)$$

After 10 periods, we get the following frequencies for the various strategies:

$$\mathbf{x}(10) = \begin{bmatrix} 40 \\ 25 \\ 35 \end{bmatrix}$$

In this case, the most used strategy is F (consisting in preserving a position). Indeed, even if only 10 parties were using this sort of strategy at time 0, at time 10 (and for all the following epochs) 40 decide to remain fixed on their positions, 35 decide to use valence issues and 25 decide to change position. Moreover, at the 10th period, the political campaign has reached an equilibrium. This can be checked easily, checking that:

$$T \cdot \mathbf{x}(10) = \mathbf{x}(10)$$

Such an equilibrium point can be also detected associating it to the equilibrium equation $\mathbf{x} = T\mathbf{x}$ a further equation stating that the total number of parties is (only) 100. We would like to be more precise. T is a *stochastic matrix*, this means that its entries are non-negative and that the columns sums are unitary. This is an example of *Markov chain*. An equilibrium point $\mathbf{x}^*$ must satisfy:

$$\mathbf{x}^* = T \cdot \mathbf{x}^*$$

or, equivalently:

$$(I - T)\mathbf{x}^* = \mathbf{0}$$

In our case, the matrix $B = I - T$ is:

$$\begin{bmatrix} 1 & 0 & 0 \\ 0 & 1 & 0 \\ 0 & 0 & 0 \end{bmatrix} - \begin{bmatrix} 0.3 & 0.55 & 0.4 \\ 0.35 & 0.45 & 0 \\ 0.35 & 0 & 0.6 \end{bmatrix} = \begin{bmatrix} 0.7 & -0.55 & -0.4 \\ -0.35 & 0.55 & 0 \\ -0.35 & 0 & -0.6 \end{bmatrix}$$

The equilibrium condition boils down to:

$$B\mathbf{x}^* = \mathbf{0} \tag{4.10.4}$$

or, more explicitly:

$$\begin{bmatrix} \mathbf{b}^1 \\ \mathbf{b}^2 \\ \mathbf{b}^3 \end{bmatrix} \mathbf{x}^* = \mathbf{0}$$

The matrix:

$$B = \begin{bmatrix} \mathbf{b}^1 \\ \mathbf{b}^2 \\ \mathbf{b}^3 \end{bmatrix}$$

is singular for an obvious fact: the columns totals of T are 1, the columns totals of I are 1 and, consequently[54]:

$$\mathbf{b}^1 + \mathbf{b}^2 + \mathbf{b}^3 = \mathbf{0} \Rightarrow \text{ any } \mathbf{b}^s \text{ is a l.c. of the others}$$
$$\mathbf{b}^s = -\sum_{k \neq s} \mathbf{b}^k$$

We can scrap an equation from (4.10.4) and replace it with the budget equation, stating the fact that the number of parties is hundred:

$$\sum_{s=1}^{100} x_s = \mathbf{u}\mathbf{x} = 100$$

being $\mathbf{u} = \begin{bmatrix} 1 & 1 & 1 \end{bmatrix}$, We construct a new matrix, say:

$$C = \begin{bmatrix} \mathbf{b}^1 \\ \mathbf{b}^2 \\ \mathbf{u} \end{bmatrix}$$

which is non-singular. The equilibrium equation becomes:

$$C\mathbf{x} = \begin{bmatrix} 0 \\ 0 \\ 100 \end{bmatrix}$$

and provides us with:

$$\mathbf{x}^* = C^{-1} \begin{bmatrix} 0 \\ 0 \\ 100 \end{bmatrix} = \begin{bmatrix} 40 \\ 25 \\ 35 \end{bmatrix}$$

---

[54]This fact holds generally. If T is a square stochastic n-matrix, then $I - T$ is singular, being I an identity n-matrix. At least one of the equilibrium equations turns out to be redundant.

4.11 Epidemics

Government interest for welfare must be able to handle difficult problems like Epidemics. We show that some interesting insights from DS-theory could be somehow relevant. Our starting point is the paper G. JAMES, N. STEELE (1990): 'Epidemics and the spread of diseases', in [6].

We will examine a sequence of models, with increasing complexity, but we will see that Maths can provide interesting insights.

Here is our plan:

- Model B (as Basic): the crude effect of an unprotected infection;
- Model R (as Recovery): which are the implications of a recovery opportunity for infected people?
- Model RI (as Recovery and Isolation): which could be the effect of some isolation decisions?
- Model RIV (as Recovery, Isolation and Vaccination): what's the effect of vaccination? In times of popular "no-Vax" people, this should be rather interesting.

4.11.1 The B Model

Take a closed population of N individuals. A part of this population, with size $i(t)$, varying in discrete time $t = 0, 1, \ldots, T$, is made by infective people.

The motion law is:

$$i(t+1) = i(t) + \beta i(t)[N - i(t)]$$

Its interpretation for our readers should be straightforward: the r.h.s. addendum

$$\beta i(t)[N - i(t)]$$

tells us that the increment in infected people, described by $\beta > 0$, depends on the possible number of (dreadful) meetings between infected and non infective people ($i(t)$ people and $N - i(t)$ people). It is a standard discrete-time logistic model.

The equilibrium equation for i^* is:

$$i^* = i^* + \beta i^* \left[N - i^*\right]$$

Apart from the trivial null solution 0, we get the equilibrium $i^* = N$.

In order to decide whether i^* is an at least local attractor, the reader knows we have a simple sufficient condition. For a DS:

$$i(t+1) = \phi[i(t)]$$

it boils down to:

$$\left| \phi' \left(i^* \right) \right| < 1 \text{ or, more explicitly } -1 < \phi' (i) < 1$$

In our case:
$$\phi' (i) = 1 + \beta N - 2\beta i$$

Hence:
$$\phi' (N) = 1 - \beta N$$

and, consequently, we get that at least local stability is guaranteed if:

$$\beta N < 2 \text{ or } \beta < \frac{2}{N}$$

This requirement is of political interest as it tells us that the sensitivity coefficient β must not be too high, especially in the case of large populations.

4.11.2 The R Model

This model introduces an additional feature w.r.t. the basic model B: during a unitary time period, some fraction γ of infected people do recover. It is convenient to assume that γ is not to high. The condition we assume is:

$$\gamma < \beta N$$

certainly satisfied for sufficiently large population sizes N.
    The new motion law is[55]:

$$i (t + 1) = i (t) + \beta i (t) [N - i (t)] - \gamma i (t)$$

The new equilibrium equation is:

$$i^* = i^* + \beta i^* \left(N - i^* \right) - \gamma i^*$$

Together with the trivial equilibrium 0 we find now:

$$i^* = N - \frac{\gamma}{\beta} = \frac{\beta N - \gamma}{\beta} > 0$$

under our assumptions.

---

[55]The quantity $\frac{\beta i(t)[N-i(t)]-\gamma i(t)}{N}$ is usually called the *incidence rate* of the disease.

Remark 4.11.1 If such equilibrium portrays the long-term behavior of the system (see hereafter the conditions guaranteeing this fact) the result tells us that the disease will however remain endemic in the population. If $\gamma < \beta N$, let $n < N$ such that $\gamma = \beta n \Rightarrow n = \gamma/\beta$ hence:

$$i^* = N - n$$

A rough conclusion stemming from this analysis is that partial recovery cannot guarantee the absence of infected people: Nature needs a strong help.

Anyway the new equilibrium point will be at least locally stable[56], but morally globally stable if[57]:

$$\beta N - 2 < \gamma < \beta N$$

Exercise 4.11.1 An exciting exercise for readers, wanting to explore wild landscapes, would consist in reshaping both the B model and the R model, removing the (intriguing) assumption that N is constant. A good option would assume:

$$N(t+1) = (1+g)N(t)$$

May be that there are no more equilibria in absolute terms, but in relative terms:

$$i(t)/N(t)?$$

4.11.3   The RI Model

We add another interesting feature: together with partial recovery, we introduce the possibility that some percentage α of infected people can be detected and isolated in order to mitigate their contribution to the disease diffusion. The new motion law will be bidimensional, as now we have 2 state variables:

- $i(t)$: the size of infected and non-isolated people at t;
- $s(t)$: the size of infected and isolated people at t.

Here are the two new evolution equations:

$$\begin{bmatrix} i(t+1) \\ s(t+1) \end{bmatrix} = \begin{bmatrix} i(t) + \beta i(t)[N - i(t) - s(t)] - \gamma i(t) - \alpha i(t) \\ (1-\gamma)s(t) + \alpha i(t) \end{bmatrix}$$

A few comments:

- what changes in the first equation is that the portion of exposed population:

---

[56] With attraction basin $(0, +\infty)$.

[57] The computations, which are quite analogous to the ones of the basic model are left to the reader as a useful exercise.

$$N - i(t) - s(t)$$

takes into account the splitting of the old $i(t)$ "infected" into the new categories: $i(t)$ "non-detected infected" and "isolated infected":

new $i(t)$ infected but non-detected

old $i(t)$ ↗

↘

isolated infected people $s(t)$

- we need an evolution rule for isolated infected: a part of them recovers and new detected entries are registered.

The equilibrium conditions are:

$$\begin{cases} i^* = i^* + \beta i^* (N - i^* - s^*) - \gamma i^* - \alpha i^* \\ s^* = (1 - \gamma) s^* + \alpha i^* \end{cases}$$

or:

$$\begin{cases} \beta i^* (N - i^* - s^*) - \gamma i^* - \alpha i^* = 0 \\ s^* = \alpha i^* / \gamma \end{cases}$$

As usual, an uninteresting equilibrium stays at 0, while the most serious one stays at:

$$\begin{cases} i^* = \beta \gamma N / \left(\beta \gamma + \alpha + \gamma^2 + \alpha \gamma\right) \\ s^* = \alpha \beta N / \left(\beta \gamma + \alpha + \gamma^2 + \alpha \gamma\right) \end{cases}$$

Several remarks could be made about this solution, but they are left to the patience and the interest of the reader. For instance:

- i^* decreases when α increases, implying that infective detection is useful to reduce the endemic part of the population;
- s^* decreases if γ increases: a better recognition system reduces the isolation needs.

4.11.4   The RIV Model

A further development of the model. We would like to understand the role of a possible vaccination against the infective disease.

Bad news about the necessary Mathematics: to model this case we will need 3 state variables:

- $i(t)$: non-detected infected people at t;
- $s(t)$: isolated infected people at t;
- $m(t)$: vaccinated and consequently immunized people.

The last variable deserves some attention. Its variation depends on the joint effect of recovery on infected people (both non-detected and isolated) and on the fact that vaccination acts on people who are both non-isolated, nor already vaccinated. We denote with v the number of vaccinated who become immune for every exposed person.

Here is the motion law:

$$\begin{bmatrix} i(t+1) \\ s(t+1) \\ m(t+1) \end{bmatrix} = \begin{bmatrix} i(t) + \beta i(t)[N - i(t) - s(t) - m(t)] - \gamma[i(t) + s(t)] \\ s(t) + \alpha i(t) - \gamma s(t) \\ m(t) + \gamma[i(t) + s(t)] + v[N - s(t) - m(t)] \end{bmatrix}$$

An equilibrium:

$$\begin{bmatrix} i^* \\ s^* \\ m^* \end{bmatrix}$$

is a constant vector collecting the size of the three groups:

$$\begin{bmatrix} i^* = \text{size of non-detected infected people} \\ s^* = \text{size of detected and isolated people} \\ m^* = \text{size of non-detected and vaccinated people} \end{bmatrix}$$

which, under the motion law remains unchanged. Such points can be found solving the equilibrium equations:

$$\begin{cases} i^* = i^* + \beta i^*(N - i^* - s^* - m^*) - \gamma(i^* + s^*) \\ s^* = (1 - \gamma)s^* + \alpha i^* \\ m^* = m^* + \gamma[i^* + s^*] + v(N - i^* - s^* - m^*) \end{cases}$$

The solution of the system provides us with a politically interesting result:

$$\begin{bmatrix} i^* \\ s^* \\ m^* \end{bmatrix} = \begin{bmatrix} 0 \\ 0 \\ N \end{bmatrix}$$

If we translate this mathematical result into normal language, its interest should become absolutely clear: the "prevalence[58] $\dfrac{i^* + s^*}{N}$ of the disease is 0" and consequently the "isolation size" s^* is, annihilated. The disease has disappeared.

---

[58] % of non-detected affected + % of isolated affected = 0.

Numerical simulations, based on realistic parameters show that, in the long-range, this is the population dynamics.

4.12 Exercises

Exercise 4.12.1 Find the general solution for the following motion laws, describing the evolution rules of continuous time dynamic systems (with $t \geq 0$, $a > 0$ and $b \in \mathbb{R}$):

(1) $x'(t) = 2t + a$ (2) $x'(t) = \dfrac{a}{t+1}$ (3) $x'(t) = 2\sqrt{t}$

(4) $x'(t) = te^{3t^2+a}$ (5) $x'(t) = \dfrac{1}{\sqrt{at+1}}$ (6) $x'(t) = 3e^{1-3t}$

(7) $x'(t) = \dfrac{t}{t^2+b}$ (8) $x'(t) = (2t+b)^2$ (9) $x'(t) = ae^{t/3+b}$

Exercise 4.12.2 Let's consider the following differential equation $x'(t) = x^2 t^2$.

(a) Classify the differential equation. If it exists, find any constant solution.
(b) Find the general solution and the particular solution considering the boundary condition $x(1) = 1$. Show that we can calculate the particular solution without finding the general one.

Exercise 4.12.3 Find the solution for the following differential equations, which describe the evolution rules of continuos time dynamic systems:

(a) $\begin{cases} x'(t) = xt \\ x(0) = 1 \end{cases}$ (b) $\begin{cases} x'(t) = 2x/t \\ x(2) = 1 \end{cases}$

(c) $\begin{cases} x'(t) = x/(t+1) \\ x(0) = 1 \end{cases}$ (d) $\begin{cases} x'(t) = e^t/x^2 \\ x(\ln 2) = 1 \end{cases}$

Exercise 4.12.4 Let's consider the following differential equation

$$x'(t) = \dfrac{1}{x(t) \cdot t}$$

(a) Classify the differential equation. If it exists, find any constant solution.
(b) Find the general solution and the particular solution considering the boundary condition $x(e) = 2$.
(c) Show how we can calculate the particular solution without finding the general one.

Exercise 4.12.5 Let's consider the following differential equation

$$x'(t) = \frac{2x(t)}{t+2}$$

with $t > -2$.

(a) Classify the differential equation. If it exists, find any constant solution.
(b) Find the general solution and the particular solution considering the boundary condition $x(0) = 4$.
(c) Show how we can calculate the particular solution without finding the general one.

Exercise 4.12.6 Consider the linear differential equation

$$\text{(a)} \begin{cases} x'(t) - x(t)/t = t^2 \\ x(1) = 1 \end{cases} \qquad \text{(b)} \begin{cases} x'(t) - x(t) \cdot t = t \\ x(0) = 2 \end{cases}$$

$$\text{(c)} \begin{cases} x'(t) - 2x(t) \cdot t = 4t \\ x(0) = -2 \end{cases} \qquad \text{(d)} \begin{cases} x'(t) - x(t) = t \\ x(0) = 2 \end{cases}$$

with $t > 0$.

(a) Find the general solution $x(t)$.
(b) Find the particular solution considering the indicated boundary conditions.

Exercise 4.12.7 Let's consider the discrete dynamic system

$$\mathbf{x}(t+1) = \begin{bmatrix} 1 & 0 \\ 0 & 2 \end{bmatrix} \cdot \mathbf{x}(t) \quad \text{and} \quad \mathbf{x}(0) = \begin{bmatrix} 1 \\ 2 \end{bmatrix}$$

(a) Calculate $\mathbf{x}(1)$, $\mathbf{x}(2)$ and $\mathbf{x}(3)$.
(b) Find the particular trajectory singled out by the boundary condition.

Exercise 4.12.8 Let's consider the discrete dynamic system

$$\mathbf{x}(t+1) = \begin{bmatrix} 0 & 1/2 \\ 1/2 & 0 \end{bmatrix} \cdot \mathbf{x}(t) + \begin{bmatrix} 1 \\ 1 \end{bmatrix}$$

(a) Find its unique equilibrium point $\mathbf{x}^*$.
(b) Given that

$$\mathbf{x}(0) = \begin{bmatrix} 0 \\ 0 \end{bmatrix}$$

calculate its position at $t = 2$.

Exercise 4.12.9 The size $S(t)$ of a population is grows with instantaneous growth rate $g(t)$.

(a) Write the first order differential equation governing such a motion.
(b) Using definite integrals provide $S(t)$ in terms of $S(0)$.
(c) Assume then that $g(t) = 9 + 4t$ and compute explicitly $S(t)$ in terms of t and of $S(0)$.
(d) Under the assumption of the preceding points (a), (b) and (c), find how much time t^* is necessary so that the size of the population is doubled.

Exercise 4.12.10 The size of a population evolves in continuous time according to the motion law

$$x'(t) = 500x^3(t)[M - x(t)]$$

where $M > 0$.

(a) Find the two equilibria $x^* < x^{**}$ for the system.
(b) Using the diagram phase technique, decide about their stability.
(c) At which position $\bar{x}$ does the speed of growth of the population attain its maximum value?

Exercise 4.12.11 A public institution has a liquid endowment of $x(0)$ at date 0. Such an amount $x(t)$ varies in discrete time for two reasons:
(i) — during each period the institution collects interest, coming from the investment of liquidity. The interest is 5% of the liquidity available at the beginning of each period. The interest is collected at the end of each period;
(ii) — at the end of each period a constant amount c is taken away to support the activity of the institution.

(a) Write the motion law in discrete time, describing the evolution of the endowment.
(b) Find the unique equilibrium point x^* for the dynamic system. Provide an interpretation for the equilibrium equation.
(c) Construct the phase diagram for the dynamic system and make a diagnosis about the stability of x^*.

Exercise 4.12.12 Consider a population of 20000 people and the possibility that an infection will spread out in it. We model the diffusion of the infection in discrete time. We denote with $x(t)$ the number of people infected at time t. The diffusion process is a standard logistic one

$$x(t+1) = f[x(t)] = x(t) + ax(t)[20000 - x(t)]$$

where a is a positive parameter.

(a) Write the equilibrium equation in x, which must be satisfied by any equilibrium point x. Show that there are two equilibria $x^* < x^{**}$ and then find them.

(b) Explicit $f(x)$, compute $f'(x)$ at a generic point and determine $f'(x^{**})$.

(c) Determine for which values of a we can be sure that x^{**} is locally stable.

Exercise 4.12.13 During some revolution, the number of riots at time t is $x(t)$. We model the evolution of riots in discrete time. The number of riots at $t + 1$ is the product of two factors:

(i) — $A[x(t)]$, which is simply proportional to $x(t)$. The proportionality factor is $\alpha > 1$.

(ii) — $B[x(t)]$, which is a linear affine function of $1/x(t)$, with unitary intercept and negative slope $-\beta$, where β is a positive constant.

(a) Construct and simplify the motion law.

(b) Construct the phase diagram and find the unique equilibrium for the system.

(c) Motivating your opinion indicate the starting numbers of riots which bring the revolution to success.

References

1. Bernheim, D.B., Whinston, M.D.: Microeconomics. Mc Graw-Hill, New York (2013)
2. Edwards, D., Hamson, M.: Guide to Mathematical Modelling. MacMillan, Cambridge (1989)
3. Granovetter, M.: Threshold models of collective behavior. Am. J. Sociol. **83**(6), 1420–1443 (1978)
4. Giordano, F.R., Fox, W.P., Horton, S.B., Weir, M.D.: Mathematical Modeling. Brooks/Cole, Belmont (2009)
5. Heywood, A.: Political Theory, an Introduction. Palgrave Mac-Millan, Basingstoke (2004)
6. Huntley, I.D., James, D.J.G. (eds.): Mathematical Modelling (A Source Book on Case Studies). Oxford University Press, Oxford (1990)
7. Lang, J.C., De Sterck, H.: The arab spring: a simple compartmental model for the dynamics of a revolution. Math. Soc. Sci. **69**, 12–21 (2014)
8. Livio, M.: The Golden Ratio. Broadway Books, New York (2002)
9. Murthy, D.N.P., Page, N.W., Rodin, E.Y. (eds.): Mathematical Modelling (A Tool for Problem Solving in Engineering Biological and Social Sciences). Pergamon Press, Oxford (1990)
10. Nikaido, H.: Convex Structures and Economic Theory. Academic Press, New York (1968)
11. Solow, R.M., Samuelson, P.A.: Balanced growth under constant return to scale. Econometrica **21**(3), 412–424 (1953)
12. Sanz, M.T., Mico', J.C., Caselles, A., Soler, D.: A stochastic model for population and well-being dynamics. J. Math. Sociol. **38**, 75–94 (2014)
13. Simon, C.P., Blume, L.: Mathematics for Economists. W.W Norton and Company Inc., New York (1994)
14. Ward, M.D.: Differential paths to parity: a study of the contemporary arms race. Am. Polit. Sci. Rev. **78**(2), 297–317 (1984)
15. Weidlich, W.: Sociodynamics - A Systematic Approach to Mathematical Modelling in the Social Sciences. Harwood Academic Publishers, Amsterdam (2000)

© Springer Nature Switzerland AG 2018
L. Peccati et al., *Maths for Social Sciences*, UNITEXT - La Matematica per il 3 + 2 113, https://doi.org/10.1007/978-3-030-02336-2

Index

A

Age-composition of a population, 246
   closed population, 247
   open population, 247
Antiderivative, 200
   calculus, 201
Arab springs, 315
Arbitrage, 11

B

Basis, 12
   fundamental, 13
   theorem, 15
Bourbaki, 127

C

Caste
   statistical distribution, 209
Central limit theorem, 212
Characteristic polynomial, 277
Cold war, 324
Colin Clark three sectors model, *see* three
    sectors model
Consensus diffusion
   Granovetter model, 310
Constrained optimization, 170
   general case, 179
   Lagrangean function, 174
   problem, 171
   set of constraints, 172
Continuous dynamic system, 249
   general solution/integral, 249
   linear, 260
     constant coefficients, 262
     endogenous vs. exogenous part, 262

     homogeneous, non homogeneous,
      262
     solution, 263
   linear autonomous
     equilibrium, 275
     equilibrium stability, 277
   linear differential equation
     integrating factor, 264
   linear differential equation of first order,
     253
   particular solution/integral, 249
   phase diagram and asymptotic behavior,
     307
   separable (differential) equation, 250
   separable equation
     definite integral, 252
Covariance, 169

D

Degrees of freedom, 181
Demand
   elastic, anelastic, rigid or inelastic, 146
   elasticity, 145
   inverse function, 148
Density
   normalization condition, 205
Density function
   frequency, 210
   probability, 210
Determinant, 41
   area interpretation, 44
   geometrical interpretation, 41
   Laplace rule, 45, 49
   Sarrus rule, 44
   volume interpretation, 45
Difference equations, 232

© Springer Nature Switzerland AG 2018
L. Peccati et al., *Maths for Social Sciences*, UNITEXT - La Matematica
per il 3 + 2 113, https://doi.org/10.1007/978-3-030-02336-2

Differential equations, 232
Discrete distributions, 209
Discrete dynamic system
    linear, 234
        constant coefficients, 236
        forcing term, 235
        general solution, 248
        homogeneous, non homogeneous, 235
            particular solution, 248
            special case, 246
    linear autonomous
        equilibrium, 276
    linear homogeneous
        solution, 237
    linear, non-homogeneous
        solution, 242
    local stability condition, 301
    phase diagram and asymptotic behavior, 295
Distribution
    Gauss(ian), Laplace, Gauss-Laplace, normal, law of errors, 211
Dynamic system, 221, 222
    autonomous, non autonomous, 230
    classic approach, 233
    continuous, 222
        numerical approach, 283
    dimension, 222
    discrete, 222
        bifurcation, 299
        boundary condition, 248
        numerical approach, 282
    equilibrium, 285
        equation, 288
        global attractor, 290
        globally stable, 290
        local attractor, 290
        locally stable, 290
        unstable, 290
    equilibrium point, 285
    extracting information from the motion law, 232
    linear autonomous
        constant coefficients, 274
    motion law, 224
    of order 2, 231
    phase diagram, 291
        equilibrium line, 292
        phase curve, 292
    state vector, 222

E

Earthquake, 32
    temporary accomodation, 32
Eigenvalue, 90
    finding, 90
Eigenvector, 90
    finding, 90
Elections, 329
Epidemics, 332
    recovery, 333
    recovery and isolation, 334
    recovery, isolation and vaccination, 335
Expectation, 205

F

Fishery, 307
Frequency density, 204
Function, 99
    concave, 140
    continuous, 110
    convex, 141
    derivable, 127
    derivative, 122
        chain rule, 135
        computation, 128
        differentiation rules, 130
        elementary, 128
        geometric interpretation, 122
        haidresser rule for product, 134
        kaki-rule, 129
        onion rule, 136
        second, 136
    differentiable, 122
    differential, 118, 122
        geometric interpretation, 122
    discontinuous, 111
    elementary, 101
    exponential, 104
        alternative notation, 107
        positivity, 106
    extrema, 137
    extremum point
        local maximum point, 138
        local minimum point, 138
    graph, 100
    increasing, decreasing, 137
    increment, 115, 119
    increment of the independent variable, 119
    inflexion points, 141
    limit, 112
    linear (affine), easy, 101

local and global behavior, 118
logarithmic, 108
    argument, 108
power, 103
quadratic, 102
ranking of infinitesimals and infinities, 115
second order derivative
    concavity/convexity, 140
smooth, 120
stationary point, 138
zeroes, 143
Function of several variables, 152
    constrained extrema, 170
        general case, 178
    constrained local maximum
        sufficient condition, 183
    contour lines, 156
    contour map, 156
    differentiable, 161
    differential, 160
    extremum
        first order condition, 164
        second order condition, 165
    geometric representation, 155
    gradient, 159
    Hessian matrix, 162, 164
        NW principal minors, 165
    level curves, 156
    partial derivatives, 159
        second mixed, 163
        second pure, 163
    saddle point, 167, 174
    Schwarz Theorem, 163
    stationary point, 160
    vector function of a vector, 153

G
Game theory, 190
Gompertz, 215
Grexit, 187, 189, 190, 301

H
Harmonic growth, 89

I
Inada, 319
Integral, 197
    Barrow's formula, 200
    by parts, 202
    by substitution, 203

calculus
    fundamental theorem, Torricelli, Barrow, 199
definite, 198
function, 198
indefinite, 201
integrand function, 198
Interval, 100
    closed, open, semi-open, 100

K
Kronecker method, 81

L
Lagrangean, 174
Lagrange multiplier
    economic interpretation, 176
    political interpretation, 177
Landau symbol o(.)
    notion, 112
    small o algebra, 114
Law of large numbers, 207
Leontief, 33
    economy, 33
    final uses, 34
    finding production vectors, 40
    intermediate uses, 34
    technical coefficients, 34
    viability (Hawkins-Simon), 51
Linear algebraic system
    complete matrix, 75
    Cramer, 72
    degrees of freedom, 72, 77
    determined, 77
    general case, 74
    impossible or overdetermined, possible, underdetermined, 72
    notion, 71
    possible, 77
        solution(s), 77
    Rouché-Capelli Theorem, 78
Linear application, 67
    theorem, 69
    Tinbergen model, 67
Logarithm
    Napierian, natural, 109
    properties, 110
Logistic model
    continuous
        phase diagram, 294
    discrete, 224
        phase diagram, 293

M

Macron, 313
Makeham, 215
Markov chain, 330
Matrix, 16
    addition, 20
    adjoint, 50
    complementary, 47
    conformability, 24
    division, 35
    empirical
        rank, 54
    Hurwitz, 278
    identity, 18, 26
    minor, 46
    multiplication, 22
        properties, 27
        vector interpretation, 30, 31
    multiplication by a scalar, 21
    operations, 18
    product, 25
    rank, 53
        determination, 54
        Kronecker method, 54, 56
        reduction method, 54
        rows and columns, 53
    rank minor, 82
    scalar, 17
    slicing, 18
    square
        algebraic complement, 47
        characteristic equation, 91
        characteristic polynomial, 91
        cofactor, 47
        complementary minor, 47
        determinant, 41
        inverse, 37
        inversion, 50
        inversion algorithm, 38
        principal diagonal, 26, 42
        principal NW minor, 46
        principal NW submatrix, 46
        secondary diagonal, 42
        singular, 37
        trace, 91
    stochastic, 330
    submatrix, 45
    transition, 225
    transposition, 19
    type of, 16
Mean
    Chisini, 241
    types, 240

Military application
    guerrilla, 322
    war, 321
Mortality rate, 215
Motion law
    continuous, 228
    discrete, 228

N

Nabla symbol ∇, 159
Network, 86
    Boolean numbers, 86
    centrality index, 87
    Freeman index, 87
Norm
    of a vector, 158
Number e, 107

O

ONG
    efficiency, 150
    efficiency & Microeconomics, 150
Optimal consumption, 177
Ordinary least squares, 167

P

Parabola
    discriminant, 102
    vertex, 102
People survival, 214
Political profit, 186
Primary sex-ratio, 260
Probability
    expectation, 60
    expected value, 60
    linear correlation coefficient, 65
    probability function, 59
    random variable
        discrete, 58
    variance, 62
Probability density, 204
Production function, 227
Production mix
    absorption function, 172
    optimal, 172
Public debt model, 270

Q

Quantity
    infinite, 112
    infinitesimal, 112

R

Random number
 discrete
  expectation, 204
Random variable
 continuous, 204
Routh–Hurwitz condition, 278

S

Samuelson's accelerator, 231
Shadow price, 176
Socio-demographic model, 254, 266, 267
Solow model, 227, 310
 demographic trap, 318
Statistical variable
 continuous
  mean, 204
 discrete
  mean, 204
Statistics
 bivariate distributions, 63
 frequency function, 59
 linear correlation coefficient, 65
 mean value, 60
 statistical variable
  discrete, 58
 variance, 62
Straight line

 intercept, 101
 slope, 101
Survival function, 214

T

Three sectors model, 225
Tinbergen model, 84

U

USA against USSR, 324

V

Variance, 169
Vector, 2, 4
 addition, 5
 components, 3
 linear combination, 8
 linear (in)dependence, 10
 null, 4
 scalar multiplication, 7
 transposition, 3
Vector function, 68
 additive, 68
 homogeneous, 68
Vote elasticity and media, 184
Votes and campaign duration, 185

Printed in the United States
By Bookmasters